METHODS IN ANCIENT WINE ARCHAEOLOGY

METHODS IN ANCIENT WINE ARCHAEOLOGY

SCIENTIFIC APPROACHES IN ROMAN CONTEXTS

Edited by Emlyn Dodd and Dimitri Van Limbergen

BLOOMSBURY ACADEMIC
LONDON • NEW YORK • OXFORD • NEW DELHI • SYDNEY

BLOOMSBURY ACADEMIC
Bloomsbury Publishing Plc
50 Bedford Square, London, WC1B 3DP, UK
1385 Broadway, New York, NY 10018, USA
29 Earlsfort Terrace, Dublin 2, Ireland

BLOOMSBURY, BLOOMSBURY ACADEMIC and the Diana logo are trademarks of Bloomsbury Publishing Plc

First published in Great Britain 2024

A catalogue record for this book is available from the British Library.

Library of Congress Cataloging-in-Publication Data
Names: Dodd, Emlyn K., editor. | Limbergen, Dimitri van, editor.
Title: Methods in ancient wine archaeology : scientific approaches in Roman contexts / edited by Emlyn Dodd and Dimitri Van Limbergen.
Description: London ; New York : Bloomsbury Academic, 2024. |
Includes bibliographical references and index.
Identifiers: LCCN 2023036816 (print) | LCCN 2023036817 (ebook) | ISBN 9781350346642 (HB) | ISBN 9781350346659 (PB) | ISBN 9781350346666 (ePDF) | ISBN 9781350346673 (eBook)
Subjects: LCSH: Wine and wine making–Mediterranean Region–History–To 1500. | Wine and wine making–Rome. | Viticulture–Rome. | Archaeology–Mediterranean Region. | Archaeology–Methodology.
Classification: LCC TP559.M47 M48 2024 (print) | LCC TP559.M47 (ebook) |
DDC 641.2/2091822—dc23/eng/20230914
LC record available at https://lccn.loc.gov/2023036816
LC ebook record available at https://lccn.loc.gov/2023036817

ISBN: HB: 978-1-3503-4664-2
ePDF: 978-1-3503-4666-6
eBook: 978-1-3503-4667-3

Typeset by RefineCatch Limited, Bungay, Suffolk

To find out more about our authors and books visit www.bloomsbury.com and sign up for our newsletters.

CONTENTS

FIGURES

TABLES

CONTRIBUTORS

Lázaro Lagóstena Barrios is Professor of Ancient History at the University of Cádiz, Spain. His main areas of interest are the economic history of the Roman Empire, but also historical socio-environmental interaction, an aspect on which he has directed several projects in recent years (Riparia I; Riparia II; Aquivergia). He is a specialist in non-invasive historical research applied to landscapes and ancient urban areas, and is responsible for the Unit of Geodetection and Georeferencing of Historical-Archaeological Heritage at the University of Cádiz.

Patrizia Basso is Professor in Classical Archaeology in the Department of Cultures and Civilizations at the University of Verona, Italy. She focuses on Roman architecture, primarily through research and fieldwork currently in Aquileia, Udine and Negrar, Verona. Since 2020 she has coordinated the project *In Veronensium mensa: Food and Wine in Ancient Verona*, which uses an interdisciplinary approach to identify the historical origins of production and consumption of wine and food around Verona from prehistoric times to the Middle Ages.

Gaetano Benčić is Curator of the Parentine Territory Museum in Poreč, Croatia. Graduating from the Faculty of Letters and Philosophy at the University of Trieste, he has since participated as a speaker and organizer in both national and international scientific conferences on issues concerning Istrian history and archaeology. He is the author of monographs and articles in both Croatian and Italian on various topics of Istrian culture, especially concerning the valley of the Mirna river and the archaeological topography of Poreč and Parentino.

Nicolas Bernigaud is an environmental and landscape archaeologist specializing in geoarchaeology, spatial archaeology and agent-based modelling. He was Senior Postdoctoral Researcher in the ERC-funded Rurland Project (2014–18), which studied the Roman countryside in northern Gaul, and the multidisciplinary RDMed project (2018–21) focused on the impact of climate change on ancient societies in southern Gaul. He currently co-directs the ANR Project MICA, which models the impact of climate change on agriculture in the Western Mediterranean and Gaul from the Bronze Age to Late Antiquity.

Frédérique Bertoncello is a CNRS Research Scientist in the Laboratory of Cultures and Environments: Prehistory, Antiquity, Middle Ages (CEPAM) at the University Côte d'Azur, France. Her research focuses on settlement and landscape dynamics during the Roman period. She has coordinated and participated in several interdisciplinary research projects studying socio-environmental interactions in ancient societies using spatial archaeology and agent-based modelling.

Alberte Bondeau is a CNRS Research Scientist at the Mediterranean Institute of Marine and Terrestrial Biodiversity and Ecology (IMBE), Aix-Marseille University, France. Her work deals with process-based agroecosystem modelling, to build tools investigating the impact of climate and land management on agricultural production. She has participated in various interdisciplinary projects analysing human–environment interactions in the Mediterranean through the combined use of archaeological data, spatial and agroecosystem modelling.

Laurent Bouby is Research Engineer at CNRS in the Institute of Evolutionary Science of Montpellier, France. His research interests focus on archaeobotany, plant resources and agricultural practices in the western Mediterranean from the Neolithic to the Roman period. He has published several papers on the history and domestication of the grapevine.

Francesco Breglia is Postdoctoral Researcher at the University of Salento, Italy. He works in the field of pre-protohistoric archaeology with a focus on archaeobotany. In recent years he has been involved in the application of geometric morphometry to the analysis of grapevine seeds to investigate the early diffusion of viticulture in the Italian peninsula during the Bronze Age.

Antony G. Brown is Professor at the University Museum, Arctic University of Norway and the Department of Geography and Environmental Sciences at the University of Southampton, UK. His research interests are in geoarchaeology, palynology and sedimentary ancient DNA (sedaDNA). He currently leads the ERC-funded TerrACE Project in Norway and the AHRC-funded Stonehenge-Avon Buried Landscapes Project in the UK.

Riccardo Carmenati is a PhD candidate in 'Education, Cultural Heritage and Territories' at the University of Macerata. He is a member of the archaeological excavation of the University of Macerata at Urbs Salvia/Villamagna (Italy), Palokastër/Antigone (Albania) and Gortina (Greece). He is also a member of the Laboratory of Archaeobotany and Palaeoecology at the University of Salento.

Diana Dobreva is Assistant Professor in Classical Archaeology in the Department of Cultures and Civilizations at the University of Verona, Italy. Her primary research aims to understand why trade networks emerge, persist, evolve, and in some cases fail. She has analysed major cultural transformations within the Roman East and West using traditional and novel approaches. Her most recent book is *Tra Oriente e Occidente: Dinamiche Commerciali in Moesia Inferior e Thracia. I dati delle anfore* (2017).

Emlyn Dodd is Lecturer in Classical Studies at the Institute of Classical Studies, University of London, UK. He was Assistant Director for Archaeology at the British School at Rome, Italy, from 2021–23, and is Honorary Fellow at Macquarie University, Australia. He is expert in Graeco-Roman wine and olive oil production, predominantly in Italy, Greece and Turkey, co-directs excavations at the Roman urban site of Falerii Novi in Central Italy, and directs a survey project investigating agricultural production in the Cyclades, Greece. Recent publications include the monograph *Roman and Late Antique Wine Production in the*

Eastern Mediterranean (2020) and 'The Archaeology of Wine Production in Roman and pre-Roman Italy' in the *American Journal of Archaeology* 126.3 (2022).

Christopher D. Dore is a Consultant with Heritage Business International in Tucson, Arizona, USA. He is also Adjunct Professor in the School of Anthropology at the University of Arizona and in the Department of Archaeology at Simon Fraser University in Burnaby, Canada. His research interests include heritage business management, Maya settlement organization, and the non-visible archaeological record as revealed through geospatial methods.

Pedro Trapero Fernández is Postdoctoral Researcher at the University of Cádiz. His work focuses on Roman landscapes and territory, especially agriculture and viticulture using non-invasive techniques and Geographical Information Systems. His most recent book is *La viticultura romana en el estuario del Guadalquivir* (2021).

Manuel Malfeito Ferreira is Associate Professor at the School of Agriculture in the Department of Natural Resources, Environment and Territory at the University of Lisbon, Portugal. His research is focused on wine microbiology, appreciation and cultural heritage. He is also a winemaker for several Portuguese companies and has written the book *Mastering the Art of Enjoying Wine, from Enology to Enosophy* (2022).

Girolamo Fiorentino is Professor in Advanced Methodologies in Archaeological Research and is responsible for the Laboratory of Archaeobotany and Palaeoecology at the University of Salento in Lecce, Italy. His research is focused on the reconstruction of human–environment interaction, through the analysis of plant macroremains in archaeological and natural sequences. He has published more than 200 papers in this field and is involved in several European research projects.

Nicolas Garnier is an analyst specialized in ancient natural products after studying chemistry, ancient history and etruscology at the École Normale Supérieure de Paris-Ulm and receiving a doctorate from the Université Paris VI-Jussieu. His research focuses on the contents of *amphorae*, including pitch, oils, waxes, grapes and wine. He has developed new approaches implementing mass spectrometry techniques for a precise identification of a wide range of materials in various types of archaeological materials.

Joël Guiot is CNRS Emeritus Research Director at the Centre of Research and Teaching of Environmental Geosciences (CEREGE), Aix-Marseille University, France. His main research interests are the impact of past, present and future climate change on Mediterranean ecosystems. His most recent project studies resilience and adaptation to droughts and extreme climatic events in the Mediterranean with a focus on the Roman Period. He has published over 200 papers in international journals, including as lead author for the IPCC special report on 1.5°C warming.

Pieter Gurdebeke is a geologist with a PhD from Ghent University (2019) who works on late Pleistocene and Holocene environmental changes and their reconstruction through bio-indicators and geochemistry, with a focus on aquatic environments in the northern hemisphere. Pieter is also interested in anthropogenic landscapes and the heritage of industries that are determined by landscape characteristics.

Mkrtich Harutyunyan is a PhD candidate at the Instituto Superior de Agronomia, University of Lisbon, Portugal, former scholar at the Calouste Gulbenkian Foundation (Lisbon), and recipient of several grants. His research interests focus on experimental archaeology, the recreation of ancient winemaking techniques and wine styles, particularly using pottery vessels, and the reconstruction of the practice of ancient viticulture based on historical sources.

Mario Indelicato is a PhD candidate at the University of Catania and Research Fellow at the Institute of Heritage Science of the National Research Council (CNR), Italy. His research interests concern the archaeology of production and archaeological methodologies with a focus on experimental archaeology applied to the study of ancient agriculture. He has published several articles on Roman viticulture and oenology.

Marco Marchesini holds a master's degree in Quaternary, Prehistory and Archaeology from the University of Ferrara and Specialisation School in Archaeological Heritage (S.I.S.B.A.) from the University of Trieste. He conducts research in palynology and palaeo-archeobotany, collaborating throughout Italy with public and private organizations, including various Italian and foreign universities, Archaeological Superintendencies, Museums, and Archaeological Societies, and is the author of 200 articles in international and national books and journals.

Silvia Marvelli holds a PhD in Science and Technology for Archaeology and Cultural Heritage from the University of Ferrara and is sector manager for the Palynology and Archaeobotany Laboratory of C.A.A. Giorgio Nicoli, Bologna, Italy. She coordinates research activities in the field of palynology and archaeobotany and collaborates with public and private institutions, various Italian and foreign universities, and superintendencies. She is the author of 170 scientific publications and the editor of sixteen volumes.

Patrick E. McGovern is the Scientific Director of the Biomolecular Archaeology Project for Cuisine, Fermented Beverages, and Health at the University of Pennsylvania Museum in Philadelphia, where he is also Adjunct Professor of Anthropology. Over the past two decades, he has pioneered the interdisciplinary field of Biomolecular Archaeology which is yielding new chapters concerning human ancestry, medical practice, and ancient cuisines and beverages. He is the author of *Ancient Wine: The Search for the Origins of Viniculture* (2003, 2006, 2019) and numerous other articles and books.

Andrew McLean is Training Fellow at the University of Edinburgh's Centre for Data Culture and Society. His research interests focus on the economy of the Roman Empire, computational modelling and the ancient Adriatic. He has published on complex methodological issues in archaeology, including his most recent article, 'Beyond Least Cost Paths: Circuit Theory, Maritime Mobility and Patterns of Urbanism in the Roman Adriatic,' in the *Journal of Archaeological Science* 138 (2022).

Ian Meadows is an independent archaeological consultant and advises multiple modern quarries on archaeology. He led excavations at Wollaston for Northamptonshire Archaeology, now the Museum of London Archaeology, UK.

Davor Munda is Archaeologist and Senior Curator at the Museum of the Poreč Territory and is author and co-author of numerous exhibitions. His areas of research include the prehistory of Poreč and its region along with ancient Roman history and culture. As archaeologist and curator, he aims to present complex archaeological material to museum visitors of all ages.

Anna Chiara Muscogiuri is an archaeologist and holds qualifications in the Specialisation School in Archaeological Heritage (S.I.S.B.A.) from the Universities of Trieste, Udine and Venezia Ca' Foscari. Since 2019, she has collaborated with the Laboratory of Palynology and Archaeobotany at C.A.A. Giorgio Nicoli, Bologna, Italy, where she carries out research in palynology and archaeobotany.

Antoni Martín i Oliveras is County Archaeologist at the Catalonian Government Culture Department and Associate Researcher at the University of Barcelona Centre for the Study of Provincial Interdependence in Classical Antiquity, Spain. His research interests focus on the archaeology and economy of Roman wine, landscape and experimental archaeology, and heritage management. Recent books include *Arqueologia del vi a l'època romana. Del cultiu al consum. Marc teòric i epistemològic* (2015) and *Productive Landscapes and Trade Networks in the Roman Empire* (2019).

Marie-Jeanne Ouriachi is Associate Professor in Roman History and Archaeology at the University of Côte d'Azur and member of the Laboratory of Cultures and Environments: Prehistory, Antiquity, Middle Ages (CEPAM), France. Her research interests include the evolution of settlement and territorial dynamics in southern Gaul during antiquity, and she is expert in spatial archaeology, spatial epigraphy and modelling, including game theory, graphs and networks, and agent-based modelling.

Alessandra Pecci is Lecturer in Archaeology at the University of Barcelona, Spain. She specializes in archaeometry and food practices, especially through the organic residue analysis of archaeological materials, mortars and plasters. She has participated in international and interdisciplinary projects in Italy, Spain, Turkey and Mexico.

Roberto Perna is Professor of Classical Archaeology in the Department of Humanities at the University of Macerata, Italy, and Director of archaeological fieldwork at Urbs Salvia and Villamagna (Italy), Antigonea, Palokaster and the Drino valley (Albania), and Gortina (Greece). Alongside archaeological–historical synthesis, his research interests focus on landscape archaeology, polyorcetics, *instrumentum domesticum*, epistemology and method, with particular attention to museums, archaeological parks and territorial and urban planning. He has recently published, with, M. Giuliodori, R. Carmenati, and J. Piccinini, *Roma e il mondo adriatico: dalla ricerca archeologica alla pianificazione del territorio* (2020–21).

Nicholas K. Rauh is Professor of Classics in the School of Languages and Cultures at Purdue University, USA. From 1996 to 2014 he directed the Rough Cilicia Archaeological Survey Project in Gazipasha Turkey. Since 2015 he has participated in the Boğsak Maritime Archaeological Survey Project in Boğsak Turkey. Most recently he has published the survey pottery datasets for both projects open access at the Purdue University Digital Research Repository (PURR).

Elisabetta Rizzoli is a Laboratory Technician at the Laboratory of Palynology and Archaeobotany of the C.A.A. Giorgio Nicoli, Bologna, Italy, and holds a master's degree in 'Sciences for the recovery and conservation of the Archaeological Heritage' from the University of Modena and Reggio Emilia, Italy. Since 2007, she has undertaken palynological, aerobiological and archaeobotanical research at a local, national, and international level. She is the author of sixty scientific publications.

Corinne Rousse is Associate Professor in Roman Archaeology at the University of Aix Marseille and a member of the Centre Camille Jullian research centre. She directs the French archaeological mission ISTRIE in Croatia with the support of the French School of Rome and the Ministry of Europe and Foreign Affairs. She is a specialist in coastal archaeology and has published on harbour archaeology, maritime villas and economy focused on the Adriatic, Italy and Gaul.

Florian Seiler was Senior Researcher at the German Archaeological Institute in Rome and Berlin. His latest research focuses on urban and landscape archaeology in Pompeii and its hinterland, from the Hellenistic period to the Roman Empire. This included excavations and multidisciplinary investigations and has resulted in publications on Roman houses, town planning and studies on the cultural landscape.

Isabel Rondán Sevilla has a PhD in Ancient History from the University of Cádiz, Spain, is member of the group HUM240 Historical Heritage of Andalusia in Antiquity, and a contracted researcher at the Heritage Geodetection Unit (IVAGRO-UCA). She specializes in the study of Roman rural settlements using non-invasive methodologies.

Simon D. Turner is Honorary Senior Research Fellow in Geography at University College, London. His research focuses on the sedimentological evidence of human activity in aquatic systems using geochemical and paleoecological techniques. His most recent research is on the legacy of the Classic Maya on soil geochemistry, metal contaminants in UK lake systems and the stratigraphic signature of unprecedented global anthropogenic change in the last seventy years.

Dimitri Van Limbergen is Postdoctoral Researcher at the Department of Archaeology, Ghent University, Belgium. His research interests include Roman archaeology and ancient economic history, with most of his work covering Italy and the Mediterranean from Republic to Empire. He is above all an expert in the archaeology of ancient wine and has published extensively on aspects of cultivation and production. His most recent book is an avant-garde edited volume on *Reframing the Roman Economy* (2022).

Andrea Zifferero is Associate Professor of Etruscology and Italic Antiquities, Museology and Management of Archaeological Heritage at the University of Siena, Italy. He is the scientific director of the Rusellae Project (Grosseto) and the Marsiliana d'Albegna Project (Manciano) and leads the ArcheoVino Project at Scansano (Grosseto). He recently published *Materiali per Marsiliana d'Albegna 1: Dagli Etruschi a Tommaso Corsini* (with Mariagrazia Celuzza, 2022).

Gianni de Zuccato is an archaeologist and has recently worked at the Soprintendenza Archeologia, Belle Arti e Paesaggio per le Province di Verona, Rovigo e Vicenza, Italy, directing excavations and research at the 'Villa dei Mosaci' in Negrar (Verona).

ACKNOWLEDGEMENTS

This book was conceived at the international conference *Vine-growing and Winemaking in the Roman World*, convened by the editors and held from 27–29 October 2021 at the Academia Belgica, British School at Rome, and Royal Dutch Institute at Rome. Even though its contents now extend far beyond the material that was presented in-person and online across these three days, without the significant financial and logistical support of those institutions, their directors, and along with the University of Padova and Maria Stella Busana it would not have been possible to embark upon this endeavour. Funding was generously provided by the Academia Belgica as a Symposium Grant and by the British School at Rome. We are also grateful for substantial support from the Structural Determinants of Economic Performance in the Roman World (SDEP) research network, led by Koen Verboven and funded by the Research Foundation Flanders, Belgium, at both the conference and through publication assistance for this volume. Finally, to the contributing authors, who have generously given their work, intellect and time to this aspirational undertaking and for which we are grateful.

ABBREVIATIONS

Where applicable, abbreviations are given in accordance with the *Oxford Classical Dictionary*, 4th edn (E. Eidinow, ed.), Oxford: Oxford University Press.

Ancient Sources

Ath., *Deip.*	Athenaeus, *Deipnosophistae*
Cassiod., *Var.*	Cassiodorus, *Variae*
Cato, *Agr.*	Cato, *De Re Rustica/De Agricultura*
Columella, *Rust.*	Columella, *De Re Rustica*
Hero, *Mech.*	Hero of Alexandria, *Mechanica*
Herodotus, *Hist.*	Herodotus, *Histories*
Hes., *Op.*	Hesiod, *Opera et Dies (Works and Days)*
Hes., *Theog.*	Hesiod, *Theogonia*
Just., *Epit.*	Justinus, *Epitome* (of Trogus)
Martial, *Ep.*	Martial, *Epigrams*
Palladius, *Op. Agr.*	Palladius, *Opus Agriculturae*
Paus.	Pausanias
Pliny, *HN*	Pliny the Elder, *Historia Naturalis (Natural History)*
SHA, *Prob.*	Scriptores Historiae Augustae, *Probus*
Strabo, *Geog.*	Strabo, *Geography*
Suet., *Aug.*	Suetonius, *Divus Augustus*
Symm., *Ep.*	Symmachus, *Epistulae*
Tac., *Agr.*	Tacitus, *Agricola*
Theoc., *Id.*	Theocritus, *Idylls*
Theophr., *Caus. pl.*	Theophrastus, *De causis plantarum*
Theophr., *Hist. pl.*	Theophrastus, *Historia plantarum*
Varro, *Rust.*	Varro, *De Re Rustica*
Virgil, *Aen.*	Virgil, *Aeneid*

Virgil, *Ecl.*	Virgil, *Eclogues*
Virgil, *G.*	Virgil, *Georgics*
Vit., *De arch.*	Vitruvius, *De architectura*

Miscellaneous

GPR	Ground-Penetrating Radar
hl	hectolitre (100 metric litres)
iug.	iugerum / iugera
ORA	Organic Residue Analysis

CHAPTER 1
SCIENTIFIC APPROACHES TO ANCIENT WINE: DEVELOPMENTS, CHALLENGES AND FUTURE PERSPECTIVES

Emlyn Dodd and Dimitri Van Limbergen

Abstract

This introductory chapter outlines the remit and purpose of this book as one designed to survey current methods and approaches, especially those building new trajectories in the study of ancient viticulture and viniculture. We demonstrate the major progress made in methods and tools during recent decades and the potential this holds to balance our knowledge of wine-related activities that are more clearly visible archaeologically and those that are ephemeral and have long gone underrepresented in scholarship. Following an extensive, methodologically themed bibliographical overview of the state of the art, we outline the structure and rationale for this volume and summarize key chapters, which are grouped thematically. Finally, we discuss future directions and challenges for the field, including limitations in perspective generated by a focus on, for example, *amphorae* or production installations, and a need to draw together and synthesize data from often-disparate scientific methodologies.

This book is designed to survey methods and approaches for the study of ancient viticulture and viniculture. It brings together an array of modern techniques that are currently used – as well as those that bear exciting future potential to be used – to untangle the complex history of wine. The origin of its contents lies in, though now extends far beyond, the conference *Vine-growing and Winemaking in the Roman World*, held from 27 to 29 October 2021 at the Academia Belgica, British School at Rome, and Royal Dutch Institute at Rome. By curating a diverse range of approaches in one volume, many of which are rarely in dialogue, we aim to provide a more accessible guide to the scientific methods that form the new bedrock of research on ancient wine, especially in Roman contexts. It is our hope that the case studies, methodologies, debates, and results within can be immediately utilized as a point of reference and inspiration by field archaeologists, scientists and researchers working across disparate disciplines, chronologies, geographies and contexts. Most of all, we aspire for this to be a starting point for future directions – a revisionist agenda for the field – and not a cul-de-sac.

Such a change of course also requires (and indeed enables) us to become more explicit about terminology. Ancient oenological studies often use the terms viniculture and viticulture interchangeably, but they in fact refer to two different things. The term *viniculture* refers to the entire process of making wine. This includes grape-growing and post-harvest processes, such as the treading and pressing of grapes, and fermentation, cellaring and ageing of must or wine. These activities and processes are generally undertaken within the setting of an estate, winery or production facility. On the other hand, the term *viticulture* refers to grapevine management, cultivation and grape production, regardless of whether wine is or is not the final product. This extends through the agricultural preparation of fields, planting and care of grapevines, and harvesting of grapes. It also includes the production of grapes, and secondary *Vitis* products (e.g. vines or leaves), for other commodities, including raisins or table grapes. This distinction is important for the field, as some archaeological datasets allow us to speak of viticulture, but not viniculture (e.g. when there is no evidence for wine production facilities).

A century of research and methodological experimentation

Since the first, almost exclusively text-based attempts in the late nineteenth to early twentieth century, the role of wine in Roman society and economy has been continuously explored using a variety of methods.[1] Studies have steadily included more archaeological data and an expanding array of analytical tools, to evolve from scrutinizing a finite number of ancient texts[2] to recovering and analysing an ever-expanding and variegated set of archaeological and scientific evidence.[3] *Amphorae*, as one of the key transport containers for wine in the Mediterranean region, have contributed vital information

[1]Manzi 1883; Beaurredon 1888; Curtel 1903; Billiard 1913; Dalmasso 1941; Mareschalchi and Dalmasso 1931; Rostovtzeff 1957; Dion 1959. In the eighteenth century, Lazzaro Spallanzani (1798) in southern Italy also utilized ethnographic observations of contemporary viticultural practice in comparison to ancient descriptions, noting substantial similarities. Also see Forbes 1993; Rossiter 1998; Frankel 1999; Lewit and Burton 2019; Dodd 2020a; Lewit 2020.

[2]E.g. Ath., *Deip.*; Cato, *Agr.*; Columella, *Rust.*; Galen, *Galeni opera omnia, Selected Works*; Hes., *Op., Theog.*; Hippocrates, *Diseases, Internal Affections, Epidemics*; Palladius, *Op. Agr.*; Pedanius Dioscorides, *The Greek Herbal*; Pliny, *HN*; Theophr., *Caus. Pl.*; Varro, *Rust.*; Virgil, *Aen., Ecl., G.* The now lost text of Mago the Carthaginian was also clearly an important influence in many later agricultural treatises, with considerable space devoted to viticulture (Dodd 2022a: 444). Many of these earlier Phoenician, Greek and Roman texts formed a foundation for the creation of Late Antique and Byzantine documents that outline wine production processes (e.g. Zeno of Verona, *Tractatus*; Symm., *Ep.*; Cassiodorus, *Expositio Psalmorum*; Pope Gregory, *Dialogi*; *The Geoponika*).

[3]Hirschfeld 1983; Amouretti and Brun 1993; Tchernia and Brun 1999; Brun 2003; Brun, Poux and Tchernia 2004; Ciacci, Rendini and Zifferero 2012; Salviat and Tchernia 2013; Thurmond 2017. Also of relevance are Forbes 1955; White 1970a, 1970b, 1975; Curtis 2001.

since the pioneering work of Dressel, fuelled by the onset of underwater archaeology.[4] This gave rise to a large corpus of ceramic and epigraphic data, refining chronologies and typologies, and later the set-up of specialist databases.[5] Spurred on by advances in quantification and content analysis, alongside developments in the study of kilns and ceramic provenance to unlock mineralogical, chemical and petrographic properties, an enormous amount of data has been generated regarding the production and distribution networks of wine and its carriage in *amphorae*.[6]

Alongside the containers in which wine was transported, another major source of information is production facilities – whether wine presses, treading floors, or storage and cellaring rooms (*cellae vinariae*). Since the seminal works of Rossiter, Frankel, and Brun,[7] the pan-Mediterranean corpus of sites with evidence of some sort of wine production has grown steadily through excavation and survey, and more focused studies have increased our knowledge on their nature, size and function.[8] Production facilities are now one of our main pieces of evidence to evaluate chronologies and scales of local and regional production.[9]

In more recent years, and in tandem with developing scrutiny of the above evidentiary forms, the study of Roman wine has grown into a vastly interdisciplinary field. It now incorporates a wide range of methods drawing from the humanities and sciences that allow us to investigate and reinvestigate evidence in a more holistic fashion. The incorporation of these new techniques has also enabled more confident development of hypotheses about the birthplace(s) and trajectories of wine, beginning in the mountainous regions of Lebanon/anti-Lebanon, the Taurus, Caucasus and Zagros.[10] Archaeobotanical methods to study seed and fruit remains suggest some degree of grape consumption in the Peloponnese and Thessaly of Greece from the early Neolithic (7000–5200 BCE), with

---

[4]See Dressel 1899; Lamboglia 1950, 1952, 1955; Benoit 1954, 1958, 1961; Zevi 1966; Tchernia 1968–70, 1969; Grace 1979; Peacock and Williams 1986.

[5]On ceramic and epigraphic data, see e.g.: Callender 1965. Databases include, for example: CEIPAC (http://ceipac.gh.ub.es/); RTAR (http://publications.univ-provence.fr/rtar/); CAEA – Rhodian Stamps (http://www.amphoralex.org/amphoras/accueil/Amphores.php); Amphorae ex Hispania (http://amphorae.icac.cat); University of Southampton ADS Catalogue (https://archaeologydataservice.ac.uk/archives/view/amphora_ahrb_2005/info_intro.cfm#4).

[6]Demesticha 2015; Ahmed 2019; Leidwanger 2020; Komar 2021. For quantification studies, see Vidal and Corredor 2018. For analyses on *amphora* contents, see Slovenec, Sondi and Crnkovic 1991; Parker 1996; Perez-Arantegui, Paz Peralta and Ortiz Palomar 1996; Remola et al. 1996; Schmitt 1998. For kiln and provenance studies, see Peacock and Williams 1986; Sondi and Slovenic 2003; Mange and Bezeczky 2006; Vitali 2007. For the use of shipwreck sites, see Parker 1990, 1992a, 1992b; Wilson 2009, 2011, 2014. See also the shipwreck database by the Oxford Roman Economy Project: http://oxrep.classics.ox.ac.uk/databases/shipwrecks_database/

[7]Rossiter 1981; Brun 1986, 2003, 2004a, 2004b; Amouretti and Brun 1993; Frankel 1999.

[8]Aydinoğlu and Şenol 2010; Peña Cervantes 2010; Van Limbergen 2011, 2017, 2019; Waliszewski 2014; Dodd 2017, 2020a, 2020b, 2020c, 2022a; Dodd, Galli and Frontoni 2023; Feige 2022; Peña Cervantes 2023. For substantial bibliography drawn from smaller-scale studies and excavations that uncovered evidence of wine production facilities in the Levant and across modern Turkey, see Dodd 2020a: 12–15.

[9]E.g. De Sena 2005; Marzano 2007, 2013; Van Limbergen 2015; Dodd 2020a.

[10]McGovern 2011, 2013; Arroyo García and Revilla 2013.

evidence of vine management in northern Greece potentially as early as the fifth millennium BCE and some sort of small-scale winemaking in the Aegean region by the late Neolithic (*c.* 4500–4000 BCE).[11] Vinicultural practices then increased through the Middle and Late Bronze Age Minoan and Mycenaean cultures and can be seen through wine pressing equipment at, for example, Monastiraki in Crete.[12] Italy appears to have experienced a similar trajectory, albeit perhaps slightly later, and archaeobotany has again played a leading role, with evidence suggesting the exploitation of grapevines from the Mesolithic and grapevine cultivation increasingly likely from the Final Bronze Age.[13] This accelerated from the tenth or ninth centuries BCE, where a variety of forms of evidence – from archaeobotanical remains, production facilities, organic residue analysis, to various implements used in the consumption and transport of wine – illustrate the embedding of wine in what would become Roman culture.[14]

Great steps forward have now been made via multidisciplinary analysis of the manufacture, distribution and use of ceramic, metal and glass containers for wine transport and consumption,[15] the study of harvest and production scenes in art and iconography,[16] and by investigating newly uncovered and legacy archaeological remains of wineries and storage facilities. One of the most prominent methods to be developed and increasingly finetuned over the last forty years is organic residue analysis (ORA), which has enabled the identification of chemical signatures that can indicate the presence of wine (or fermented grape juice) in archaeological materials, including pottery and production facilities.[17] ORA has since helped to clarify whether wine, oil or a mixture of commodities was stored in various containers and produced using certain structures.[18] It also has the potential to detect a wide range of non-grape additives found in ancient wines.[19] Through this, it builds our understanding of mixed beverages (e.g. Etruscan and Greek 'grogs'), which may have been precursors to more pure grape wine, as well as creating a nuanced scientific dataset of flavoured wines to be used in comparison with the substantial ancient literature that provides descriptions for various types of Greek and Roman wine (e.g. Columella, *Rust.* 12.27; Palladius, *Op. Agr.* 11.14-19;

---

11 Valamoti 1998, 2004, 2009, 2015; Valamoti et al. 2007, 2015, 2018, 2020; Garnier and Valamoti 2016; Pagnoux et al. 2021.

12 Fiorentino and Solinas 2006; Sarpaki 2012.

13 Although increasingly early evidence may push this into the Neolithic – see overview in Dodd 2022a.

14 Dodd 2022a: 448–51.

15 On the analysis of glass in particular, see Evershed 2008; Ribechini et al. 2008a, 2008b; Craig, Saul and Spiteri 2020.

16 Bielefeld 1997; Balmelle and Brun 2005; Di Pasquale 2010.

17 Evershed 1993; McGovern et al. 1996, 2004; Singleton 1996; Garnier 2003; Garnier et al. 2003; McGovern 2003, 2013; Pollard and Heron 2008. Examples of its use on ceramics include Condamin et al. 1976; Formenti, Hensard and Tchernia 1978; Rotshild-Boros 1981; Garnier 2007; Pecci 2007; Jerkovíc et al. 2011; Zlateva and Rangelov 2015; Pecci et al. 2017, 2021. The use of organic residue analysis on production facilities is more recent, cf. Middleton et al. 2010; Pecci, Cau and Garnier 2013; Pecci et al. 2013.

18 Romanus et al. 2009; Garnier et al. 2011; Pereira and Silvino 2015; Woodworth et al. 2015.

19 Cf. McGovern et al. 2013 for the presence of rosemary, basil, thyme and pine resin in Etruscan wines.

Pliny, *HN* 14.19). The detection of wine residues in this manner, however, brings its own methodological challenges (in part due to the complexity of the product) and none of the current methods produces unambiguous results.[20] Furthermore, although slowly improving, much of this data comes from periods and societies outside the Roman world[21] and best practice, systematic and integrated residue analysis remains a long-term goal.[22]

Our vinicultural analytical repertoire has also grown through the use of ancient DNA (aDNA) analysis and the archaeobotanical[23] study of grape pip morphometrics – both of which have advanced significantly in recent decades.[24] This is supplemented by key advances in data recording and synthesis alongside specific lab-based techniques (e.g. isotopic analysis).[25] These approaches have furnished insights into the processes of grapevine diffusion, interrelationships between cultivars, gradual domestication of the grapevine, and the spread of winemaking across a broad temporal and spatial range.[26] While bearing immense promise, uptake of some of these analyses within 'Roman' Mediterranean archaeological projects has been slow, and they are historically better applied to pre- and protohistoric sites.[27] As with ORA, the success of biomolecular and morphometric studies depends heavily on favourable preservation conditions as well as controlled and rigorous sampling protocols. High degrees of genomic and physical preservation only occur under waterlogged or desiccated conditions – even charring, which is the most ubiquitous form of preservation, causes deformation due to carbonization.[28]

A more recent development is the introduction of spatial analysis and computational modelling to simulate aspects of vine-growing, such as land suitability, yield and the

---

[20]Evershed 2008; Bernal Casasola 2015; Garnier 2015; Pecci 2018; Drieu et al. 2020; Garnier and Pecci 2021; McGovern et al. 2021.

[21]E.g. Guasch-Jané et al. 2004, 2006; Garnier and Valamoti 2016.

[22]Allevato et al. 2012; Arobba et al. 2014. See the contributions of McGovern and Pecci in this volume.

[23]The study of plant remains from archaeological contexts – referred to as palaeoethnobotany in some North American studies (Lodwick and Rowan 2022: 593).

[24]On aDNA, see Vouillamoz et al. 2007a, 2007b; Vouillamoz and Grando 2006; Wales et al. 2015, 2016; Firmati et al. 2021. Also see the contributions of Zifferero and Basso and Dobreva in this volume. On grape pip morphometrics, see Mangafa and Kotsakis 1996; Perret 1997; Terral et al. 2010; Bouby et al. 2013; Orrù et al. 2013. On the development of archaeobotany more broadly, see Lodwick and Rowan 2022.

[25]Marston, Warinner and D'Alpoim Guedes 2014; Lodwick and Rowan 2022: 598.

[26]Figueiral et al. 2010; Terral et al. 2010; Bouby et al. 2013; Aversano et al. 2017; De Grummond 2018; Guasch-Jané 2019; De Lorenzis et al. 2019, 2020; Fuks et al. 2020; Peña Cervantes 2020; Raimondi et al. 2020; Bonhomme et al. 2021; Del Mastro et al. 2021; Pagnoux et al. 2021; Pérez-Jordà et al. 2021; Trentacoste and Lodwick 2023. See contributions by McGovern and Zifferero in this volume.

[27]Costantini and Costantini Biasini 1999; Menozzi et al. 2002; Mercuri et al. 2006, 2021; Delpino 2007; Mariotti Lippi et al. 2012; Marvelli et al. 2013; Cardarelli et al. 2015; Gismondi et al. 2016; Lentjes 2016; Lentjes and Semerari 2016; Depalmas et al. 2020; Motta and Beydler 2020; Acconcia and Piergrossi 2021; Magris et al. 2021; Marchesini, Marvelli and Rizzoli 2021; Mercati et al. 2021; Moricca et al. 2021; Pini et al. 2021.

[28]Smith and Jones 1990; Ramos-Madrigal et al. 2019; Landa et al. 2021; Lodwick and Rowan 2022: 594–6.

effects of climate.[29] The potential benefits of predictive models are substantial, offering hypothetical scenarios that can enhance our knowledge of viticultural strategies, rural landscape use, and ancient trade and market economies.[30] However, efforts to date remain almost entirely based on guidelines in the ancient farming manuals – itself problematic in many ways – along with modern (thus anachronistic) conceptions of vineyard land selection and yield.[31] Studies with an overly quantitative focus are often also limited by exponentially increasing factors of error each time a new calculation is layered based on hypothetical yields, planting densities, or land sizes. Finally, various aspects of Roman oenology are being explored through experimental archaeology, including the reconstruction of vine-growing and winemaking techniques, and the reproduction of wine taste profiles.[32]

Introducing the present volume

The preceding overview clearly illustrates the enormous extent to which the analytical study of ancient wine, including the methods to extract information from archaeological material, has progressed over the past few decades. This book builds on this trajectory by foregrounding an array of scientific and archaeological techniques now used to study ancient vine-growing and winemaking. The following chapters cover a wide temporal and geographical range, though they are restricted mostly – though not exclusively – to the Mediterranean region and chronologies broadly associated with Roman cultures. The book is arranged into thematic sections, covering biomolecular archaeology and chemical analysis, archaeobotany and palynology, vineyard and landscape archaeology, and computational modelling and experimental archaeology. We acknowledge that there is significant potential overlap between these sections and, indeed, encourage such fluidity, which fosters exactly the multiproxy, interdisciplinary trajectory that is most promising to unravel wine's complex territorial identities and histories.

Following this introduction, Chapter 2 sets the stage by providing context for the modern scientific and archaeological study of wine. Patrick McGovern builds on his long history of fieldwork and laboratory analysis by exploring debates behind vinicultural trajectories including how wine became the dominant beverage of Greek and Roman culture. His particular area of expertise in the field of ORA is brought to the fore, recommending integration of the wide range of analytical techniques now at our disposal and considered and contextualized interpretation of data.

[29]Poglajen and Zerjal 2007; Stubert et al. 2020; Verhagen and Whitley 2020; Bernigaud, Bondeau and Guiot 2021; Bernigaud et al., forthcoming. Also see Bernigaud et al., Martín i Oliveras, and Mclean in this volume.

[30]Goodchild 2009, 2013.

[31]Diederich 2005; Hine 2011; Trevizam 2013; Witcher 2016; Van Limbergen 2022.

[32]Tchernia 1998; Tchernia and Brun 1999; Indelicato, Malfitana and Cacciaguerra 2017; Boulay 2018; Caillaud 2018; Bouvier 2000; Feier et al. 2019; Indelicato 2020a, 2020b. Also see Harutyunyan et al. in this volume.

In Chapter 3, Dimitri Van Limbergen and Pieter Gurdebeke offer a potential research framework through which all kinds of datasets and methods discussed in the volume can be integrated to unravel terroir profiles in ancient viniculture. They introduce the research pathway of 'palaeo-terroir' in Roman wine archaeology as a way to 1) extract more holistic interpretations from the ancient source material; 2) untangle the natural and human ecosystems of which our artefacts and structures were part; and 3) provide renewed guidance to the field in the coming decades.

Biomolecular Archaeology and Chemical Analysis

Chapter 4 highlights aDNA data as one of the most exciting new contributions to this field, with Andrea Zifferero giving particular focus to the Albegna Valley region in Tuscany. The importance of integrated approaches combining archaeobotanical and archaeological study is stressed, which has led to new insights regarding the introduction of Greek cultivars to the Italian settlements of this region in antiquity. Chapters 5 builds from this, while also supplementing McGovern's contribution in Chapter 2, to discuss recent results from ORA undertaken at sites across the Roman Mediterranean. Alessandra Pecci outlines the validity and uses of ORA, with data consolidated by comparative modern experimental testing. Of particular importance are studies that have successfully sampled and tested production facilities, helping to discern those that produced wine, oil or other commodities. The case for this practice is strengthened by Corinne Rousse and collaborators in Chapter 6, who provide a key case study in the use of ORA at a villa site in Istria (Croatia). They illustrate the potential of ORA, not only for deepening our understanding of ancient production facilities, but also – if applied systematically – to shift entire regional agricultural and economic trajectories.

However, as groundbreaking as these analyses are, there remain issues of use and production that require deeper reflection. For example, how many times (or how often) must a facility be used for sufficient trace material to be left for detection in ORA? And, if installations were used for multiple commodities and we detect, for example, both olive oil and wine residues, how might this impact our interpretation of the facility, including the frequency of use for each commodity over its lifespan? Critical reflections such as these are crucial if we want to further the analytical potential of ORA and advance our insights into the spatiotemporal use of production installations.

Archaeobotany and Palynology

Chapter 7 surveys the various types of archaeobotanical evidence that can be utilized to discern aspects of ancient vine-growing and winemaking. Patrizia Basso and Diana Dobreva then build on this by exploring a case study of the Valpolicella area in Italy, where a multidisciplinary project has produced a wealth of archaeobotanical data. They build a strong case for the need for continuous interaction between archaeology, archaeobotany, chemistry and archaeometry to illuminate some of the remaining unresolved queries regarding Roman viticulture and wine production.

In Chapter 8, the authors show how grape pips, pollen, wooden remains of grapevines, and other organic elements (e.g. pomace), can be preserved, recovered and synthesized with additional forms of archaeological and literary evidence to indicate cultivation in the Roman period. Perhaps most importantly, and in the context of northern Italy, they confront the challenging question of how one uses this evidence and at what scale it must be found to discern viticulture for wine production rather than other purposes.

Riccardo Carmenati and colleagues then turn to central Adriatic Italy in Chapter 9 to discuss the application of morphometry to a rich carpological assemblage of 163 grape pips, dating between Late Antiquity and the Early Middle Ages. They clearly illustrate the potential of biometric analyses to reveal surprising results, with most pips belonging to the wild subspecies range. These preliminary data complement other recent work[33] and highlight the important role of wild grapes in wine production and the slow and potentially fluctuating process of grape domestication, including the possibility of reversion closer to wild or feral states during certain socio-political moments.

Chapter 10 highlights palynology as a technique particularly useful in atypical viticultural regions. Antony Brown and collaborators explore the growing body of scientific data now suggesting that viticulture – and perhaps wine production – was practiced on the British Isles following the Roman invasion. The authors also explore the development of new techniques, such as sedaDNA, to recover *Vitis* DNA in the European Alps, as well as the possibility to detect and trace grapevine cultivation through pollen dimorphism – which, if undertaken in tandem with morphometric and DNA analysis, bears real potential to shed light on the evolution of wild and domesticated grapevines.

Vineyard and Landscape Archaeology

Vineyard archaeology is a comparatively less-developed area and one in need of focused attention and advancement. Pedro Trapero Fernández and colleagues in Chapter 11 discuss a series of techniques applicable to the study of viticultural landscapes, such as GIS modelling of conditioning factors and the use of non-invasive geophysical technologies. As they illustrate, the application of these techniques is crucial for the future of vineyard archaeology, but much remains to be done in terms of tracing, analytical refinement and dating to ensure that methods are rigorous and data robust.

In Chapter 12, Florian Seiler turns to the region around Pompeii – a region with exceptional levels of preservation and thus opportunity, but one which simultaneously bears potential to dangerously skew our understanding and apply geographically and temporally specific interpretation too broadly. Seiler does an excellent job of recognizing this as he discusses a recent excavation project at Scafati near the River Sarno, which is the first example of a comprehensive study of viticulture in an Italian alluvial plain

---

[33]E.g. Bouby et al. 2013, 2014.

setting. The multidisciplinary analysis and micro-excavations presented in this chapter enrich our knowledge of Pompeian viticulture, showing how grapevines were successfully grown on the plain outside the city at a large scale.

Christopher Dore and Nicholas Rauh take a slightly different approach in Chapter 13 to illustrate how an assessment of contemporary, modern landscape vegetation can potentially build our understanding of the situation in antiquity. They discuss an innovative methodology using an analysis of multispectral satellite imagery of the western Rough Cilician region of Türkiye. The results of these analytical techniques, which have rarely been tested elsewhere, revealed extensive wild grapevine growth hidden throughout maquis scrub of remote mountainous landscapes in south coastal Türkiye. Dore and Rauh suggest that this is a residual image of the Roman landscape, which has been largely preserved due to limited post-antique habitation until recently and reinforces the ability of the grapevine to thrive in this regional environment, as the ancient literature implies.

Modelling and Experimental Archaeology

The final section focuses on the development of predictive modelling tools and experimental archaeology to produce reconstructions of vine-growing and winemaking. In Chapter 14, Antoni Martín i Oliveras attempts to quantify Roman wine production in the *Ager Barcinonensis*. By applying different methods of calculation, he assesses and quantifies production processes and yield ratios, and, by integrating the number of known winemaking facilities in a particular region, obtains estimated scales of vineyards managed and vines grown on a certain property. This is extended in Chapter 15, where Nicolas Bernigaud and colleagues discuss a range of methods to reconstruct the geography of wine production and model the yields and profitability of vineyards in Roman Gaul and the wider Roman Empire. They address the possibilities and constraints of spatial analysis methods, climate vegetation models and agent-based modelling, and raise an important question of how we might address conflicts between modelling results and archaeological data. The chapter also begins to probe interconnections with climate change, including diachronic comparisons between eastern and western Mediterranean regions.

Andrew McLean turns to the Adriatic and South Etrurian regions in Chapter 16 to quantify agricultural and landscape suitability. His analyses illuminate how different factors – for example, slope, aspect and proximity to infrastructure – affected the distribution of Roman rural sites. The outcome demonstrates, amongst other things, that preferences for slope are fairly consistent across the two regions, but that it appears to have been more important in South Etruria than the Adriatic, where the changeable Bora and Mistral winds made preference for direct southerly slopes less apparent. Finally, Chapter 17 explores the potential for experimental archaeology to deepen our understanding of Roman viti- and viniculture. The authors address the general value of the field to the study of human nutrition, before discussing its application in ancient wine studies by using some specific examples for the Roman world.

Future trajectories

The chapters in this book demonstrate the major progress made by studies in ancient wine over recent decades. The methods and tools discussed here have the potential to greatly advance our understanding of ancient viticulture and viniculture in the Mediterranean and beyond. Many of them have already begun to balance our knowledge of wine-related activities that are more clearly visible archaeologically and those that are ephemeral and have long been underrepresented in scholarship. This maturing of the discipline is undeniable, and a crucial development in itself, but it is also necessary to point out a number of significant challenges that still lay ahead.

First, the historic and current focus on *amphorae* and presses – if logical and necessary – generates major gaps in our reconstructions. The durable nature of ceramic *amphorae* creates an underrepresentation of less detectable containers, such as wooden barrels,[34] and, similarly, easily detectable press facilities are almost always those constructed from stone or brick, leaving those made from wood often unacknowledged.[35] Such a focus also creates bias in our interpretation as more easily detectable remains highlight specific spheres of operation. *Amphorae* were best used for transport over water in ships and barges, but less so for overland transport, where barrels and skins were preferred.[36] Similarly, larger-scale winemaking is clearly visible in stone and brick facilities whereas smaller-scale production may have used less-permanent spaces constructed from wood.[37] Other interpretative issues also exist. *Amphorae* were used for a wide range of other products,[38] and presses were used to process not just grapes for wine but also olives for oil (and likely other products) and their often-fragmentary preservation makes it difficult to discern exact uses.[39] Both are also now known to feature heavily in processes of reuse and recycling.[40] Lastly, despite great progress, many *amphora* chrono-typologies remain unsatisfactory,[41] while exhaustive and integrated press databases, typologies and chronologies do not exist for most of the Mediterranean region.[42] Segregated research will, therefore, almost always result in faulty, or at least incomplete and unbalanced, interpretations.

Second, *amphorae* and presses provide excellent information on production, trade and consumption, but say little or nothing about aspects of vine-growing and grape

---

[34]Baratta 1997; Desbat 1997; Marlière 2001, 2002; Mille and Rollet 2020; Sands and Marlière 2020.
[35]Liverani 1987; Brown et al. 2001; Busana 2002, 2003; Bacchetta 2003; Pesavento Mattioli 2003; Rosada and Turchetto 2008; Van Limbergen and De Clercq 2021.
[36]Panella 1989; Bevan 2014.
[37]Brun 1986; Foxhall 2007; Dodd 2020a: 82, 113, 116; 2022a; 2022b: 164; 2022c.
[38]Bernal Casasola 2015; Pecci et al. 2017.
[39]Rossiter 1981; Brun 1993b; Curtis 2001, 2008.
[40]E.g. Pecci et al. 2017. Also see Pecci and Rousse et al. in this volume.
[41]For example, around the Adriatic: cf. Lindhagen 2009, 2013, 2020; Lipovac Vrkljan 2009; Carre et al. 2014; Lipovac Vrkljan et al. 2017; Lipovac Vrkljan and Konestra 2019.
[42]One exception is the limited press database developed for the Oxford Roman Economy Project (http://oxrep.classics.ox.ac.uk/databases/olive_oil_and_wine_presses_database/).

cultivation, especially relative to environment and climate. This is not helped by the fact that vineyard archaeology as a subfield remains underdeveloped and almost all our current data is the result of chance finds.[43] Many sites are partially excavated or roughly dated, lack modern scientific analysis, and vast areas of the Roman viticultural world remain unexplored. To remedy this, we must develop more robust interdisciplinary, especially non-invasive, strategies to investigate traces of ancient vineyards. Alongside this, despite growing interest in palaeoclimate studies,[44] our current proxies remain too coarse to adequately explore the relationship between climate and viticulture. To do so on a regional scale, we must invest in the capture of records with high spatiotemporal resolution and wide geographical applicability.

Third, a trajectory is now developing of more circumscribed research (in content and method), which fosters increasingly cocooned or siloed studies. This is a risk inherent to any increasingly specialized field that is generating huge amounts of data and opening new possibilities, but it has left the discipline somewhat disoriented with our knowledge at times fragmented and disparate. Research is undoubtedly pursuing important pathways, but to really enhance our understanding future trajectories need to form research programmes that act in concert with one another. If we continue along the current course, research trajectories will become increasingly unbalanced and lacunae ever more obvious. It will also be difficult to answer critical new research questions. Were the production facilities routinely detected in archaeological fieldwork used for wine, olive oil, both, or even other commodities, including *garum*? Were domesticated grapes, wild species, or a blend used to make wine? What form(s) did vineyards take and how did they respond to – and in return impact – their environments? How might we begin to model ancient viticulture to gain an understanding of landscape suitability, quantification and palaeo-terroir? Given the wide ranging, high-resolution scientific toolkit now at our disposal, we can no longer be content relegating these inquiries to prioritize more traditional analysis.

Some of the contributions in this book address these challenges explicitly and propose initial pathways forward.[45] While many cannot be answered through traditional methods of analysis, they also cannot be solved by using one method alone, whether new or old. It is now clear that to truly unravel the history of wine, we must use multidisciplinary approaches that foster a holistic perspective. Therefore, finally, this volume is also a call for concerted collaboration in data collection and analysis but also in research foci and strategies. Computational landscape modelling must be used alongside palaeoclimate studies, archaeobotany and DNA analyses to illuminate links between vineyard structuring, climate, environment and the selection of grape cultivars. Investigations into

[43]E.g. Brun and Pommerpuy 1992; Brown and Meadows 2000; Brown et al. 2001; Brun and Laubenheimer 2001; Marinval et al. 2002; Toupet and Lemaître 2003a, 2003b; Turner and Brown 2004; Boissinot 2009; Volpe 2009; Poux et al. 2011.

[44]McCormick et al. 2012; Harper and McCormick 2019.

[45]Complementing the contents of this volume, other aspects of these issues are also addressed in Van Limbergen, Dodd and Busana, forthcoming.

regional *amphora* production and winery facilities need to be integrated with other archaeological and natural science datasets. Perhaps most importantly, these studies need to be done in a systematic way, with results and methods made easily accessible and databases generated to foster temporal and spatial comparison and thus improve our ability to trace viticultural trajectories across the *longue durée*. By bringing together often-disparate scientific methodologies in ancient wine research, this book aims to set a new benchmark from which such a future can be launched and is a first attempt to unify and align over two centuries of expertise generated by researchers within this field.

Such a future also includes methods and approaches other than those treated in this volume. While we have tried to bring together the most vibrant and exciting fields in the archaeological study of ancient wine today, other promising research pathways and methodologies remain less explored. One example is the heuristic value of comparative historical analysis and ethnographic analogy. This bears great potential to increase our understanding of regional vine-growing and winemaking practices, explaining archaeological datasets and patterns, but it remains ill-tested and under-studied. The use of palaeoclimate proxies to understand the physical characteristics and chronological trends of ancient viticulture in relation to climate and environment is also in its infancy. Crucially, specific aspects of winemaking infrastructure await more systematic scrutiny and typological re-establishment, such as detailed geographical inventories of presses and their layouts, the various press technologies used, or the still under-studied integration of earthenware *dolia* and *pithoi* in Greek and Roman contexts, including their evolution and technical developments.[46] Only by committing persistently to collective, interdisciplinary scientific analysis can we shape this future.

Finally, this volume also illustrates the geographically skewed nature of scientific analysis in Roman wine research. Almost all the chapters focus on Italy and the central–western Mediterranean, except for scientific analysis of earlier cultural material (e.g. Neolithic, Bronze Age, Phoenician) from areas around the eastern Mediterranean, Middle East and Fertile Crescent (cf. Chapter 2). Indeed, while the bibliography of this chapter illustrates that scientific study of vinicultural features is increasing in Hellenistic and Roman Greece, the Levant and Black Sea region, generally it remains vastly underutilized in the eastern and southern Mediterranean despite the importance of wine production in certain periods. Future efforts must consciously be made to embed analysis within projects in these regions to increase our analytical datasets and minimize interpretive bias.

---

[46] cf. Cheung 2021a, 2021b; Carroll 2022.

CHAPTER 2
ANCIENT VINICULTURE: A MULTIDISCIPLINARY HOLISTIC PERSPECTIVE

Patrick E. McGovern[1]

Abstract

With ever-improving archaeological and scientific techniques and methodologies over the past half century, our understanding of ancient Roman viniculture has greatly advanced. A multidisciplinary holistic approach has proved essential in developing working hypotheses and testing them. This approach is illustrated by a review of how viticulture, winemaking, and the 'wine culture' moved from the Near East to the western Mediterranean where they were consolidated under the Roman empire. Organic Residue Analysis, a representative scientific approach, is also reviewed. Technological comparisons between East and West (e.g. stomping vats, winepresses, fermentation in pottery jars, finishing processes and additives) show the conservative character of Roman viniculture, which was spread throughout the empire. Rome's influence lasted for centuries until the present. Experimental archaeology enables us to reimagine what Roman wine tasted like by testing the many variables involved in its production and pointing the way forward for future investigation of ancient viniculture.

Introduction

Before the advent of scientific archaeology,[2] our knowledge of Near Eastern and Roman viniculture was largely limited to texts written by classical authors.[3] Those writers in turn depended on Greek and Phoenician sources, especially the agricultural compendium by the third to second century BCE Punic author, Mago the Carthaginian. His twenty-eight

[1]I thank Emlyn Dodd, Dimitri Van Limbergen and Maria Stella Busana for their invitation to present a keynote address at the conference in Rome, as well as the Belgian, British and Dutch institutes for their support. I learned much in the process of being brought up-to-speed on ancient Roman viniculture in my quest to understand its past and advance its future investigation. I also benefited greatly from my colleagues world-wide, whose knowledge and assistance have been invaluable over the years.

[2]McGovern 1995, 2019a.

[3]cf. White 1970a.

books in Punic (Phoenician) detailed many Levantine vinicultural techniques, such as raisined wine, that were adopted by the Romans. Scholars also culled ancient artwork, especially Greek vase paintings (Figure 2.1) and Etruscan and Roman frescos, for information on viniculture. Egyptian tomb frescos and reliefs (Figure 2.2) could be compared. Because they were produced as much as 3,000 years earlier, they showed how Roman viniculture likely was ultimately rooted in Levantine technology, as was the earlier Egyptian industry.

Artefacts and ecofacts (e.g. grape seeds) might provide other relevant data. Their usefulness, however, was undermined by their general lack of archaeological context and dating. Stratigraphic excavation was not yet practiced, and the art market was not regulated. A relic could well have an uncertain history in its use and how it was conserved. Traditional Italian winemaking, which had likely resisted change since Roman times, provided another avenue for understanding ancient viniculture. The drawback was

Figure 2.1 Dionysus sailing the Mediterranean, after he has miraculously grown a grapevine up his ship's mast and transformed his attackers into dolphins, by Exekias inside a *kylix* of the mid-sixth century BCE. Courtesy of the Staatliche Antiksammlungen und Glyptothek, Munich, no. 8729.

Figure 2.2 Winemaking scene from the Theban tomb of the scribe Nakht, *c.* 1400 BCE. Courtesy of The Metropolitan Museum of Art, Rogers Fund, 1915 [15.5.19e].

discerning the differences between past and present. Together, texts, artwork, artefacts/ecofacts and ethnography provided a first approximation to ancient Roman viniculture. But a more scientific methodology was needed to prevent inaccuracies.

To start with, it should be appreciated that archaeological science is a historical ('soft') science, like geology and cosmology.[4] Unlike the 'hard' sciences (e.g. chemistry and physics), experiments cannot be repeated in the laboratory. Past events are unique, and they are preserved to a limited degree. We infer a working hypothesis from the available data, assuming regularities in nature and human culture. Consequences, if the hypothesis is true, are then deduced and tested and credibility grows with repeated confirmation and ever-increasing levels of probability.

Scientific archaeology is extremely multidisciplinary, drawing upon the hard and soft sciences as well as the humanities. Attention needs to be given to all the available evidence, to interpret it in the most meaningful and holistic way possible. Ancient viniculture, so central to so many cultures, offers the prospect of achieving this goal as few other human activities can.

Principal scientific disciplines

The following approaches are essential in reconstructing ancient viniculture for a specific region and time period:

1. **Precise dating** of the pertinent archaeological contexts and artefacts/ecofacts, especially by radiocarbon determinations.

[4] See McGovern and Hall 2015.

2. Comprehensive **surveying** for ancient vineyards and winemaking facilities by geophysical prospection (e.g. magnetometry and ground-penetrating radar), drones and satellite imagery. Follow-up with test excavations and analyses of recovered materials.
3. Reconstruction of the **ancient environment**, **climate**, **biome** and **microbiome** by excavation and coring, followed by geological, archaeobotanical, archaeozoological, and physicochemical analyses (e.g. soil pH). Apply modern precipitation levels and temperatures as proxy measurements.
4. **Landscape archaeology** to establish relationships between ancient vinicultural features within their physical setting throughout the year, including soil composition, altitude, sun exposure, water availability, humidity, and wind direction and velocity.
5. **Site excavation** by the best stratigraphic archaeological methods available, with special attention to geomorphology, pedology and hydrology. Vinicultural samples for scientific analyses handled with gloves, avoiding contaminants (e.g. plasticizers) and properly stored.
6. **Taphonomy** of vinicultural remains, inorganic and organic, assessed by analyses and known physicochemical properties. Soil samples collected from the same contexts are essential as controls. Careful observation of archaeological contexts for other sources of degradation and disturbance (e.g. insects).
7. **Archaeobotany and archaeozoology**, using standard techniques and flotation of macro-remains, ideally coupled with high-power microscopic observations (e.g. starch, pollen and phytoliths).
8. **Molecular Archaeology**:
    a. **Organic Residue Analysis (ORA)** of putative wine vessels and installations by applying many chemical techniques to identify biomarkers ('finger-print' compounds) that derive from the original natural products.
    b. **DNA analyses** to determine the genetic sequences of the wild Eurasian species (*Vitis vinifera* ssp. *sylvestris*) and its domesticated varietals (*Vitis vinifera* ssp. *vinifera*). Also sequence and identify other microorganisms in the soil that are important in wine fermentation (e.g. yeast) along with any additives (other fruits, herbs and natural products) mixed with the wine.
    c. **Inorganic analyses** of putative wine vessels and artefacts made of pottery, metal, glass/faience, etc. by microscopic examination, chemical analyses (e.g. instrumental neutron activation analysis (NAA), petrography, etc.), to determine their contributions to the detection of grape/wine and vinicultural processes within the archaeological and environmental context (including soil).
9. **Experimental archaeology** employed as proxy for laboratory replication of results to support a hypothesis, based on the available archaeological, botanical, textual, artistic, ethnographic and other scientific evidence. Numerous interacting variables are tested for their effects on ancient vineyard design, winemaking process and wine flavour.

Many of these approaches are applied to specific case studies in the following chapters of this book. When taken together, a holistic, albeit still partial, picture of the beginnings and subsequent development of Roman viniculture can be seen to be emerging. In no small measure, the integration of multidisciplinary techniques accounts for this signal achievement. Much is still left to do, and many more scientifically excavated sites are needed to firm up working hypotheses and round out the picture. Because of the limited archaeological and scientific database that is available to work from, controversies about interpretations and the best methods to use will continue to persist.

Organic residue analysis

One key example of how the application of any scientific technique to ancient vinicultural hypotheses needs to be constantly assessed and refined is provided by an evaluation of recent developments in my principal area of expertise: ORA.

ORA is one of the most promising tools in our scientific repertoire, with great potential in reconstructing ancient viniculture across regions. This discipline has played a pivotal role in the larger field of biomolecular archaeology to identify ancient grape and wine remains. Great strides forward have been made in the use of chemical biomarkers to identify wine, additives and preservatives, as well as methods to elucidate how wine was processed and aged, perhaps even whether it went through gradual fermentation or was intentionally heated and smoked to speed up this process.

Progress over the past several decades has been especially impressive for the Italian peninsula and the emergence and consolidation of Rome and its empire. Pompeii and Herculaneum, whose inorganic remains were preserved *in situ* under volcanic debris, were informative but only up to a point. Organic compounds of archaeobotanical remains were incinerated, but the physical forms of fruits (e.g. grapes) were retained. Artistic evidence, such as frescos, showed how wine was served in home and tavern, and how it was important and integrated into the social, religious and economic life of the city. There are few archaeological sites that have contributed so much from so many perspectives to a detailed understanding of ancient Roman vinicultural expertise and wine culture. Yet, there was much more to know that only better-preserved organic materials from other archaeological sites, together with modern surveying and excavation methods, could reveal. Archaeologists in collaboration with scientific specialists have shown just how much information can be gleaned to illuminate how vineyards were laid out, what specific types of winepresses were employed, what grape cultivars might have been cultivated, and how the grape must was vinified in *dolia defossa* and specially designed cellaring facilities.[5]

[5]Dodd 2022a.

As part of a holistic multidisciplinary approach, ORA has been increasingly used to analyse contemporaneous organic remains from excavated sites in the central and western Mediterranean.[6] However, the results are only as good as the samples to be tested, which are ideally well-provenanced, well-dated, well-preserved, and were not contaminated after their deposition (e.g. by ground water) or post-excavation in the lab or storage facility (e.g. by human handling without gloves or conservation materials). Specific chemical extraction and analytical methods must also be scrutinized for their data precision and accuracy and overall effectiveness in solving important archaeological and historical questions.

Findings should not be accepted at face value and presented as assured results until they have been critically evaluated. For example, chemical evidence has been used to bolster theories of Bronze Age cultivation of the native wild Eurasian grape vine in the western Mediterranean, advocating its possible domestication at that time in contrast to traditionally dominant 'East to West' diffusionist models in which the Eurasian grape was domesticated in a unique event in the mountainous Near East at an earlier period. While there is increasing evidence that the grape was exploited in pre-Roman Italy (especially, in native 'grogs'), there is yet insufficient evidence to argue for extensive cultivation in the Italian peninsula, let alone domestication (i.e. planting and exploitation of the hermaphroditic vine cloned by vegetative horticultural methods) before the arrival of the Iron Age Phoenicians and Greeks. Late Bronze Age interactions with Canaanites and Mycenaeans also appear to have had a comparatively minimal impact, at least as regards the introduction of the domesticated vine. The much earlier domestication in the Near East represents a truly revolutionary process, yet unparalleled for the Eurasian grapevine across its broad geographic range from the Iberian Peninsula to central Asia or among the many other species of *Vitis* worldwide.

Very recently, the chemical evidence for ancient Mediterranean and Near Eastern viniculture, including ancient Italian and Roman investigations, has come under serious scrutiny and scepticism.[7] The critics are generally tied to what might be denoted as the 'Gas Chromatographic-Mass Spectrometric (GC-MS) lipid school' of ORA, based on the primary use of one chemical technique.[8] As already pointed out, however, the very limited archaeological evidence for ancient grape or wine should be assessed from a broader scientific and historical perspective, otherwise we run the risk of becoming overly dismissive of data that are crucial to developing plausible working hypotheses. Such scepticism might then be extended out into the secondary literature, as has already begun to occur in this instance,[9] which might impede the long-term development of the field.[10] It is useful here, therefore, to contrast a variety of ORA techniques to judge how

[6]Indeed, Pecci describes many of these in Chapter 5 of the present volume.

[7]cf. Drieu et al. 2020, 2021; Briggs et al. 2021; Whelton et al. 2021.

[8]Evershed 2008.

[9]cf. Craig 2021.

[10]This has happened in the past with other applications of the natural and social sciences to archaeology (e.g. radiocarbon dating before it was put on a firmer basis by tree-ring calibration).

the discipline might best proceed into the future in carrying out analyses of ancient wine and grape remains.

First and foremost, absolute chemical certainty should not – and cannot – be the goal, especially when based on a single chemical technique. The shortcomings of such an approach have been laid out previously.[11] The upshot of this critique is that a careful weighing of the available evidence from multiple disciplines is required, combining and balancing natural scientific and socio-cultural approaches. Working hypotheses can then be induced and tested deductively.

As regards ORA's role in this process, special note should be given to how tartaric acid (TarA) is analysed, since it is the characteristic organic acid of wine, occurring in large amounts only in the Eurasian grapes of the Near East and Mediterranean regions. Of course, it cannot be used exclusively as a biomarker to detect wine, since it occurs to a limited extent in other natural products of the region such as the pomegranate (*c.* eighteen times less than the Eurasian grape) and is also produced in small amounts by soil microorganisms. Due regard also needs to be given to whether a specific natural product with higher amounts of the acid, such as baobab fruit from Africa, *Viburnum* fruit from the Russian plain[12] or hawthorn fruit from east Asia, might have been imported into the region in the period under investigation.

The possibility has been raised recently that small organic acids, such as TarA, are particularly susceptible to being degraded or contaminated by environmental factors as opposed to small lipid compounds, such as palmitic and stearic fatty acids.[13] In fact, the latter are less soluble in water than small organic acids. But what might be gained by less groundwater leaching and contamination of fatty acids is negatively offset by the weaker binding properties of palmitic and stearic acids (pKa's of 4.75) in a pottery matrix, the usual material of the vessels being tested by ORA, as compared to the enhanced ionic binding to the aluminosilicate structure of clay by a stronger acid like tartaric (pKa 2.98 at 25°C for its first acidic function; note that pKa is a measure of the dissociation of a proton from an acid compound in aqueous solution). The amphiphilic nature of fatty acids also subjects them to movement in both hydrophilic and lipophilic materials, such as soils and especially in poorly provenanced and mixed archaeological contexts.

It should also be noted that the analysis of precipitated tartrate salts, including potassium bitartrate and calcium tartrate, at the bottom interiors of pottery vessels offers excellent prospects of detecting TarA as they are better protected from environmental factors and usually comprise larger amounts of the acid.[14] If the ancient pottery vessel being tested once contained a liquid, this effect will be further enhanced by the absorption of TarA into the pottery matrix where conversion to the insoluble salt can also occur.

[11]McGovern and Hall 2015 contra Evershed 2008; Roffet-Salque et al. 2013.

[12]Drieu et al. 2020.

[13]Whelton et al. 2021.

[14]Calcium tartrate, which often forms in the calcareous geological regimes of the Near East and Mediterranean, has a solubility of only 100 mg/L in water at 20°C – markedly less from that of the acid (1.3 kg/L) and much more comparable to the solubility of palmitic acid (7.2 mg/L).

Especially in dry climates and in well-defined and protected archaeological contexts such as a tomb or well-sealed floor of a building, TarA in a wine vessel will then be better absorbed, bound by ionic forces to the pottery matrix, retained, and preserved over time than will palmitic and stearic acids. A similar argument can be made for the accumulation and preservation of malate and other organic acid salts.

Moreover, if the sample to be tested comes from a narrow-mouthed pottery vessel or, even better, one that was stoppered at some time during its history – as was almost always the case for storage and transport containers like *amphorae* – then the prospects of retention of TarA are further improved.[15] Note, however, that associated archaeological evidence is usually required to make the case for this working hypothesis, such as the excavation of the stoppers in the immediate vicinity of the tested whole vessels and ascertaining that the stoppers have identical diameters to the mouths of the vessels. If possible, it is also advisable to test any visible deposits on the interiors of vessels that might represent wine residues. Careful visual inspection and low-power microscopic examination should be standard practice.

Since the goal of any ORA investigation is to obtain reliable results that bear meaningfully on important cultural, technological and/or historical developments, the all-important starting point should be a careful assessment of the archaeological context of the proposed samples to be tested for their potential in developing and testing working hypotheses.[16] The specific archaeological contexts that the samples came from – whether intact tomb, *in situ* primary settlement feature, or a secondary or tertiary disturbed context such as a refuse pit or building refurbishing – need to be fully detailed and published. Non-primary archaeological contexts are especially problematic, and they should be excluded from any ORA study.

Unfortunately, many GC-MS lipid studies come up short by either not providing such information or not clearly articulating their archaeological rationale for their choice of samples.[17] Because these studies can be based on poorly provenanced sherds and rarely include whole vessels, one cannot independently judge how meaningful the samples are and to what extent supposed biomarkers have been contaminated and/or degraded by groundwater, soil chemistry, microbial activity and other factors. Although most of these studies involve large assemblages of samples, it is then difficult to evaluate the statistical significance of the data in testing and corroborating a working hypothesis.

Soil samples collected near the pottery vessels to be tested, which are contemporaneous with the latter, are the most effective way to monitor the loss and/or gain by groundwater movement, contamination, and/or degradation of the biomarker compounds, especially for longer time spans. Instead of contemporaneous soil samples as the most effective proxy, one study opts for an absolute amount of extracted residue – 5 μg lipid per gram

[15]For examples of unfired raw-clay stoppered vessels that tested positive for TarA from sites throughout the ancient Near Eastern and Mediterranean worlds and dating back to the early Neolithic period, see McGovern 2019a with references to published technical articles.
[16]McGovern and Hall 2015.
[17]E.g. Drieu et al. 2021.

potsherd – 'for a valid interpretation'.[18] But such a relatively small amount is no guarantee of this, as already discussed.[19] Much higher amounts, combined with contemporaneous soil controls, are best and should be emphasized.

As our recent critique has also made clear,[20] Drieu et al. (2020) wrongly assumed that the ancient samples extracted by the Lamuela-Raventós group were done with potassium hydroxide (KOH) before testing for TarA/tartrate.[21] In fact, the Lamuela-Raventós group used only 0.1 per cent formic acid in water/methanol (80:20 vol/vol) in their extractions, which was adequate for detecting TarA/tartrate if well-preserved in sufficient amounts. If KOH had been part of their extraction, it would certainly precipitate out the nearly insoluble potassium bitartrate, which explains the < 0.1 per cent recovery yield when a pure tartaric acid standard was run.[22] Secondarily and separately, the Lamuela-Raventós group did extract their samples using an alkaline KOH fusion reaction, to analyse for syringic acid derived from the red pigment, malvidin. The mistake of using KOH in the initial extraction has been made elsewhere and continues to be employed.[23]

Previous studies give the misleading impression that the full range of published extraction methods were evaluated by Drieu et al. (2020), and only that of Garnier and Valamoti proved to be adequate.[24] Whelton et al. state that 'the difficulties and pitfalls in attempting the detection of wine through ORA of specific "biomarkers" have recently been comprehensively reviewed and discussed in Drieu et al.' (p. 4). They have, however, overstated their case, since the latter study did not evaluate the specific extraction method used in conjunction with the now state-of-the-art method employing liquid chromatography high resolution Orbitrap mass spectrometry (LC-HRMS) or tandem LC-MS/MS.[25] The best extraction method for detecting TarA by this technique was determined to be an ammonium hydroxide solution in water/methanol, followed by ultra-purification protocols. Associated control soils from the specified archaeological contexts went through the same procedure. This approach, which was apparently not available for testing by the GC-MS lipid group, is currently the most sensitive and accurate chemical method for the detection of TarA/tartrate.

Tandem LC-MS/MS was the method of choice in analysing early Neolithic wine from Georgia in the South Caucasus.[26] It represents a model study of the superior capabilities of this method in which a multidisciplinary approach incorporating ORA is integrated with other pertinent archaeological, archaeobotanical, climatic and scientific data. This is

---

[18]Whelton et al. 2021: 9.

[19]Indeed, the arbitrariness of the cut-off point has been revealed by scholars who advocate a much lower limit of only >1 µg/g of pottery (cf. Drieu et al. 2020). Reduced further to >0.1 µg/g in Briggs et al. 2021.

[20]McGovern et al. 2021.

[21]Beginning with Guasch-Jané et al. 2004.

[22]cf. Drieu et al. 2020.

[23]cf. Pecci, Cau and Garnier 2013; Pecci et al. 2020; Drieu et al. 2021; Briggs et al. 2021.

[24]Whelton et al. 2021.

[25]cf. McGovern, Mirzoian and Hall 2009; McGovern et al. 2013, 2017.

[26]McGovern et al. 2017.

not to say that the Garnier-Valamoti extraction method, which avoids using KOH and is well-tailored to GC-MS, cannot be equally effective in identifying ancient grape products, especially wine, at sites throughout Italy and elsewhere, thereby shedding important new light on vinicultural development over millennia.

A more recent paper additionally claims to have discovered a 'robust analytical approach' for identifying ancient Eurasian grape (*Vitis vinifera*) products, including wine, 'with a high degree of confidence'.[27] It argues that if a pottery sherd recovered from the soil of an ancient archaeological site has a tartaric acid (TarA) content exceeding 35 per cent compared to the sum of its malic acid (MA) and TarA, its original vessel contents very likely derive from grape; if less than 35 per cent, the likelihood is nil. In advance of a more detailed critique, the following points should be considered:

1. The MA content of fully matured wild *Vitis* species generally exceeds that of TarA, often by as much as three times.[28] Wild and modern domesticated varietals from the 2022 vintage in Georgia are now being analysed to test that likelihood for *Vitis vinifera*. If true, the cut-off point for the TarA percentage will be considerably decreased, certainly below 35 per cent.
2. To enable consumption, humans would have been motivated to lower the MA content, a quantitatively variable genetic trait, by clonal selection. The composition of the wild grape was adapted to birds, which can tolerate higher acid contents than humans.[29] Depending upon the historical course of viticultural practice, a cut-off above 35 per cent for the TarA percentage might apply to recent European cultivars but not to ancient cultivars on Sicily or elsewhere.[30] The latter probably have considerably lower MA contents than ancient ones.
3. The validity of the TarA–MA ratio likely also depends greatly on the specific grape cultivar, its altitude, whether the grapes were harvested prior to or at full maturation, groundwater movement and/or contamination, soil type and pH, the microbiome (e.g. MA is generally more susceptible to attack by microorganisms than TarA), possible malolactic fermentation whether accidental or intentional, dilution effects due to an increase in size of the domesticated grape, and the relative initial absorption and retention of both acids into the pottery matrix over time.

Such factors, taken together and especially for the distant past, could result in widely varying relative percentages of TarA to MA, essentially invalidating any exact cut-off point.

---

[27]Drieu et al. 2021.
[28]Kliewer 1967; Burzynski-Chang et al. 2019.
[29]Waterhouse, Sacks and Jeffery 2016; Schneider and Troxell 2018.
[30]Drieu et al. 2021.

Beginning and spread of viniculture and 'wine culture'

The preceding review of ORA, as applied to ancient viniculture, highlights how careful appraisals of other archaeological, botanical, historical, textual, artistic, ethnographic and scientific approaches, following a similarly rigorous approach, are needed to advance the field. In what follows, I concentrate on the most basic hypothesis of all: how was viniculture as a holistic entity first established on Roman territory? An understanding of its beginnings will serve as a reference point from which to judge how conservative Roman winemaking might have been or whether it improvised and improved upon the technology. An assessment can also be made of its influence elsewhere as the empire expanded.

To answer this fundamental question, we begin in the Near East. Briefly stated, my working hypothesis posits that the wild Eurasian grapevine (*Vitis vinifera* ssp. *sylvestris*) was first taken into cultivation and domesticated at a single time and place in the mountainous Near East. On current evidence, the general area can be delimited to the mountainous South Caucasus and nearby regions of the eastern Taurus and northwestern Zagros Mountains.[31]

We now know that grape domestication occurred at least by *c.* 6000 BCE during the early Neolithic period. Illustrative of the holistic multidisciplinary approach advocated here, an international project arrived at this conclusion by integrating the results from multiple scientific disciplines, including archaeological excavation, spatial analysis, radiocarbon dating, archaeobotany (low-power microscopy, and starch and pollen studies), biomolecular archaeology (ORA and DNA analysis), ancient climatic reconstruction (from which the Winkler Index for viable winemaking could be calculated) and artefactual studies.[32] Because the domesticated grape (*V. vinifera* ssp. *vinifera*) is hermaphroditic (i.e. male and female sexual organs occur together in the flowers of the same plant unlike the wild species in which male and female are separate plants), it is more easily pollinated. Consequently, it produces more fruit for larger-scale and more efficient wine production.

From these beginnings, a 'wine culture' was eventually consolidated in the mountainous Near East. Such a culture, as exemplified by ancient and modern Georgia in the Caucasus and Mediterranean Italy, is one in which wine becomes the centre of human activities from birth to death, for rites of passage, religious celebrations and offerings, economic activity, political negotiations, feasts with song, dance, and other entertainment like the Roman *convivium*, or simply an everyday meal, possibly accentuated with toasting. Once established, wine culture can be hypothesized to have gradually radiated out in time and space to become a dominant socio-economic, religious, artistic and medical force throughout the Near East and Mediterranean region. This development entailed transplanting the domesticated grapevine and the adoption of vinicultural techniques by other peoples, often with the advice of those who were already adept winemakers.

---

[31]See McGovern 2019a, especially the updated Afterword, which should be consulted for more detailed discussion and bibliography on this topic and others, addressed below.
[32]McGovern, et al. 2017.

For our purposes in tracking the course of wine culture from 'East to West' across the Mediterranean, its arrival in the northern Levant represents a crucial point of departure. Likely adopted there by at least 5000 BCE, it took longer to reach the southern Levant where the wild vine did not grow. Beginning around 4000 BCE, the Jordan Valley and environs were planted with the domesticated vine and began producing wine. Later in the fourth millennium BCE, the probable ancestors of the historic Bronze Age Canaanites of the southern Levant started exporting wine to nearby Egypt, which was also devoid of grapes. Most remarkable is the tomb of the Dynasty 0 pharaoh Scorpion I at Abydos on the middle Nile River, which was stocked for eternity with some 700 jars (4,500 litres) of wine as early as *c.* 3150 BCE. This finding, too, was based on biomolecular archaeological and archaeobotanical evidence.[33]

However, since an Egyptian winemaking industry was yet to be established, from where did the jars and their wine come? Clues to their origin were their unusual shapes, decorations and manufacturing details, which were not local to Egypt. The best parallels came from sites in the Jordan Valley, near the Dead Sea, and in the vicinity of Gaza along the Mediterranean, the closest settled area in the southern Levant to Egypt. Since stylistic comparisons can be misleading, an inorganic chemical technique was applied: NAA. By measuring as many as thirty-five chemical elements, especially the rare earths, and applying powerful statistical techniques, it was shown to a 99 per cent probability that the wine jars from the Scorpion I tomb were made of the same clays as those found in the Jordan Valley, the Hill Country of the West Bank and Transjordan, and the Gaza region.[34]

Correspondence between the stylistic and chemical findings could hardly be coincidental. At this time – the Early Bronze Age I in the southern Levant – most pottery is argued to have been locally produced in small workshops, as was further corroborated by the NAA evidence for the exploitation of local Levantine clay beds. The most parsimonious and straight-forward conclusion was suggested: where the pottery was made was most likely near where the wine inside the jars was made.

With the pharaohs perhaps setting the precedent of importing wine from Canaan, a native Egyptian industry was established in the Nile Delta *de novo* by Dynasty 1, *c.* 3000 BCE. Canaanite winemakers thus likely played a major role in overseeing and advising on vinicultural decisions and techniques. The advanced stage of Canaanite technology is evident from Egyptian tomb frescos and reliefs (Figure 2.2), the earliest known depictions in the world, which began to be produced in the third millennium BCE. We see how the vines were deliberately trained on trellises and pergolas, to protect the grapes from the potentially damaging sun and for easier cultivation and harvesting. Small treading facilities, well attested throughout the Levant,[35] are shown with red grape must gushing

---

[33]McGovern, Mirzoian, and Hall 2009; McGovern 2022.
[34]McGovern 2001.
[35]Frankel 1999; Fischer-Genz 2016; Orsingher, et al. 2020.

from spouts. Stoppered *amphorae*, which evolved from the Canaanite Jar[36] to become the container *par excellence* for ageing and transporting wine, are also illustrated. They bear detailed hieroglyphic wine labels noting the year of the pharaoh and the winery's location, akin to the information provided by the later *dipinti* on Roman *amphorae*.

With this success behind them, the Canaanites from their city-states along the Levantine coast – Byblos, Tyre and Sidon – became superb shipbuilders and seafaring entrepreneurs. In their state-of-the-art 'Byblos ships' made of cedar of Lebanon, they ventured ever farther out into the Mediterranean. According to my working hypothesis, they applied a similar formula wherever they went: import wine and other luxury goods, entice the rulers with the wine culture by presenting them with specialty wine sets, and establish a native vinicultural industry by transplanting the domesticated Eurasian grapevine and establishing a wine industry.

Canaanite seafaring prowess is exemplified by the fourteenth century BCE Uluburun shipwreck, which thus far is the earliest one found in the Mediterranean and has been convincingly argued by its excavators to be a Canaanite ship.[37] Scientific testing demonstrated that the vast majority of the approximately 150 *amphorae* on board contained a terebinth-tree resinated wine.[38] Several jars were filled only with the resin, to use as a sealant or for adding more to the wine as a preservative on its journey. Similar ships were probably already travelling to the nearest island, Cyprus, by the mid-third millennium BCE, moving on to Crete by around 2200 BCE, and to the Greek Mainland probably shortly thereafter.[39]

During the early first millennium BCE, the Phoenicians ventured farther into the western Mediterranean where they founded colonies in Africa, on strategic islands including Malta, Sicily and Sardinia, and along the Spanish coast, even out into the Atlantic to Huelva, near Cádiz.[40] Archaeological remains illustrate that they likely promoted wine culture to local cultures, as they did in Egypt, and fostered the production of wine locally. Widespread Phoenician, and later Greek, activity gave rise to new domesticated grape varietals due to the secondary crossing of the Near Eastern grapevine with local wild grapevines. In the wake of Phoenician trade and colonization, native fermented beverages, especially mixed fermented beverages ('grogs'), such as the Greek *kykeon*, with their numerous ingredients, were marginalized, modified, and displaced by grape wine.[41]

Probably by the tenth to ninth centuries BCE, when their influence was already felt in Spain,[42] the Phoenicians might well have made contact with people in Etruria in central Italy. Archaeology shows the adoption of many eastern Mediterranean technologies,

---

[36]McGovern 2019a: 110.
[37]See McGovern and Hall 2015 with references to published reports.
[38]McGovern and Hall 2015.
[39]Hadjisavvas 2020.
[40]Botto 2013.
[41]McGovern 2009.
[42]cf. Ramón 2010; Botto 2013; Peña Cervantes 2020.

including glassmaking, metalworking, shipbuilding, ivory carving, Royal Purple dyeing, and preeminently wine consumption in special drinking sets and production.[43] These developments followed probable native exploitation of wild grapevines already growing on the peninsula and making a fermented beverage perhaps as early as the Bronze Age (*c.* 1500–1400 BCE), possibly stimulated by contacts with traders from the Aegean region.[44] The Etruscans can be hypothesized to have carried this tradition a step further by drawing upon Phoenician expertise in viniculture.

A prime piece of evidence for this working hypothesis, in my opinion, is that the Etruscan *amphora*, which was eventually adapted by the Etruscans as their container for exporting wine to southern France,[45] is more closely modelled after the Phoenician *amphora* than the Greek one, represented by Massiote and Greek types of the same period (Figure 2.3). While problematic to confirm, similarities in container styles, especially in the early stages of their adoption, might well imply similar contents – in this instance, wine. A comparable case can be made for the Phoenician *amphora* and its culture having influenced the development of the *amphora* and winemaking in their Iberian colonies. The pivotal role of Mago the Carthaginian in providing a foundation for later Roman texts might provide further evidence of this influence.

From Etruria, wine culture passed over into the Roman world, with added Greek elements which increasingly penetrated northwards from *Oenotria* (and Magna Graecia more broadly) at the southern tip of the Italian peninsula.

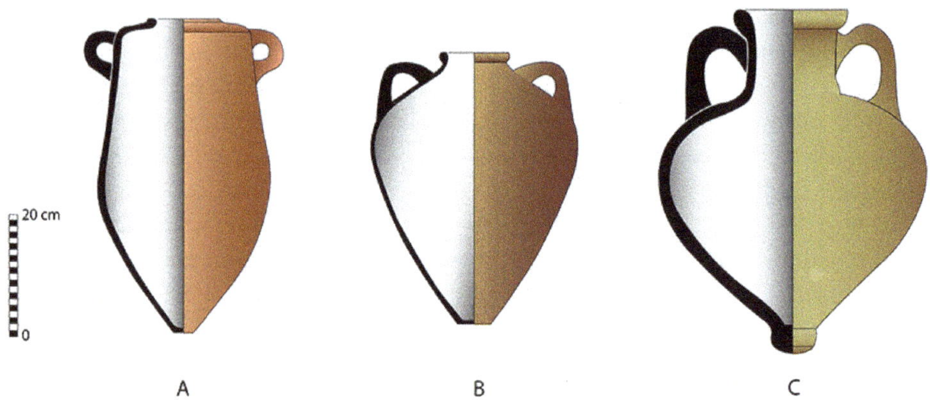

Figure 2.3 Early *amphora* types from the western Mediterranean: (A) Phoenician type A-PUN, *c.* 700–600 BCE; (B) Etruscan type A-ETR 4, *c.* 626–575 BCE; (C) Massaliote type A-MAS 1, *c.* 550–475 BCE. Drawings courtesy of B. P. Luley.

[43] Del Mastro et al. 2017, 2021.
[44] Dodd 2022a: 446–50.
[45] McGovern et al. 2013.

Technological comparisons between the East and West

Specific technological features of the Roman winemaking industry support this general hypothesis that viniculture flowed from 'East to West'. If we examine the Roman vinicultural *chaîne opératoire*, it is remarkable how similar some treading vats, trellising methods, vinification techniques, and wine additives are to those in the East at much earlier periods (cf. Figure 2.4). The Levant had hillside stomping platforms with settling basins perhaps from prehistoric times; Rome had *palmenti*, which span a long prior history.[46] The Near East had buried vinification jars at Chalcolithic Areni in Armenia[47] and Bronze Age Byblos in Lebanon;[48] Rome had similarly subterranean *dolia defossa*. In both regions, again beginning earlier in the East, grapevines were variously grown up trees as was their natural habit, supported on linear trellises and pergolas above ground (Figure 2.4), trained to run along the ground, or pruned into low bushes.[49] Tools that originated in the East (e.g. the pruning hook) were likely carried on ships as trade goods[50] and adopted in Etruria and Rome.[51] Herbal and tree resin ingredients, which were

Figure 2.4 Black-figured vase by the Athenian Amasis Painter (sixth century BCE), recovered from the Etruscan site of Vulci, Italy. It is the earliest depiction in the Greek world of the vinicultural *chaîne opératoire*. Photograph by P. Neckermann, courtesy of the Martin von Wagner Museum, University of Würzburg (redrawn by B. P. Luley).

46 Dodd 2022a.
47 Areshian et al. 2012.
48 Brun 2004a: 59.
49 White 1970a: Figure 2.
50 Delpino 2012: Figure 7.
51 Zifferero 2012; De Grummond 2018.

chemically identified in wine vessels throughout the Near East beginning in the Neolithic period, were also regularly added by vintners to later Roman and Greek wines (cf. Pliny, *HN* 14.19, Columella, *Rust.* 12.27; Palladius, *Op. Agr.* 11.14–19).

By contrast, the differences that exist between the vinicultural technologies, East and West, are relatively minor. For example, while the archaeological remains of facilities show that Roman winemakers, especially in Italy, perhaps carried out their primary fermentation in open-air collection vats, probably followed soon after by continuing the process in partially or fully underground *dolia* so as to free up the collection vat,[52] the Levantine custom was generally to do these operations sequentially in multiple tanks and smaller *amphorae* and jars, such as *pithoi*, fully above ground (Figure 2.2). Both regions, however, completed their secondary fermentations and ageing in above-ground pottery vessels or those slightly embedded into a surface. The minor differences remain to be explained. The Roman (Italian) method was rarely adopted in the Levant, except at King Herod's winter palace whose *dolia* were probably imported from a pottery workshop in Italy.[53]

Additives that went into the wine, which were very important in pharmacopoeias both East and West, also varied only slightly, depending mainly on access to local and imported herbs, spices and tree resins. Diluting wine with water, however, was less common in the East where peoples preferred it neat.[54] Adding seawater to wine, as a form of 'amelioration', was the mark of the best Greek imported sweet whites from the Aegean, especially from Kos (Pliny, *HN* 14.9). The practice probably came via Greek colonization in Italy, since the sweet Levantine wines, which were accorded high praise by classical authors (e.g. Bybline: Hes., *Op.* 589; Euripides, *Ion* 1177; Theoc., *Id.* 14.15; Archestratus of Gela, fr. 59.5; Ath., *Deip.* 1.29B), were not described by them as having been similarly treated.

While the origins and diachronic distribution patterns of mechanical winepresses remain difficult to untangle, they may again point to the Near East; for example, the lever-and-fulcrum concept was already used in Egypt as early as the third millennium BCE for an irrigation device, the *shaduf*, as depicted in Old Kingdom artwork. The ancient Egyptians, and presumably the Canaanites, also had sack winepresses for extracting the maximum amount of juice from the grapes from an early time. Textile sacks were stretched between two poles by teams of as many as eight men and a more mechanical version was constructed within wooden frames, which was operated by several men rotating a stick to twist the sack.[55] It is likely depicted in one of the earliest hieroglyphs of the First Dynasty, appropriately designating a vine, vineyard, or wine.[56]

The earliest winepress that combined the lever-and-fulcrum and the sack concepts was the lever-and-weights type, numerous examples of which were excavated at third

---

[52] Dodd 2020a, 2022a.

[53] Porat et al. 2018.

[54] McGovern 2019a: 235.

[55] James 1995: Figures 13.6–13.10.

[56] James 1995: 200–1, Figure 13.3, Hieroglyph A.

millennium BCE Ugarit in the northern Levantine. Based on current evidence, Brun argues for their exclusive use for olive oil extraction, but Zamora makes a reasonable case for concomitant extraction of grape juice.[57] The lever-and-weights winepress was definitely well established by the second millennium BCE throughout the Levant, and this type continued in use for millennia until the present.[58]

Such advances provided a foundation for other winepress types described by the Roman authors and whose features are corroborated by excavated archaeological remains and artwork. They included the lever-and-drum (or winch), lever-and-screw, and the single and double direct-screw sub-types, among others.[59] Although their origins are uncertain,[60] they might well have been developed in the central or western Mediterranean where they are best represented. The increasing use of the lever-and-screw and the direct-screw types in the East from the fourth century CE might also suggest that these types were first developed in the West and later adopted in the East.[61]

Many other steps in the *chaîne opératoire* might be similarly explored from the perspective of the East–West hypothesis. Among many topics worthy of investigation are the specific origins of wine grapes across regions with a focus on white varietals which are said to have been especially appreciated by the Romans,[62] vineyard layouts, irrigation methods, pruning and grafting of the vines, plant rotation and fertilizing, tools such as large funnels and hoes, and the inception and development of *passito* wine (*passum* in Latin) made from raisins for which Mago the Carthaginian first provided a recipe. One's success in pursuing other relevant topics is sometimes limited by the available evidence. For example, ancient vineyards in the Near East surprisingly have yet to be located for any period. Even where Roman imperial activity is well-documented (e.g. Herodium), the location of the vineyards associated with the winemaking facility are unknown. Surveys and mapping by satellite and aerial reconnaissance, geophysical prospection and excavation are needed, along the lines of model studies carried out in Sicily, the Crimea and around Pompeii.[63]

Conclusion and the future

My goal in this review of ancient viniculture from a scientific perspective has been multifold. I have articulated a working hypothesis of how this highly consequential technology and its associated wine culture probably arose in the Near East and spread

[57]Brun 2004a: 50–8; Zamora 2000.
[58]Lewit and Burton 2019.
[59]Brun 2003, 2004a, 2004b; Frankel 2020.
[60]cf. Lewit 2020.
[61]Lewit and Burton 2019.
[62]Tchernia 1990: 65; Komar 2020.
[63]Sicily: Boissinot 2009; Crimea: Smekalova et al. 2016; Pompeii: Jashemski 1979, 1993. See also the contribution by Seiler in the present volume.

westward across the Mediterranean to Rome. There it blossomed and lay the foundation for winemaking, not just elsewhere in the empire but around the world in the centuries to come, up until today. I stress that this hypothesis is by no means 'proven'. Rather, a multidisciplinary holistic approach, suitable for resolving and testing important questions for a historical science such as archaeology, should lead the way in directing future research. The challenge is to bring to bear the most appropriate and advanced methodologies in the natural and social sciences in concert with those of the humanities.

I believe that the primary 'East to West' hypothesis has thus far been borne out. Rome received its vinicultural knowledge and expertise, together with the transplanted domesticated Eurasian grape, ultimately from the Levant and region of the fertile crescent. For 'Roman' viniculture specifically, the main channel for the adoption of related technology was the Phoenicians via the Etruscans who, in turn, passed it on to the Romans. Greece played a later auxiliary, albeit crucial, role in the process. Roman farmers, winemakers, artisans, engineers, and other craftspeople (e.g., potters and metalworkers) undoubtedly made their own contributions to various stages in the process (whether through irrigation, crop husbandry or mechanical pressing methods), but they do not presently appear to have contributed any advance that was revolutionary and might have had far-reaching effects on the long-standing practice of viniculture.

Any influence going from West to East from the empire as it expanded was comparatively minimal. Local viticulture and winemaking in the Levant and further east continued along the same indigenous lines that it had followed for millennia. Some improvements were no doubt made as a result of new developments in the central and western Mediterranean, such as better cultivation techniques and more diverse pressing operations, but those were at the margins, either too difficult to discern in the archaeological record or requiring more investigation. Only when a ruler such as Herod the Great deliberately imitated Roman practice and custom by probably importing *dolia* made in Italy or using preferred botanicals from there or elsewhere can one say that a more extensive transfer of technology might have occurred.

Much more remains to be done. For example, experimental archaeology, as a proxy for laboratory replication of results to support a hypothesis, can be employed to evaluate the numerous, interacting variables that might be proposed for an ancient vineyard design, winemaking apparatus, or finishing process.[64] Depending on the success or failure of the replicative experiment, a working hypothesis can be gradually refined.[65] The experimental archaeology approach to ancient winemaking can take many forms. For example, you might integrate the available archaeological and scientific evidence for viniculture from a specific region and period with an ancient recipe for a specific wine according to contemporaneous texts and artwork. This is what Jean-Pierre Brun and André Tchernia, the archaeologists, and Hervé Durand, the vintner, did at the Mas des

[64]E.g. Indelicato 2020a, 2020b; Harutyunyan and Malfeito-Ferreira 2022; Harutyunyan et al. 2022. Also see the contribution of Harutyunyan et al. in this volume.

[65]See McGovern 2017a regarding mixed fermented beverage recreations, including grape wine.

Tourelles winery, along the southern Rhone River at Beaucaire. where a Gallo-Roman pottery workshop and winery once stood next to a vineyard.[66]

They based their approach on their extensive ancient vinicultural knowledge and on detailed recipes of the Roman writer Columella (4 BCE–*c.* 70 CE). A massive wine press and Roman cellar were reconstructed following Brun's recommendations, and Roman *dolia* and *amphorae* made according to Tchernia's specifications. Durand chose to grow a recently hybridized grape, Villard Blanc, in ancient Roman fashion on pergolas, single-poles and trees. The grapes were stomped and pressed, juice collected in open concrete vats, and then transferred to and fermented in pitch-lined *dolia*, partly buried in the ground. The finishing touches were to add fenugreek, *defrutum* syrup (grape juice boiled down to half its volume), seawater and lime, with quince, iris, lilac, and other ingredients experimented with in later vintages. The final product, named Turriculae, was aged in *amphorae* in the cellar.

I did a tasting of Turriculae with Tchernia and Durand in October 2002 and was duly impressed by the deep yellow colour and its sherry-like nutty quality, accented by herbs, such as the fenugreek with its own sherry-like caramelized character. In addition to their later experiments, they might also have tried basil, thyme and/or rosemary which were chemically detected in early Etruscan wine.[67] Roman wine, with its many additives, represents the apotheosis of this approach. The assumption of such experimental re-creations of an ancient wine is that human senses and the 'mind-altering' experience caused by alcohol have remained largely constant over thousands of years despite changes in individual and societal tastes. That's a reasonable expectation, even for Palaeolithic hominins with brain structures probably similar to our own.[68]

Ideally, any tasting component of a re-created beverage should be coupled with scientific measurements. Besides compiling the judgements of the best vintners and sommeliers, their assessments need to be calibrated with Electronic-Nose GC-MS in which the taster provides descriptors of the volatile aromatics as they come off the chromatograph and then go into the spectrometer for chemical identification. Much of what we call taste is gleaned from our olfactory sense and similarities and differences for human and machine may turn out to be quite revealing. For example, knowing a wine's colour ahead of time can severely prejudice your evaluation of it, even for the expert taster.[69] Once the different taste variables have been described and identified from one another as much as possible, they can be correlated with the available archaeological, textual, artistic and other scientific data to help in their interpretation and refine additional experiments. Thus, a holistic multidisciplinary scientific approach, again, comes to the fore.

An important investigation still to be done is to compare the olfactory properties of wines vinified for long periods on their lees of skins, seeds and stems in underground

[66]Tchernia and Brun 1999.
[67]McGovern et al. 2013; note the re-creation of an Etruscan 'grog' in McGovern 2017a.
[68]McGovern 2009: 7–8, 23; McGovern 2019c.
[69]Castrioto-Scanderberg et al. 2005.

pottery jars, like the Georgian *qvevri*, with those made in oak barrels.[70] Both Old and New World winemakers have begun avidly experimenting with fermentation above or below ground in pottery replicas of ancient vessels, sometimes importing them from Georgia and sometimes producing them out of local clays and in keeping with distinctive indigenous styles.[71]

From a chemical vantage point, wine made and aged in oak, especially toasted barrels, can be expected to be more resinous in character. Using oak in winemaking is a kind of nod to ancient, resinated wines. Woody flavours and aromas are imparted by lactones, produced from the lipids in oak. Its tannins readily hydrolyse to form acetal, which may yield a pleasing but vegetal taste. When heat is applied to the wood, its lignins break down into compounds such as vanillin and other phenolic aldehydes, which are often lamented in over-oaked wine. Oak's high carbohydrate content is converted into caramelization products by heating. By contrast, pottery and other ceramics as inorganic materials are relatively neutral chemically. But they can contribute a different range of flavours and aromas from those of oak, owing to the porosity and ionic structure of specific tempered clays with metal ions that undergo oxidation-reduction, polymerization and enzymatic reactions, many of which are yet to be characterized. Stainless steel and most other metals have little or no effect on the final product.

Anyone who has tasted an *amphora* or *qvevri* wine will have been struck by how unusual its flavours and aromas are compared to the standard wines made according to Western 'industrial' standards. They are worlds apart. You might be captivated by the new or nauseated by the old method. Yeast and grape variety can add to your confusion, especially if the cultivar is not one of the twenty-five or so varieties, usually of French derivation, that are most commonly planted. If the yeast and other microorganisms have taken up residence in the pottery pores and cannot be easily washed out between fermentations, then they may contribute their own blend of indigenous flavours (and perhaps off flavours) to the mix.

Because of the millennia-long development of viniculture in the East, which then spread westwards, we now have 5,000 or more grape cultivars, with their seemingly infinite varietal range of tastes. They all ultimately derive from the Eurasian species (*Vitis vinifera*) of the mountainous Near East transplanted or crossed again and again as the wine culture moved from 'East to West,' and then spread out to the rest of the world. While our understanding of these developments is still limited, our increasing ability to deploy archaeological, historical and scientific techniques, based on a holistic multidisciplinary approach, promises to fill in the gaps in our knowledge. This volume represents a major step forward in that direction.

---

[70]cf. Grigalashvili and Khomasuridze 2021.

[71]Barisashvili 2017; Glonti and Glonti 2013.

CHAPTER 3
INTRODUCING PALAEO-TERROIR: A FRAMEWORK TO STUDY THE GEOGRAPHY OF WINE IN THE ROMAN WORLD

Dimitri Van Limbergen and Pieter Gurdebeke[1]

Abstract

This chapter introduces the research approach of 'palaeo-terroir' in Roman wine archaeology as a novel, interdisciplinary way to explore the relationship between man and nature in vine-growing and winemaking. In this first-time discussion of the concept, we focus on the potential of ancient literature and archaeology to inform us on palaeo-terroir aspects, and then on the use of geophysical, archaeobotanical and palaeoclimate methods and datasets for building site-based and regional wine geographies in the Roman world. In doing so, we argue for the potential of palaeo-terroir to groundbreakingly deepen and upgrade our reconstructions of ancient wine landscapes and economies.

Introducing 'terroir' in Roman wine studies

Wine is a complex product. Throughout its multi-stage production, wine is shaped continually by a multi-layered and multi-scale concoction of natural and human components that in regionally distinct ways interact to create unique wine landscapes and economies. Vine-growing is a highly elastic and versatile branch of agriculture, outlined to a large extent by the vine's extreme sensitivity to (changes and variations in) environment and climate. Indeed, natural parameters influence entire wine economic structures, from physical territories over the organization of farms and production lines to whole industries. They dictate vineyard site selection, layout and management, as well as the choice of grape cultivars, and so condition the human expenditure of resources, time and energy in the set-up (e.g. training systems) and maintenance (e.g. drainage and

[1]The concept and methodology of palaeo-terroir were originally conceived by Dimitri Van Limbergen, who is responsible for the majority of the text. Aspects of geology and pedology in particular were further refined in collaboration with Pieter Gurdebeke. This paper has benefitted much from long and fruitful discussions with Philippe De Smedt, Wieke De Neef and Lieven Verdonck (on geophysics), Andrea Borsato, Silvia Frisia and Andrea Columbu (on palaeoclimate and speleothems), Koen Deforce (on archaeobotany in general, and pollen and phytoliths in particular), and Davide Cristoferi (on comparative history), whose kindness in sharing knowledge and expertise is immensely appreciated.

irrigation infrastructure) of vineyards. As such, this dynamic and regionally variegated interaction between nature and man strongly reverberates in land use and productivity patterns, and much regulates wine economies.[2] Especially in the pre-industrial past, wine had an important role in mobilizing and organizing local communities. Vine-growing strategies contributed to the moulding of land configuration practices and tenurial systems, and vice versa, while winemaking customs had a decisive influence on the size, design and location of production infrastructure. Wine as a food product is also a true chameleon, subject to frequent change in quantity, quality and its sensorial features (look, taste, smell), driven in part by socio-cultural dynamics, dietary habits, commerce and trade.

On a larger scale, in an expanding and consolidating Roman world – in which wine became a core drink, food and cultural product – the geography and the course of production were further regulated by mechanisms of inter-regional connectivity, socio-material acculturation and glocalization, power politics, government policies and ensuing economic developments.[3] The effects of these processes get even more pronounced when considering the primacy of sea transport, and so the strong clustering of people and activities on the coast, thus strongly encouraging the littoralization of economic, agricultural and viticultural activities in the ancient Mediterranean.[4]

In sum, wine landscapes as constructed realities reflect deep cultural resonances of the relationship between humans and the places they inhabit. In contemporary oenology, this dynamic relationship is referred to as 'terroir'. The term was officially established in the French Renaissance, meaning 'land' or 'soil', and in oenology embodies the notion of a deep link between wine characteristics and the physical features of a vineyard.[5] Terroir as a popular explanatory label for wine quality has become a heavily contested term in modern viticulture.[6] In its original sense, however, terroir unambiguously refers to the impact of environment and climate on vineyard layout, vine-growing practices and grape phenology.[7] In addition, in archaeology, we perceive terroir configurations as profoundly reflecting economic, cultural and political spheres and developments on various interacting levels, which can be explored through the cultural material aspects of past landscapes.[8]

This paper connects to this vision and introduces 'palaeo-terroir' in Roman wine archaeology as a framework to investigate how wine agricultural systems and landscapes were shaped by the firm interaction between nature and man. In what follows, we argue not only for a key role of palaeo-terroir in correctly and fully understanding Roman

---

[2]Jones and Davis 2000; Jones et al. 2005; Jones 2007; Jackson 2008; Bucelli and Costantini 2009; Jones and Webb 2010; Malheiro et al. 2010; Santos et al. 2012; Van Leeuwen and Darriet 2016.

[3]Costantini et al. 2012, 2016; Conca Messina et al. 2019–20.

[4]Braudel 1949; Horden and Purcell 2000; Broodbank 2013; Manning 2018.

[5]Turner and Greasy 2003; Parker 2008; Robinson 2015; Johnson and Robinson 2019.

[6]Winkler et al. 1992; Goode 2006; Dougherty 2012.

[7]Laville 1990.

[8]Cf. The geographical study of contemporary viticulture (Dougherty 2012).

wine economies, but also for its major potential in aligning and steering current and future research, in terms of geography, datasets and methods, as such establishing a much-needed revisionist agenda for Roman wine studies.

Perceiving palaeo-terroir in literature and archaeology

The basic role of soil and landscape in vine-growing was already recognized by the philosopher Theophrastus in the fourth to third century BCE,[9] but the notion of a close relationship between wine, vine and environment found its first full expression in the farming manuals written between the second century BCE and the first century CE (e.g. Cato, *Agr.*; Varro, *Rust.*; Columella, *Rust.*; Pliny, *HN*; Virgil, *G.*). Their accounts – accompanied by a rich collection of iconographic evidence[10] – reveal a wide variety of vine-training systems (Varro, *Rust.* 1.8; Columella, *Rust.* 4.12; 16.3–4; 17; 19; 26; Pliny *HN*, 17.35.164–166) (Figure 3.1),[11] clearly adapted to various topographical and environmental circumstances, as well as to the use of different cultivars (Pliny, *HN* 14.4).[12] They also attest to the existence of both monoculture and polyculture cultivation practices (Col. *Rust.* 3.5; Pliny, *HN* 17.35.171) – as corroborated by archaeology[13] – often linked to specific landscape settings and/or socio-economic needs.[14] In a few exceptional cases, combining this information with archaeological and comparative data allows for capturing proper palaeo-terroir traits. So, for example, while many of the flatter, low-lying coastal areas of Tyrrhenian and (northern) Adriatic Italy were perhaps not ideal in terms of land capabilities for viticulture, the economic advantages of these locations were undeniable, and so a frequent attraction pole for investment and anthropic modifications. Current research on the *arbustum* in Roman Italy is, we believe, a case in point. Indeed, it is becoming increasingly clear that these plantations – made up of cereal fields that were subdivided by long rows of trees serving as living supports for vines – were the most favoured vine-growing systems of the coastal (and inland) plains and wetlands, perfectly adapted to deal with (or at least mitigate) the detrimental effects of flat and watery environments (according to modern standards not suitable for viticulture), and sustained by an ingenious system of land drainage and division schemes of the type we often refer to as Roman centuriation (Figure 3.2).[15]

In most cases, however, the interpretation of these texts – empirical, anecdotal and much constructed from the situation around Rome and central/northern Italy – remains

---

[9]Theophr., *Caus. pl.*

[10]E.g. Bielefeld 1997; Balmelle and Brun 2005; Di Pasquale 2010.

[11]White 1970a: 231–6 still offers the best overview.

[12]Van Limbergen 2015.

[13]E.g. Boissinot 2009; Volpe 2009.

[14]E.g. Van Limbergen et al. 2017; Van Limbergen 2022.

[15]Van Limbergen 2020b.

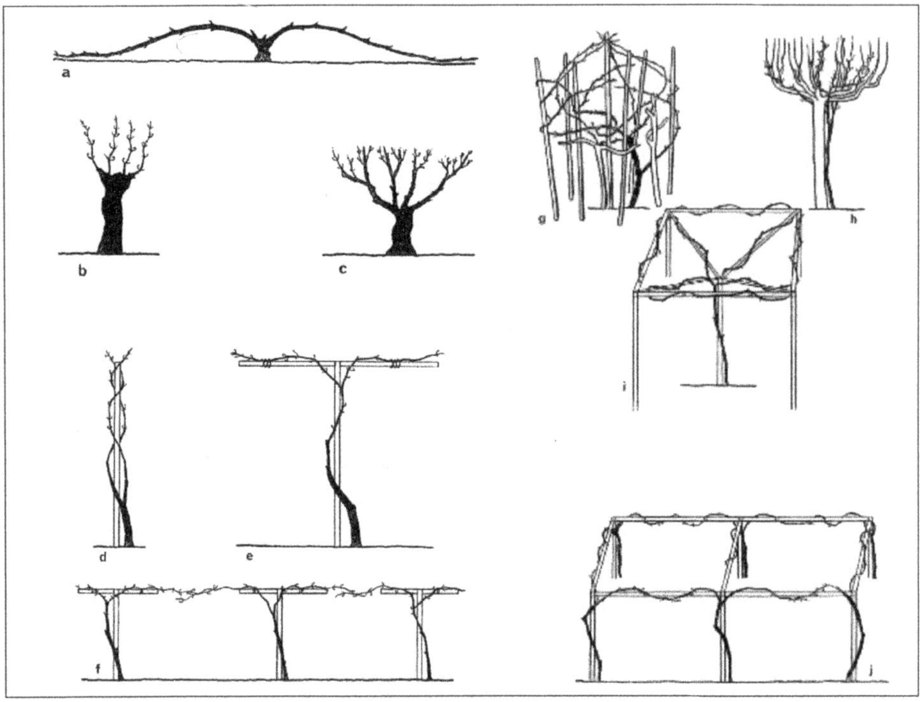

Figure 3.1 Roman vine-growing systems, as described in ancient agronomic literature: a) trailing vine (*vitis prostrata*); b) branched vine (*vitis bracchiata*); c) headed vine (*vitis capitata*); d) single-pole vine; e) yoked vine (*vitis iugata*); f) continuous yoked vine; g) palisaded vine (*vitis characata*); h) tree-wedded vine (*arbustum*); i) roofed vine (*vitis compluviata*); j) pergola. After White 1970a: 234, Figure 2.

problematic in many ways[16] and their understanding much biased by modern anachronistic views on viticulture.[17] So, despite their fundamental importance as sources of information, on their own they can only offer sketchy and distorted visions of vine land use and wine terroirs in the Roman world. Archaeological datasets hold much more potential to explore the various aspects and expressions of palaeo-terroir at work, but – as we discuss below – current knowledge needs to be much better and differently organized, and explored in more interdisciplinary and systematic ways, to realize this paradigm shift in Roman wine studies.

Figure 3.3 presents a schematic view of the various factors and processes at play in palaeo-terroir dynamics to show how they modelled wine geographies in the Roman world throughout Republic and Empire, and by different levels of micro- to meta-influence. This elucidates the complexity of such configurations, but also the role that archaeology can have in reconstructing their various constituting elements and

[16]Diederich 2005; Hine 2011; Trevizam 2013; Witcher 2016.
[17]Van Limbergen 2022.

Figure 3.2 Remnants of alberata/piantata vineyards (the eighteenth- and nineteenth-century descendants of the Roman *arbustum*) near Ocriculum (Otricoli, Umbria, Italy). Photographs by D. Van Limbergen.

mechanisms. The most direct palaeo-terroir proxies on micro- and meso-level consist of wine production facilities (treading floors, presses, vinification/fermentation features, storage rooms with *dolia*, etc.) and vineyard traces. Some of the problems associated with these kinds of evidence have been highlighted elsewhere in this volume,[18] others are elaborated below. On macro-terroir levels, concerning the interplay with local, regional, and inter-regional factors of trade and consumption, a big role is reserved for transport *amphorae*, which help in reconstructing (above all) long-distance commercial networks.[19]

[18] See the contribution by Dodd and Van Limbergen in this volume.
[19] E.g. Van Limbergen 2018.

Archaeological town remains and rural settlement patterns, however, are other important datasets, as they can offer some structural and demographic guidance (through archaeological reconstructions of population scenarios) to explore the role of domestic (urban and rural) consumption in regional wine trajectories, and in particular their impact on the vitality of export activities.[20] Finally, on meta-levels, wider pan-Mediterranean and pan-Roman institutions and infrastructures come into play to explain specific and general wine developments against the background of processes of expansion, integration, and transformation from Republic to Empire.

Before we proceed to outlining some of the developments in data and methods required to properly build palaeo-terroir profiles – in archaeology but especially in palaeoenvironmental and palaeoclimate aspects of terroir (Figure 3.3, right) – and to enable the use of palaeo-terroir as a research agenda, it is useful to briefly discuss the reflection of terroir dynamics in archaeological datasets in two examples, as to illustrate the massive potential for new and more grounded interpretations.

First, the layout and technological characteristics of press installations were clearly linked to regional environmental circumstances, and this heavily influences their traceability and reconstruction potential. So, in the North Syrian Limestone Massif, the thousands of presses scattered throughout the barren landscape and cut out in the rock and/or built with local limestone blocks, are a clear expression of the local geological conditions, as was the massive use of stone rollers in this area (and other parts of the Mediterranean where stone was amply available on-site or in the immediate vicinity) (Figure 3.4).[21] This contrasts markedly, for example, with the northern Adriatic, an area famous for wine according to the ancient texts, but where archaeology has largely failed to record the presence of winemaking facilities. Part of the problem may indeed be related to the organic nature of the latter installations, which were more easily made from perishable wood in these densely forested areas.[22] A similar argument has been made for the Var region in southern France.[23]

Second, one of the most characteristic elements in Roman vinification, if still massively underexplored, was the use of earthenware jars (*dolia*) for fermentation, ageing and storage. Wine cellars filled with these large globular pots are ubiquitous in Italy and much of the Roman world (Figure 3.5).[24] Recent archaeometric[25] and comparative oenological work[26] has revealed how these containers not only were the product of specialized ateliers, made with specific clays by skilled artisans, but also that their shape, structural porosity, and position in the ground (often semi-buried) contributed decisively to wine quality and its visual and sensory features. This strengthens the case for a deliberate use of these jars

[20]E.g. Van Limbergen and Vermeulen 2021.
[21]Van Limbergen 2017.
[22]Liverani 1987; Busana 2002, 2003; Bacchetta 2003; Pesavento Mattioli 2003; Rosada and Turchetto 2008.
[23]Brun 1986.
[24]E.g. Feige 2022.
[25]Manca et al. 2016; Carrato 2017a; Trojsi 2017; Carrato et al. 2019; Carroll 2022; Montana et al. 2021.
[26]Van Limbergen and Komar 2023, forthcoming.

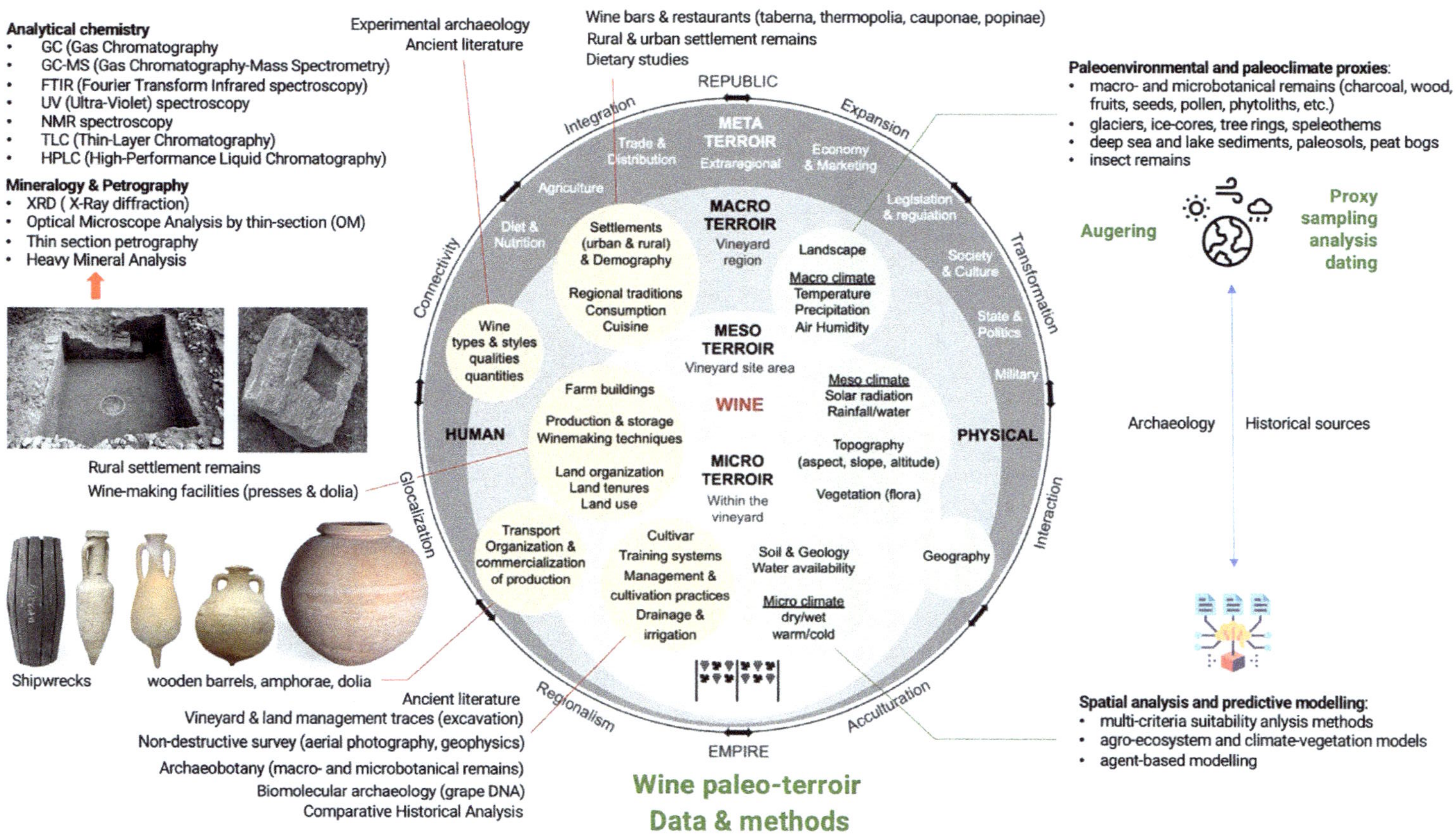

Figure 3.3 Paleo-terroir scheme: data and methods for the Roman world. Diagram by D. Van Limbergen.

Figure 3.4 Wine press with stone rollers in the village of Behyo, north Syria. After De Voguë 1865–77, pl. 113.

Figure 3.5 The *dolia defossa* wine cellar of Villa Regina at Boscoreale, Italy. Photograph by E. Dodd, courtesy of the Ministero della Cultura – Parco Archaeologico di Pompei.

in Roman winemaking, aimed at obtaining quality wines with specific characteristics. Further in-depth scrutiny of the *dolia* phenomenon in antiquity – through, for example, quantitative and shape/setting analyses, archaeometry, organic residue analysis and ethnoarchaeology – would greatly advance our knowledge on ancient wine types and styles and elucidate a crucial piece of the terroir puzzle.

Developing palaeo-terroir as a research agenda: The work ahead

The apprehension that the building of past wine landscapes and economies was fundamentally structured by the constant interplay between man and nature makes the unravelling of palaeo-terroir dynamics key to understanding Roman (and ancient in general) wine economies. It also makes us realize how little we currently capture of their core mechanisms – apart from a few notable exceptions – through archaeology and ancient texts. Developing the territorial and geographical study of Roman wine in the coming years starts above all with a mental shift, by seeing the world of Roman wine through the lens of palaeo-terroir. The proper building of palaeo-terroir profiles as interpretative frameworks, however, requires a multidisciplinary research agenda that concerts the merging of archaeology and history with earth and environmental sciences. Such an endeavour should of course integrate all methods and approaches discussed in this volume, but also systematically include palaeoenvironment and palaeoclimate studies, as well as comparative history and ethnography. Establishing palaeo-terroir as a research pathway also demands the development of new approaches and research lines on a regional scale, and often a renewed systematization and optimalization of archaeological datasets on a pan-Roman level, primarily winemaking facilities (e.g. presses, treading facilities, wine cellars). Below, we outline some of the main actions that are required to start realizing this next paradigm shift in Roman wine archaeology.

Data systematization: Presses

One of the most traceable and informative – if yet underexploited – proxies for reconstructing wine terroirs come in the form of winemaking facilities such as grape treading floors, presses and (*dolia*) wine cellars.[27] These facilities are an ideal proxy for studying the diachronic evolution and nature of wine production in the ancient world. Archaeology has already hinted at the research potential of these facilities, as regional studies and field surveys have shown both the extent of the archaeological remains detectable even without excavations and the density of pressing installations in areas intensively devoted to viticulture in antiquity.[28] Still, ancient presses, despite being one of

[27]Rossiter 1981; Amouretti and Brun 1993; Brun 2004a, 2004b, 2005.

[28]Peña Cervantes 2010; Van Limbergen 2011, 2017, 2019; Dodd 2020a, 2020b, 2022a; Feige 2022.

the most valuable and representative indicators of wine production, remain massively underexploited as proxies for capturing the nature, scale and organization of local wine production through time and space, and for deepening our understanding of change, diversity and innovation in ancient wine technology.[29]

Incomplete and unbalanced inventories, along with a chronic and consistent lack of expert analysis, profoundly hamper the huge potential offered by presses for making broader learned assumptions on the nature and the development of the ancient wine economy. One major issue is the limited uniformity in typological classification and a lack of detailed systematic geographical and chronological overviews. Another is the recurrent difficulty of discerning between wine and olive oil presses, either because of fragmentary archaeological knowledge or a lack of in-depth, comparative cross-regional studies.[30] Finally, there is the sheer quantity of the archaeological data available, which has in the past obstructed the creation of much-needed systematic overviews.[31] The building of a new, palaeo-terroir-framed database for wine (and olive oil) presses, designed to deal with the kinds of questions and issues raised above, is a key step towards implementing a more territorial and geography-inspired research strategy in ancient wine archaeology. Such a huge undertaking is necessarily a long-term research community effort, and important steps are currently being taken for a multi-institutional-led pilot project.

Vineyard archaeology: Geophysical prospection, pollen and phytoliths

As crucial as they are, presses (and *amphorae* for that matter) cannot tell us how and under which conditions the vine-growing activities that nurtured their building (and making) took place. Vineyard studies are the only way to properly disclose the territorial expression of ancient wine economies, but vineyard archaeology remains unequally developed across the Roman world – current data are largely the result of chance finds in (central) Italy and France and are often difficult to interpret because of their fragmentary nature[32] – and a standard non-invasive research strategy able to systematically investigate their traces in the landscape is lacking. Such strategies need to be devised explicitly to gain accurate and deep understanding of ancient vine-growing and their environments on site- and region-level.

---

[29]Some recent studies have started to address these issues for specific regions, e.g., Marzano 2013; Van Limbergen 2015; Dodd 2022a.

[30]Brun 1993; Curtis 2001.

[31]The basic wine and olive oil press database developed by the Oxford Roman Economy Project, while pioneering in its nature and aims, originally included only sites with archaeological evidence for two or more presses, ignoring innumerable single press installations, and then only for parts of Italy and Spain. Later versions included more areas of the Roman world, but none of them exhaustively (pers. comm. A. Marzano) (http://oxrep.classics.ox.ac.uk/databases/olive_oil_and_wine_presses_database/).

[32]E.g. Brun and Pommerpuy1992; Brown and Meadows 2000; Brown et al. 2001; Brun and Laubenheimer 2001; Marinval et al. 2002; Toupet and Lemaître 2003a, 2003b; Turner and Brown 2004; Boissinot 2009; Volpe 2009; Poux et al. 2011.

The most promising approach to this purpose, which is discussed here, would be to scan large areas around Roman farms for traces of ancient plantations by geophysical prospection, thus producing high-resolution spatial data that can then be integrated with focused coring and pollen sampling (palynology), controlled and complemented by test-pit excavations for further archaeobotanical (especially phytoliths, see below) and material culture sampling and dating.

Geophysical prospection

Of all sensing techniques – including remote LIDAR, aerial photography and satellite imagery – geophysics provide the best chance to detect past ephemeral traces with low topographic expression, and thus to trace the subsurface remains of ancient vineyards. Archaeological geophysics is an exhaustive term to denote a series of non-invasive, ground-based survey methods able to detect buried features based on their contrasting physical properties in relation to the surrounding soil.[33] Geophysics and archaeology share a long history,[34] and over the years the use of near-surface scanning techniques in archaeological prospection has developed from a mostly auxiliary method for excavation to a proper investigation method able to address a wide range of topics in both urban and rural contexts in new, non-intrusive ways.[35] Geophysical techniques dramatically improved in the 1990s, and especially in recent years the field has seen major advances, with instrumentation continuously improving in sensitivity, spatial resolution and data acquisition speed, and new multi-sensor arrays that allow for covering rapidly increasingly large tracts of land. These developments have significantly improved our ability to investigate various, even ephemeral traces of human activities over large surfaces and in hard-to-access terrain.[36] Through purposely integrated and interdisciplinary approaches, this now offers unprecedented possibilities for the understanding of past vineyard- and other agricultural-landscapes.

In Roman archaeology, geophysical prospection has yet mainly been used to investigate 'positive' structures such as walls, floors, and other building features.[37] They can, however, also be employed to study 'negative' structures like pits, holes and trenches,[38] and this has been shown extensively in prehistoric archaeology.[39] It is clear from literature and archaeology that Roman vintners planted their vines in either circular or quadrangular pits (*scrobes*), elongated furrows (*alvei*) or trenches (*sulci*), and these features are theoretically, based on contrasts in soil properties, detectable in geophysical imagery.[40] The same goes for associated land structuring traces, such as posthole-fills of wooden vine-training systems (described in literature) or drainage

[33]Scollar et al. 1990; Clark 1996; Gaffney and Gater 2003; Linford 2006.
[34]Wynn 1986; Barker 1998.
[35]Campana 2016; Deiana et al. 2018.
[36]Orengo et al. 2020; De Smedt et al. 2021.
[37]E.g. Vermeulen et al. 2012; Verdonck et al. 2020; Hoffelinck and Vermeulen 2021.
[38]Bitella et al. 2015.
[39]E.g. De Neef 2020; De Smedt et al. 2021; Hegyi et al. 2021.
[40]Tol et al. 2020.

channels in wetter areas. A particularly pertinent matter is the use of vine agroforestry systems in which vines were trained on rows of trees; a system amply discussed in ancient agronomic literature for central and northern Italy, but yet virtually absent in archaeology.[41] While the application of geophysical techniques to target rural dwellings and production sites, in particular *amphora* kilns and their associated buildings, is now fairly common,[42] relatively few geophysical surveys have shown how buried traces of agricultural fields can be revealed and the use of such methods to map traces of ancient viticulture and arboriculture is incipient. A few cases exist for Italy[43] and the Crimea,[44] but the latter are of ancient Greek date and exceptionally well-traceable by their surrounding stone walls, while the Italian data remain poorly exploited and published. Neither the methodology nor the massive informative potential of geophysics for exploring ancient vineyard sites have thus been properly and systematically dealt with.

It is therefore no exaggeration to state that the contribution by Trapero et al. to this volume represents a proper milestone in this field. The reader is referred to their chapter for a discussion on the use of geophysics in the sandy, gravelly and calcareous soils of Baetica. Here, we offer some general considerations and recommendations on the use of three major geophysical survey methods: 1) electromagnetic induction (EMI), 2) magnetometry (MAG) and 3) ground-penetrating radar (GPR). These guidelines take into account the effects of various soils and geologies, the ability of the techniques to produce complementary spatial data layers of Roman vine-growing features – thus generating enhanced levels of interpretation (different methods detect different features) – and the need to map the palaeo-landscape as a guide to subsequent palaeobotanical and archaeological sampling, so to optimize our chances of obtaining high-quality environmental and dating information.

Frequency-domain (FD) EMI sensors measure simultaneously the apparent electrical conductivity (EC_a) and magnetic susceptibility (MS_a) of the soil. This provides EMI with unique advantages compared to 'single-characteristic' geophysical methods such as GPR and magnetometry.[45] The technique relies on an EMI active sensor coil transmitting a primary electromagnetic field into the soil and receiving a secondary field through induction, which is measured by a second passive coil.[46] EC_a values are the outcome of variations in physico-chemical properties of the soil, such as clay content and mineralogy, and are proportional to the soil's moisture content, while MS_a is linked to the characteristics of magnetic minerals.[47] This makes EMI ideal to generate detailed information on soil properties, such as water content, texture, compaction, and differences in lithology, and to explore soil-landscape architectures.[48] For the same

[41]Van Limbergen 2020b.
[42]Welc et al. 2017, 2019, 2020.
[43]Broise and Jolivet 1995, 1996; Boissinot 2009; Gallo et al. 2009; Dodd 2022.
[44]Smekalova et al. 2016.
[45]Won et al. 1997.
[46]Doolittle et al. 2014; Calamita et al. 2015.
[47]Grisso et al. 2009; Al-Gaadi 2012; Tang et al. 2018.
[48]Reynolds 2011.

reasons, EMI has major potential in (landscape) archaeology,[49] even more so because at low but acceptable resolutions it enables to cover large areas quickly, but its application in the (Roman) Mediterranean remains limited.[50] We would advise EMI to be used in the first place 1) to carry out large-scale exploratory scanning operations with mobile sleds on targeted sites to demarcate zones of major interest quickly and efficiently that can then be examined at higher resolutions, with either EMI, MAG or GPR; and 2) to pinpoint the best spots in the landscape for coring and pollen sampling.

MAG measures small and local variations/anomalies in the magnetic field of the earth. These are caused by the presence of buried features that have different magnetic properties in comparison with the surrounding soil. This contrast in magnetization can be 'remanent' or 'induced' in nature. Thermoremanent magnetization comes from structures or materials that have been magnetized permanently through exposure to very high temperatures, such as pottery kilns, ovens and furnaces, or walls built with fired materials such as tiles and bricks. In contrast, induced magnetization refers to circumstances in which magnetization is created when an external magnetic field (e.g. that of the earth) is applied, which locally enhances said field. The magnetic susceptibility of a material determines how strong the enhancement will be. This is the kind of magnetization that shows in organic fills (e.g. through soil chemical properties, bacterial processes) such as pit-, ditch- or building-fills.[51] MAG can register a wide range of anthropomorphic features in great detail, and by mounting multiple probes onto mobile carts it is a very cost-effective way to survey large areas quickly. This makes MAG another main scanning method, especially in case of complex and widespread vineyard or plantation features. A joint use of MAG and EMI in specific zones can be particularly useful in cases of insufficient magnetic contrast. Despite its advantages and widespread applicability, MAG is highly susceptible to modern magnetic disturbances and therefore less effective in areas close to modern buildings and other constructions with large amounts of metal in them or which transmit electrical currents (e.g. transmission towers, pipes, cables). On fields close to modern roads it is affected by the magnetic effects of passing traffic.[52] Such biases can be remedied by incorporating GPR.

The GPR is an instrument that transmits radio-frequency electromagnetic pulses (created by a transmitter antenna) into the ground. These pulses travel through the soil, are reflected off interfaces between layers or structures with contrasting physical properties and received back at the surface by a receiver antenna. The energy that is not reflected, penetrates further into the soil, and can be reflected by deeper discontinuities, until all energy is absorbed by the soil. So, the presence of void spaces and buried features with different properties than the surrounding soil will cause sequential reflections whose amplitude and time delay can be measured by the instrument. The most important factor that governs the occurrence of reflections, is the varying degree of water content

---

[49] Conyers et al. 2008.
[50] De Smedt et al. 2013.
[51] Aspinall et al. 2008; Armstrong and Kalayci 2015; Fassbinder 2016.
[52] Schmidt 2009.

between the various layers and structures in the ground.[53] The most important advantage of GPR is fine spatial discrimination potential (high resolution) and, by measuring the time interval within which a wave returns at the surface, the potential to gain direct insight into vertical variations (depth position of buried features). Important drawbacks to consider are the limited penetration depth in soils with a high clay content[54] and the fact that the technique is more demanding in time and data processing compared to MAG and EMI. For these reasons, we consider GPR best used in specific target areas to provide information complementary with features detected by EMI and MAG.

Palynology

A second approach that needs to be further developed in tandem with geophysical survey, and indeed standardly integrated into non-invasive approaches, is the use of palynology to locate and characterize ancient vineyard sites and their environments. Palynology is the analysis of subfossil pollen grains and spores in sedimentary archives and is a technique most suited to reconstruct changes in past vegetation and human land use.[55] Direct evidence of past viticulture may be recorded as the presence of *Vitis* pollen in subsurface sediment records.[56] The use of vine pollen data in archaeology has long been hindered by uncertainty on *Vitis* pollen dispersal patterns, and so doubts on the representativity of its (generally low) concentrations in pollen diagrams. This has blurred the link betwixt rates of occurrence of *Vitis* pollen and the presence (or absence) of ancient (and modern) vineyards.[57] However, recent studies of pollen deposition near modern vineyards have clarified this – i.e. *Vitis* being a naturally weak pollinator with a low potential of atmospheric transport[58] – and have thus established thresholds for *Vitis* pollen percentages that can be applied as baseline information for locating past vineyard sites.[59] The work by Tony Brown was fundamental within these developments and his chapter with Ian Meadows and Simon Turner in this volume shows what can be done now in this field, with special relevance to the data from the British Isles.

Besides direct evidence for the presence of vine-growing, pollen (and non-pollen) records can also give us proxies on wider land use (change) patterns, and in this way document processes of vegetation modification and/or clearance, the (dis)appearance of certain species in relation to viticulture,[60] or vineyard management practices. They can, for example, shed light on the role of mixed vine-growing, such as Pliny's *porculetum* (vines and cereals) (*HN* 17.35.171) or the aforementioned *arbustum* (vines trained on

---

[53]Leckebusch 2003; Conyers 2012; Manataki et al. 2015.
[54]Vermeulen et al. 2017.
[55]Bottema 1999.
[56]Arobba et al. 2005, 2010, 2013; Mariotti Lippi et al. 2007; Calzolari 2008; Marchesini and Marvelli 2009; Cremonini et al. 2013; Rinaldi et al; 2013; Bosi et al. 2015a, 2015b, 2017, 2020; Bosi and Marchesini 2017.
[57]Bottema and Woldring 1990; Atherden et al. 1993; Boissinot 1998; Boissinot and Borchier 1998; Bouby and Marinval 2001; Zech-Matterne and Bouby 2011.
[58]Kimura et al. 1998; Gauthier and Joly 2003.
[59]Brown et al. 2001; Turner and Brown 2004; Kvavadze et al. 2010; Naqinezhad et al. 2018.
[60]Lechterback and Jensen 2020.

trees), amply discussed in ancient literature but poorly explored in archaeology.[61] Quantitative pollen-based reconstructions of past vegetations – long problematic because of complex and cumbersome statistical calibration models[62] – are becoming increasingly feasible due to major advances in computational simulation techniques.[63] In fact, palaeobotanical studies for earlier periods have demonstrated how such data hold exciting promise for tracing and framing the natural and anthropic landscape settings of past agriculture by developing a systematic research approach.[64] We need much more efforts in exploring the possibilities for palynology to identify and reconstruct the natural and man-manipulated habitats in which past vine-growing took place, as done in an exemplary way by Brown and colleagues.

One potential drawback is the need for ideal preservation circumstances. Vineyard plots themselves have little potential of conserving a temporal record of past viticulture history, even more so if they have been cultivated ever since. When ancient, buried soils are preserved, they may still deliver environmental information on the specific time slices in which these soils were formed.[65] Still, to have useful pollen records, one needs depocenters in which sediments have accumulated over longer periods, and thus formed archives of environmental history. Systems such as alluvial, lacustrine and coastal marine sediments, swamps and peat bogs are potential pollen archives, especially when anoxic conditions prevail.[66]

Within integrated approaches, suitable target areas for coring should be selected based on the geophysical results, which are able to pinpoint locations in the landscape with good natural conditions for pollen preservation. Corings can be performed by hand augering, with samples collected at regular intervals in the coring profiles. Manual coring is a relatively simple and cost-effective method for obtaining sediment sequence information, and with only a few well-chosen boreholes a lot of information on diachronic landscape cover can be retrieved. Indeed, besides pollen, sediment records may also hold other vegetational and environmental data by means of non-pollen palynomorphs[67] and archaeobotanical macro-remains (charcoal, wood, fruits, seeds), which help to further refine interpretations and dating.

Phytoliths

One underdeveloped line of research with big potential is phytolith analysis. Phytoliths are micro-sized silica (SiO_2) or calcium oxalate (CaC_2O_4) particles that collect naturally in sedimentary deposits after their release under plant decomposition.[68] Plants absorb

[61] Van Limbergen 2020b.
[62] Davis 1963; Andersen 1970; Parsons and Prentice 1981; Prentice and Parsons 1983; Prentice 1985.
[63] Sugita 1994; Middleton and Bunting 2004; Gaillard et al. 2008; Hjelle et al. 2015.
[64] Stevenson and Moore 1998; Magri and Sadori 1999; Andrieu-Ponel et al. 2000; Bottema and Sarpaki 2003; Guido et al. 2004; Mariotti Lippi et al. 2007.
[65] Moore et al. 1991; Bakels 2020; Doorenbosch 2013; Kaal et al. 2013.
[66] Bottema 1999.
[67] E.g. Marret et al. 2021.
[68] Albert et al. 2008.

silicium and calcium by their roots, and through complex and little understood biochemical pathways, phytoliths are formed which reflect cell or tissue morphology of the plant. Their various morphotypes can so be used to identify specific plant taxa.[69] Phytolith analysis can as such be a powerful tool to collect information on on-site vegetation and human ecology.[70] Compared to pollen, they have the advantage of being deposited *in situ*, and in general being more resistant to environmental degradation because of their inorganic nature.[71] The potential of phytoliths to study past viticulture, however, remains unexplored. Current barriers are *Vitis* producing only calcium oxalate phytoliths, which preserve much less than silica phytoliths, and their yet ill-studied productivity patterns,[72] which limit their current diagnostic utility.[73] To advance their use in producing reliable and high-resolution data on past vineyards, targeting phytolith sampling from small test-pit excavations (of vineyard traces detected by geophysics) is needed, integrated with the taphonomic study of phytoliths from modern *Vitis* plants.

Regional palaeoclimate reconstructions through speleothems

A proper revisionist agenda in Roman wine archaeology cannot exist without the concomitant and systematic exploration of deep-insight proxies into palaeoenvironment and climate in relation to viticulture. As discussed above, the aimed inclusion of archaeobotanical records undoubtedly offers the best chances for tracing the natural and man-made manipulated habitats in which past vine-growing took place. Similar exciting opportunities to augment our understanding of the intrinsic link between viticulture and climate in the past lie in the use of speleothems (stalagmites, stalactites, flowstones). Of all available palaeoclimate proxies – that is, indirect evidence of climatic patterns in the past, such as glacier extent, ice cores, tree rings, deep sea and lake sediments, peat bogs and pollen records – speleothems provide in both spatial and temporal terms the best chance for advancing our knowledge on the deep relationship between climate and ancient viticulture. Speleothems are defined as reprecipitated (secondary) mineral deposits formed from dripping (e.g. stalagmites, stalactites) or flowing water in caves. Our interest focuses on those made of calcium carbonate minerals (calcite, aragonite) through rainfall infiltration within the soil and the rock. Stalagmites are particularly well-suited as they provide geochemical and physical proxies reflecting surface rainfall patterns and their changes through time (season, years, decades), and as such offer a gateway to key insights into regional and local (changes in) hydroclimate and vegetation. They are also of salient interest to past viticulture in that they offer insight into temperature patterns as well. Equally important is their unique age control, at times with higher resolution than achievable for ice cores, within the last ~500,000 years.[74] Indeed, an

[69] Mulholland and Rapp 1992.
[70] Fredlund and Tieszen 1997; Stromberg 2002, 2004; Cabanes 2020.
[71] Grave and Kealhofer 1999; Piperno 2006a.
[72] Tsartsidou et al. 2007.
[73] Tyree 1994.
[74] Cheng et al. 2020; Columbu et al. 2020.

important asset of speleothems is that they can be dated with exceptionally high precision by using U-Th techniques.[75] Within the timeframe of this paper, 2σ uncertainties of less than ten years are easily attainable.[76] The accuracy of dating is further attained by counting visible (petrographic) and chemical (including organic) annual laminae between radiometrically dated layers.[77] Caves create highly advantageous circumstances for palaeoclimate reconstruction, as they form well-enclosed systems, well-protected from exogenic impacts, and therefore characterized by constant temperatures that mirror (and change accordingly to) the annual average surface ones.[78] These assets of speleothems for palaeoclimatology have been recognized since the 1960s,[79] but their systematic exploitation is a more recent phenomenon[80] and – aided by some major analytical advances[81] – they now hold great promise to steer the future of studies in past climate and viticulture. This is in particular true because: 1) temperature is central to all aspects of viticulture, in particular vine-growing suitability (as in determining its geographical limits)[82] and grapevine phenology,[83] and only speleothems offer a suitable spatiotemporal archive (for example, by climate-driven geochemical and physical proxies and high chronological resolution, so potentially enabling the detection of even minimal changes) to examine this association; and 2) caves are widespread in the Mediterranean and Western Europe (and present in regions that may lack other well-dated and continuous high-resolution records),[84] often close to wine producing regions, meaning that their analysis can be applied widely. These are excellent settings to include speleothems as a standard tool to gauge the climate component in palaeo-terroir reconstruction.

The key for successful speleothem data sampling plans is the collection of timeseries both from within (or close to) the study areas and the wider regions/territories. This strategy is necessary because the formation of speleothems is influenced by large-scale (hemispheric, regional) rainfall and temperature (changes), local morphology (with orographic effects), geology and groundwater chemistry within a particular cave.[85] We therefore need records sampled from different locations representative of the whole latitudinal expanse of the climate zone of interest and of diverse terrain settings (coastal, plains, hills, mid-montane) to obtain accurate records ('replicated signal') of regional climate (change) during Roman times. All samples should be subjected to multi-proxy analysis, first of stable isotope ratios ($\delta^{18}O$ and $\delta^{13}C$),[86] and then petrography fluorescence

[75]Drysdale et al. 2012.
[76]Wendt et al. 2021.
[77]Baker et al. 2021; Faraji et al. 2021.
[78]Fairchild and Baker 2012.
[79]Broecker et al. 1960; Hendy and Wilson 1968; Hendy 1971.
[80]Spötl and Mattey 2012; Fairchild and Baker 2012.
[81]Wong and Breecker 2015.
[82]Jones 2007; Jones and Webb 2010; Jones et al. 2012; Nesbitt et al. 2018; Cardell et al. 2019; Van Limbergen and De Clercq 2021.
[83]Jones and Davis 2000; Jones et al. 2005; Gladstones 2011; Van Limbergen and De Clercq 2021.
[84]Comas-Bru et al. 2020.
[85]Finné et al. 2014.
[86]McDermott 2004; McDermott et al. 2005.

microscopy[87] and trace element analyses.[88] Additional U-Th dating is needed, as well as physical and/or chemical lamina counting. The combined analysis of physical and chemical annual laminae provides much-needed information to disentangle temperature and rainfall signals at seasonal scale, which is what is needed to understand what parameters influenced the most viticulture in a region. Dating accuracy can be improved by integrating U-Th age models built with open-source algorithms (e.g. StalAge and/or COPRA) and annual lamina counting, allowing to create as low as possible uncertainties in the climate proxy timeseries.[89] Finally, to mitigate the risks of heuristic bias, the speleothem results need to be integrated with pollen data and other (dendrological and sedimentological) climate proxy data.[90] In return, speleothems can help to assess the human impact on our pollen signals.[91]

Comparative historical analysis

A final, completely different, trajectory for terroir reconstruction is the use of comparative historical analysis to build a heuristic script of potential vine-growing methods, locales and scenarios on regional levels.[92] Such interpretative pathways are possible because within a certain territorial setting the natural rules of terroir imply that – even if influenced by a wide range of cultural and societal factors – viticulture choices operate within a finite optional range set by a region's unique landscape configuration. This opens the door to the creation of vine-growing biographies with strong analogical possibilities that can deepen our understanding of long-term land use for viticulture, strengthen our interpretations of both the ancient source material and our natural science datasets, and increase our comprehension of Greco-Roman viticulture and assess the legacy effects of past viticulture activity on historical and current vine land use.

For data from late periods to be useful as regressive explanatory tools, however, some conditions need to be fulfilled. First, they need to be as regional as possible, with sources that have direct, or at least consequentially indirect, relevance for the study areas. Second, the periods from which we choose to collect the data need to approximate the farming conditions of the Greco-Roman era, which implies to consider periods that were characterized by low-technology (at least compared to modern standards) and labour-based farming practices, while at the same time still approaching the expertise of Roman farming that is reflected in the ancient textual sources. Third, we would also like for these data to approach some form of sampling validity, that is, based on verifiable topographical evidence that can be transferred into a GIS environment (in case of visual media) and including numeral or statistical records that can be modelled over a certain period (in

---

[87]Baker et al. 2021; Faraji et al. 2021.
[88]Verheyden 2004; Jochum et al. 2012; Desmarchelier et al. 2006.
[89]Scholz and Hoffman 2011; Breitenbach et al. 2012.
[90]Brewer et al. 2013; Chevalier et al. 2020.
[91]Ding et al. 2017.
[92]For a recent effort regarding Roman wine yields, see Van Limbergen 2022.

case of texts). To this regard, a most productive strategy would be to focus on the sixteenth to nineteenth centuries. This period saw only minimal advances in mechanization, and industrial-like farming techniques still had little impact on agriculture and food production.[93] At the same time, this era saw a renewed interest in agronomic knowledge, and as a result the redaction of a growing number of agricultural and viticultural treatises and agrarian censuses. In addition, the era is evidenced well graphically because of a major upsurge in cartographic and cadastral documentation, due in large part to large-scale military operations and governmental imperialistic policies, and important advances in land measurement and survey techniques.[94]

Potential sources are cartographic and cadastral sources (secular/ecclesiastical) and agronomic treatises, complemented with supportive iconography (drawings, paintings, illustrated calendars with agricultural cycles, photographs) and illuminating texts (primary and secondary), as well as further contextualized through academic and grey literature (e.g. publications by local historical and cultural societies). Cadastral maps are particularly suited for this purpose, as they are (according to pre-modern standards) high-resolution plans, based on proper field survey and often with land use information down to parcel level. For Italy in particular, the same is true for *cabrei* – the Italian predecessor of the cadaster which, despite their far more impressionistic character and only sporadic inclusions of illustrations, often hold decisive clues for the sixteenth to eighteenth centuries.[95] Also, the possibility of 'dealing' with these early cadastral and land division documents within a GIS environment have in the past two decades opened many new perspectives to scholars of ancient topography. Software developments now make it possible to georeference those cadasters that have a geodetic system (generally true for the nineteenth-century specimens) and the levels thus obtained can be used to georeference less accurate cadastral sketches and *cabrei* (usually so for eighteenth-century cartographic data).[96] Terroir parameters that could potentially be explored through these media include grape type/variety, vineyard location in the landscape, vine-training systems, vineyard management infrastructure (e.g. supports, drainage, irrigation) and yield potential.

Conclusion

In this paper, we have introduced the research approach of palaeo-terroir as a new and integrated way to analyse the natural and human ecosystems of Roman and ancient viticulture and wine production. By outlining its basic heuristic framework and discussing some of its most promising methods and datasets, we have tried to show the

[93]E.g. Pinto et al. 2002; Cianferoni et al. 2002; Eterovich and Spalatin 1964, 1970.
[94]E.g. Cosgrove 1992; Biggs 1999; Akerman 2009.
[95]E.g. Ginori 1978; Aimar et al. 2021.
[96]Corsi and Vermeulen 2007; Favretto 2012; Arriaga-Varela and Takahashi 2019.

massive potential of this research pathway for better and more historically grounded reconstructions of Roman (and past) wine landscapes and economies, able to groundbreakingly inform us on the interaction between man and its environment on a regional level in Roman agriculture, daily life and society. Also, by providing concrete guidelines for the use of reproducible spatial, palaeobotanical and palaeoclimate datasets, we hope to have generated a strong stimulus for others to start applying palaeo-terroir models and ideas in old and new research endeavours. There obviously lies a long, collaborative road ahead, but therein lies genuine potential to start building a proper atlas of wine regions in the ancient world.

PART I

BIOMOLECULAR ARCHAEOLOGY AND CHEMICAL ANALYSIS

CHAPTER 4
ARCHAEOLOGY AND GRAPE DNA

Andrea Zifferero[1]

Abstract

This paper reviews different approaches to detect the domestication of the grapevine. Evidence of ancient grapevine varieties may survive in today's archaeological landscape and to detect this the research projects VINUM and ArcheoVino jointly employed botanical, archaeobotanical, biomolecular and archaeological analyses. These data established the genetic relationships between current grapevine varieties and wild grapevine populations growing close to Etruscan and Roman sites along the central Tyrrhenian area. The chronology of occupation helped to determine the correct time of domestication of *Vitis vinifera* ssp. *sylvestris* into *Vitis vinifera* ssp. *sativa* around these sites. Data revealed that cultivars originally produced by sophisticated viticulture in Greek colonies of Southern Italy and Sicily were introduced into districts marked by intensive viticulture and wine trade activities probably from the eighth and seventh century BCE onwards in the Albegna Valley (Grosseto). Here, the results obtained by the ArcheoVino Project were used to create an experimental vineyard close to the Etruscan site Ghiaccio Forte made by local ancient varieties.

Research on ancient grapevines: A methodological problem

Investigations into the archaeology of vines and wine were initiated in 2004 by the University of Siena (Department of Historical Sciences and Cultural Heritage), in collaboration with the University of Milan (Department of Biology – Section of General Botany and Department of Plant Production) and the National Association of Wine Cities, starting from a question that has always accompanied fieldwork and subsequent laboratory analysis: is it possible that residual portions of an archaeological landscape can survive in living form to the present day?

This question marked the beginning of a long journey between different disciplines spanning archaeology, botany and molecular biology, which continues now with new

[1]My contribution to the Vine-growing and Winemaking in the Roman World conference moves from my esteem for Jean-Pierre Brun and admiration for his work on the archaeology of wine and oil, begun in the distant days of the Scansano (2005) and San Quirico d'Orcia (2007) conferences.

partners and other ways of supporting research. The aim was to deepen our knowledge of wild grapevine (*Vitis vinifera* ssp. *sylvestris*) populations distributed around archaeological sites (especially of Etruscan and Roman age), to clarify their evolutionary stages determined by their close relationship with the communities that cultivated them.[2]

The VINUM Project started from the assumption that those vines were the remnants of ancient vineyards, reabsorbed by the surrounding environment with the abandonment of the site, to revert to the wild form.[3] What part, however, did the wild subspecies play in the vineyards of the Etruscan and Roman periods, and how much did the processes of its domestication influence the development of the typical varieties of the cultivated form (so-called cultivars or vines), the expression of today's *Vitis vinifera* ssp. *sativa*? Finally, what were the mechanisms and variety of circulation in the ancient Mediterranean, and how can their profile and historical context be accurately reconstructed, combining botanical and archaeological data?

The experience gained with the VINUM Project was essential to introduce operational protocols suitable for the study of the wild vines surrounding the archaeological sites and, above all, new research objectives for the subsequent ArcheoVino (cf. below) and Eleiva Projects – the latter to investigate the origins of olive (*Olea europapea* L.) cultivation. Among the most important of these were the identification and active conservation of the rural landscape in the mid-Tyrrhenian area and the preservation and enhancement of surviving forms of ancient biodiversity.[4]

The study of grapevine DNA, fine-tuned since the early years of this century, has increased our possibilities of understanding the exact genetic profile of individual cultivars and the kinship relationships between present-day vines, while at the same time allowing a detailed map of the differences between wild and cultivated vines to be generated. Prior to a genetic approach, the question of the domestication of wild grapevines, which grow wild in the Mediterranean, sub-Mediterranean and Caucasian ranges, with significant expansions toward the Pontic and Caspian regions, had been addressed in the innovative studies of Gaetano Forni, with the elaboration of a map of the so-called primary centres of domestication, built on a diffusionist model that recognized the progressive development of viticulture from East to West.[5] The stages of this process, which began between the sixth and fifth millennia BCE in the South Caucasus and the Near East area and ended practically in the Roman period, were fixed on a wide-meshed chronological grid, based for the most part on the morphometric characters of grape seeds from archaeological excavations (Figure 4.1).[6]

---

[2]Research data devoted to grapevines and wine are collected in Ciacci and Zifferero 2005; Ciacci et al. 2012; Ciacci and Giannace 2012; Firmati et al. 2015.

[3]*VINUM: a project for the recognition of wild vines in the archaeological landscape of Tuscany and northern Latium*, was funded in the three-year period 2004–6 by the Monte dei Paschi di Siena Foundation and was conducted with the organizational and logistical support of the National Association of Wine Cities (Siena).

[4]Ciacci et al. 2012: 485–722; on the *Eleiva* project, cf. *San Quirico d'Orcia* 2010.

[5]Also see McGovern in this volume.

[6]Forni 2012a, with previous bibliography; a synthesis can be found in Forni 2012b.

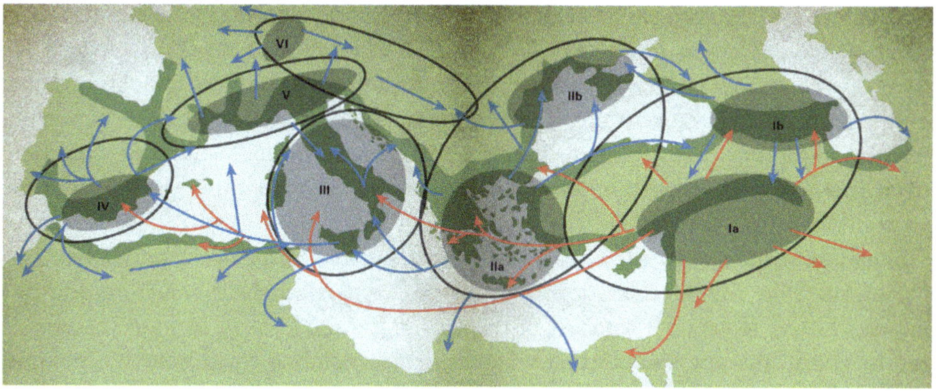

Figure 4.1 Map indicating paradomestication areas and primary domestication centres (shaded) of the wild grapevine (*Vitis vinifera* ssp. *sylvestris*) in the Mediterranean and Eurasian areas. Red arrows indicate certain routes of domestication spread documented by archaeobotanical studies; blue arrows indicate probable additional routes. Numerals indicate the progressive chronology of the spread of domestication: Ia/Ib (sixth to fifth millennium BCE), IIa/IIb (fourth to third millennium BCE), III/V (second to first millennium BCE), VI (Roman era). Map by A. Zifferero, redrawn after Forni 2012a.

It was thus understood how, already from the early stages of the Bronze Age in the protohistoric cultures of mainland and island Italy, morphological and dimensional differences in grape seeds expressed a viticulture still linked to the use of wild grapevines. This was indicated by small piriform grape seeds with an unpronounced beak. On the other hand, a more advanced viticulture was emerging, signalled by piriform grape seeds of more conspicuous size and longer beaks, indicative of purposeful cultivation, which selected plants with respect to cluster size and sugar concentration in the berries.[7]

Following this, the development of DNA polymorphism analysis derived from SSR (Simple Sequence Repeats) and SNP (Single Nucleotide Polymorphism) markers provided a new highly accurate tool for reconstructing domestication processes, ensuring more confident and targeted results than the traditional observation of phenotypes in ampelographic studies and, in general, the methods applied by archaeobotany.[8] The analysis of DNA polymorphism involved sampling parts of the inflorescences and leaves of the grapevine. The progressive and final step of this approach was the sequencing of aDNA (ancient DNA) extracted from carpological remains and, in general, from organic remains from archaeological excavation and preserved under special anaerobic conditions.

[7]On grapevine carpological remains, see Mori Secci 2005; domestication centres are covered by Forni 2012a, with updates on the distribution and chronology of grapevine remains (mainly grape seeds, but also shoots and stems) found in peninsular Italy and the large Tyrrhenian islands in Acconcia and Piergrossi 2021.

[8]Scienza and Failla 2016; for the contribution of molecular biology to the study of grapevines, see Imazio and Labra 2005 and now, generally, Scienza and Imazio 2018. On the use of SNP markers cf. De Lorenzis et al. 2019, 2020.

The VINUM Project, and later the ArcheoVino Project, had a solid methodological and theoretical basis, according to which plants located close to the archaeological site have different genetic profiles from those of more distant plants, induced by domestication. This process could also occur through introgression (i.e. the introduction of new genes into the germplasm of plants, thanks to the usual mechanisms of fertilization), resulting from the arrival of allochthonous varieties. Associated with this principle was the circumstance that the genetic characters of the vegetation, fixed in the period of occupation of the site by the resident community, had a chronology ascertainable by juxtaposing the biomolecular data with the archaeological data. Therefore, the two projects helped to set up a database of the wild grapevine genome in some districts of south-central Tuscany and northern Latium, corresponding roughly to the political territories of the Etruscan cities of Populonia and Cerveteri. Special attention was also paid to the Albegna Valley, which had played a significant role in Vulci's viticulture in the Etruscan period and continued to produce large quantities of wine in the Roman period.[9]

Analyses of the phenotypes (particularly leaf and bunch morphology) and genotypes of wild grapevine populations observed and sampled in the vicinity of archaeological sites have made evident two main trends: 1) the pronounced morphological and genetic differences with wild grapevine populations in locations farther from the site and therefore less subject to or completely unrelated to ancient forms of domestication; and 2) strong genetic similarities between present-day and native grape varieties, cultivated today in the areas investigated.[10]

Further interaction between scholars from different disciplines, however, stimulated the investigation of cultivation techniques, known mainly from literary sources, in an attempt to reconstruct the shape and articulation of ancient vineyards.[11] This led to the identification of presumed vineyards relicts from the Etruscan and Roman eras, surviving in the wild state of *lambruscaia*, in some areas of the Tyrrhenian area.[12] These areas have a high preservation level due to reduced anthropogenic impact, offering vines tall trees such as oaks, maples and elms, which ensured their survival by providing them with a living brace around which to cling. The continuity between these fossil vines and the archaeological site was sometimes related to the presence of wine press bases, carved

[9]Ciacci et al. 2012: 485–722; on wine production and Etruscan transport *amphora* kilns in the Albegna Valley cf. Zifferero 2017a and Rojo Muñoz 2021. For a preview on the results of archaeometric analysis of *amphorae*, see Fornacelli et al. 2022. Local Roman-age production is discussed in *Ravenna* 2007.

[10]Scali et al. 2018.

[11]Boissinot 2009.

[12]The term *lambruscaia*, now attested in the Tuscan-Latial Maremma area investigated by the VINUM and ArcheoVino Projects, originates from the Latin term *labrusca/lambrusca*: it is first used by Virgil (*Eclogae* V: 4–7), to describe the wild vine that grows on the edge of cultivated fields. The name has been preserved in the current usage of some areas of Piedmont, Lombardy and Emilia-Romagna, of ancient Ligurian and Etruscan culture, to qualify the wild vine but also some varieties of vines (including *Lambruschi*), which have a very close genetic kinship with the wild form. For the history of the term *labrusca/lambrusca*, see the still fundamental Sereni 1964: 151–71. On the relationship between the wild vine and the varieties of *Lambruschi*, cultivated mainly in the Modena and Reggio Emilia areas, see Scienza and Imazio 2018: 1–27.

Figure 4.2 *Lambruscaia* with wild vines on live supports at the Protohistoric, Etruscan and Roman site of Tufarelle (Allumiere, Italy), in the ancient territory of Cerveteri. Photograph by A. Zifferero.

from blocks of local rock, as has been recorded several times in the Monti della Tolfa (Rome) district. This area has also been explored for the notable presence of wild olive trees (i.e. olive trees turned back to their wild form), in the vicinity of Roman sites with olive crushing wheels and oil press bases (Figure 4.2).[13]

The tree-wedded vine, called *arbustum* in the Latin sources, is the typical expression of Etruscan viticulture. In the protohistoric period it proliferated itself in cleared fields near watercourses and at the edges of cultivated areas, only to be transferred within

[13]On wine and oil presses documented in the Monti della Tolfa area, inland from Civitavecchia, see Vallelonga and Zifferero 2010; Ciacci et al. 2012: 485–630; Allegrezza 2017.

cultivated fields to form rows consisting of aligned trees, each with the function of supporting two plants with its own stem and controlled development of branches. This is the kind of vineyard that originated with the organized landscape of the countryside of Etruria proper between the last quarter of the seventh and early sixth centuries BCE. It spread afterwards to the areas of colonial expansion, to the north in Etruria *padana* and to the south in Etruria *campana*.[14]

Today, the remnants of this ancient system can be observed in Emilia-Romagna, where it has taken the names of *piantata bolognese* and *piantata modenese-reggiana*, in Tuscany (in the Arno Valley and Casentino, in the province of Arezzo) and in northern Campania (in the province of Caserta), with the respective names of *alberata* and *alberata aversana*. This kind of vine-growing system evolved into the Roman *vinea*, where the supporting trees and the vines leaning against them were arranged differently, but always following the planting proper to the *arbustum*, which required long pruning of the shoots. Sources trace these methods of viticulture and plant pruning, along with the provisions on the ritual use of wine, to the reign of Numa Pompilius (Figure 4.3).[15]

Greek grape cultivars in the Tyrrhenian area?

One of the central uncertainties in the history of pre-Roman and Roman viticulture is the spread of grape varieties of Greek origin in the Mediterranean area. In archaeological circles, the topic is usually associated with the wine trade and focuses in particular on the production and distribution of transport *amphorae*.[16] Among the scholars who have addressed it, the botanist Attilio Scienza has proposed a model that reconstructs the varietal circulation of vines in the ancient Mediterranean, using the genetic approach and recognizing two forms of diffusion. The first is related to the direct transfer of grape varieties from the Greek motherland through citizens who, moving to the new *apoikìa*, planted them in the vineyards of the *chora*. The second is related to the emporic system, where the Mediterranean ports of trade acted as centres for the accumulation of grapevine germplasm and as propellers of grape varieties through the normal channels of the wine trade, now traceable by the distribution of transport *amphorae*.[17]

A grape variety limited to the area investigated by the VINUM and ArcheoVino Projects, whose Greek origin has been hypothesized, is *Ansonica/Inzolia*. This cultivar

---

[14]Ciacci and Zifferero 2007a: 265–8.

[15]In general, on the history of the *piantata* and *alberata*, see Sereni 1964; 1987: 40–3; on the Emilian *piantata*, cf. in particular Antonini and Marchesini 2017; Zifferero 2017b: 279–82; on the *alberata aversana*, Di Pasquale et al. 2012. Roman associated viticultural techniques are discussed in Braconi 2012, with further remarks in Zifferero 2021.

[16]Still relevant on the relationship between viticulture, *amphora* production and wine circulation are Van Der Mersch 1996; Sourisseau 2011. For the reconstruction of wine-growing contexts in Etruscan Campania and the integrated approach between literary sources and archaeological data, see, among recent contributions, Bellelli 2018 with insights into *amphora* production and wine trade in the lower Tyrrhenian area.

[17]Scienza 2004.

Figure 4.3 The survival of Etruscan viticulture in the Italian landscape seen through the Aversana planting before winter pruning near Casal di Principe (Caserta, Italy). Photograph by G. Di Pasquale.

has white grapes and is attested along the coastal strip of Tuscany and on the islands of the Tuscan Archipelago but cultivated particularly in the Albegna Valley (Grosseto) and on Giglio Island. At Giglio, the vine usually grows leaning against a cane pole, which keeps the plants low thanks to very short pruning to protect them from sea winds - a traditional cultivation technique on a dead support of single stakes or canes, reliably Greek in origin (Figure 4.4).[18]

[18]Bandinelli and Di Collalto 1997; Brandaglia 1997, 2001; Scienza 2006: 88–9; Ciacci and Zifferero 2007b; *Scansano* 2007: 259–63; on the Greek technique of viticulture using single poles, stakes tied to espaliers or canes and its survival in southern Italy, see Forni 1999; Del Lungo 2016.

Figure 4.4 Vineyard of *Ansonica* with planting supported by reed stakes on Giglio Island, Italy. After Brandaglia 2001.

Analysis of *Ansonica* germplasm then revealed strong genetic similarity with the Greek varieties *Roditis* and *Sideritis*, which are still widespread in southern Greece, the Peloponnese and the island of Euboea, along with substantial similarity to the *Inzolia* variety, grown in eastern and southern Sicily.[19] By observing the distribution of this vine and comparing the eastern and western sectors of the Mediterranean, it is likely that *Ansonica*, and consequently *Roditis* and *Sideritis*, are the by-product of the contacts of the Euboic navigators with Sicily and the Etruscan area and constitute, at least for Sicily, the result of the colonization that took place on the Ionian and Tyrrhenian coasts in the eighth century BCE. Particularly interesting is the spread of the vine in the Albegna Valley, the district under the control of Vulci, the Etruscan city where regular contacts with Euboea are evidenced by the spread of Euboic pottery and the activity of local workshops that produced Etruscan-geometric pottery under the influence of resident Greek artisans.[20]

Genetics have furthered our ability to trace the articulated evolution of cultivars, investigating the relationships and degree of kinship between present-day vines and that between present-day vines and wild vines by comparing the main cultivated varieties in

---

[19]Labra et al. 1999, with previous bibliography; on the Greek grape varieties *Roditis* and *Sideritis* (or *Sidiritis*), cf. Del Zan 2004: 628–30 and 635–6.

[20]On the argument in general Naso 2012.

the Mediterranean area. In fact, evolutionary processes depend on their movement following people and communities to territories other than their original ones and on introgression phenomena with wild vines and other native varieties.[21] Comparison between the grape varieties grown in the districts that hosted the colonial phenomena of Sicily and Magna Graecia (especially Campania and Calabria) and the varieties found in Greece today, expresses a high level of genetic variability, due to the complexity of evolutionary processes. According to the most recent research, Sicily and Magna Graecia are to be considered areas of acclimatization of Greek viticulture, during its spread towards the Western Mediterranean. This area would have given rise to the *Sangiovese* vine, whose arrival in the Albegna Valley could date back to the Etruscan period (cf. infra).[22]

Although genetics has contributed greatly to the classification of viticultural varieties, an obvious limitation of this tool, when applied to research on ancient varietal circulation, is the lack of precise chronological anchors that can contribute to the reconstruction of a historical framework. In other words, the framework provided by the archaeological context from which the biomolecular data comes is essential for its proper use. As an example, the progressive formation of the *Marzemino* grape variety, from its origin on the southern shores of the Black Sea to its arrival in the Po Valley, was traced by geneticists, at first, to Euboic navigators and commercial carriers and to the colonial phenomena of Doric expansion on the shores of the Adriatic Sea, only to completely change the historical context, attributing the variety's birth to the trades of Venice between the Middle Ages and the modern age.[23]

The recurrent absence of an adequate historical and archaeological framework in the application of the so-called hard sciences such as archaeometry in the humanities often compromises the value and accuracy of the final interpretation. To this regard, I refer to the important works of Stefano Del Lungo, who has reconstructed in an exemplary way the wine production of ancient *Oenotria* (coinciding for the most part with the area of the present Basilicata region), reinserting it in the correct historical dimension through the use of literary sources combined with archaeological, topographical and epigraphic data, without neglecting the contribution of genetic data.[24]

The ArcheoVino Project at Scansano: Towards a park of ancient viticulture

The clear productive nature of the Albegna Valley motivated the initiation of research dedicated to the local wild grapevine, which came about in the wake of the VINUM

---

[21]Scienza 2004, 2010; *Castiglione d'Orcia* 2011; Failla 2014; De Lorenzis et al. 2019.

[22]*Firenze* 2006; Vouillamoz et al. 2007a, 2008; Di Vecchi Staraz et al. 2007; Scienza and Imazio 2018: 159–81; on the genetic difference of present-day vines from ancient varieties cf. Scienza and Failla 2016; the introduction of *Sangiovese* and *Ciliegiolo* in the Etruscan period is supported in Zifferero 2016, 2017a.

[23]On the *Marzemino* grape variety, cf. Scienza and Failla 1996: 221–30 and more recently Scienza and Imazio 2018: 17–50.

[24]Del Lungo 2016: 37–46; 2017.

Project.[25] ArcheoVino has also been directed to observe the phenotypes and genotypes of *Vitis vinifera* ssp. *sylvestris* in the vicinity of the Etruscan settlement of Ghiaccio Forte (Scansano) and other rural sites from the Etruscan and Roman periods. It starts from the hypothesis that the grapevine populations now widespread in Maremma may have affinities with plants cultivated in ancient times and that they have re-naturalized (i.e. rewilded) with time, thus becoming part of the spontaneous vegetation of the area (Figures 4.5 and 4.6).[26]

The research carried out in the field with the census and sampling of wild grapevine populations clearly indicated the domestic character of the local populations, subjected to strong anthropogenic pressure for cultivation purposes. Also of great interest are the cases of plants marked by genetic traits close to the *Sangiovese* and *Canaiolo nero* (but also *Ciliegiolo*) grape varieties, identified in the Ghiaccio Forte context near Etruscan and Roman production sites.[27] These plants support the hypothesis that the middle and lower Albegna Valley was a centre of accumulation of germplasm from outside, poured into the local populations by forms of introgression, especially with the beginning of Etruscan viticulture.

A dissemination phase of the results of ArcheoVino included the design of a Park of Ancient Viticulture, of which the experimental vineyard described here represents the first significant piece. The park sits within the category of so-called 'light parks' – a form of conservation and enhancement of archaeological, environmental and landscape resources already included within existing routes (in this specific case, the Colli di Maremma Wine and Flavors Route).[28] Therefore, it was possible to prepare its master plan as part of a technical-scientific collaboration agreement, initiated in 2014 between the University of Siena and the Municipality of Scansano, divided into three phases of interventions.

The actions contained in the first phase allowed for the planting of an experimental vineyard in 2015 and the planning of its connection with the Archaeological and Vine and Wine Museum of Scansano, to be carried out on site with signage and panels and on the web through the portals of the Municipality of Scansano, the Consorzio di Tutela del Morellino di Scansano and finally the Maremma Museum Network. The first phase also includes the design of itineraries of viticultural biodiversity between the valleys of the Albegna river and the Sanguinaio ditch (the *Via delle Lambruscaie*), along with the preparation of the pathways, with the creation of the information system and related

---

[25]The *ArcheoVino Project: from the wild vine to the reconstruction of the Etruscan vineyard*, was developed between 2006 and 2008 by the Department of Historical Sciences and Cultural Heritage of the University of Siena, in collaboration with the Department of Life Sciences of the same University and the National Association of Wine Cities, with funding from the Region of Tuscany and the Municipality of Scansano. It still continues today with funds from the Municipality of Scansano.

[26]Ciacci et al. 2012: 631–722; an outline of the site of Ghiaccio Forte (frequented as a sanctuary from the last years of the seventh century BCE) and the surrounding area, is in Firmati 2012, 2013.

[27]Vignani et al. 2012; Scali et al. 2018.

[28]Fabbrini 2003; Zifferero 2012; Firmati et al. 2015.

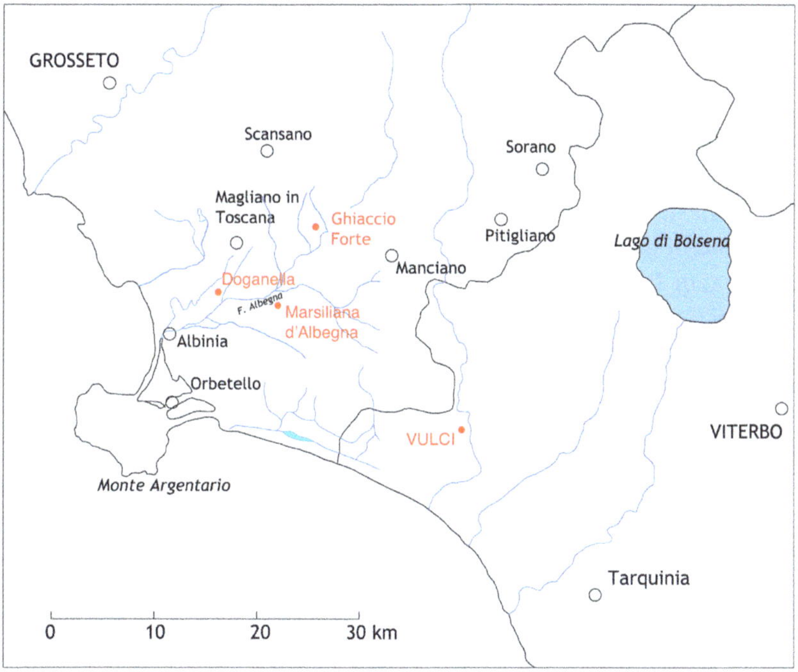

Figure 4.5 Map of the Albegna Valley (Grosseto, Italy) indicating the sites mentioned in the text. Drawing by M. Firmati and A. Zifferero.

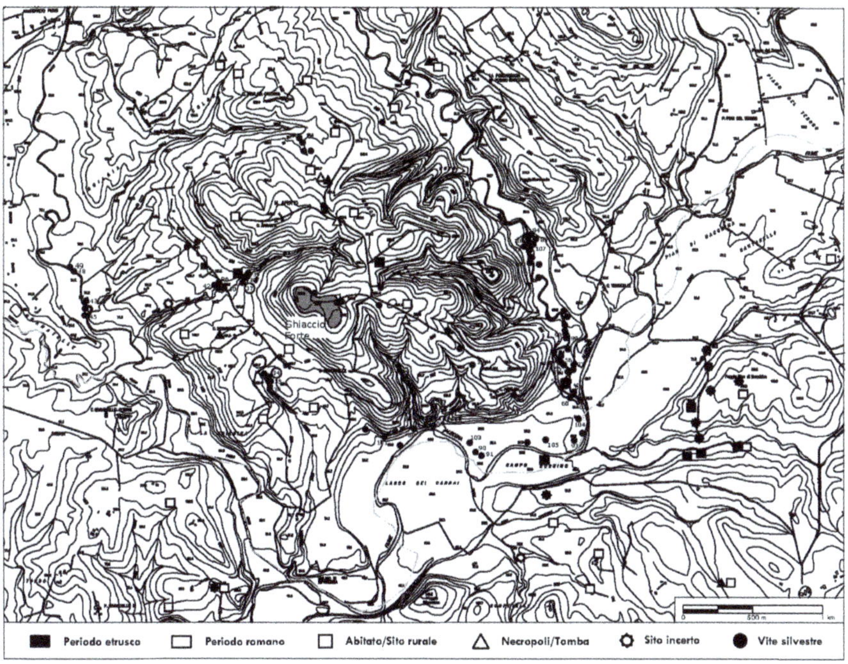

Figure 4.6 Map of the area explored by the ArcheoVino Project between the Albegna and Sanguinaio valleys in Italy, showing the main Etruscan and Roman sites and the location of wild grapevine populations. Map by A. Zifferero.

signage. The basic idea is the development of a loop trail around the village of Ghiaccio Forte, linking the experimental vineyard with the nearby Etruscan necropolis of Poggio Marcuccio and the wild grapevine populations distributed in the area. The implementation of the trail could bring the Sanguinaio Valley, due to its high level of environmental conservation and low anthropogenic impact, under protective care, in this case motivated by the survival of ancient biodiversity, as is suggested by the regional legislation on protected areas of Tuscany.

The second phase will involve, in the vicinity of Ghiaccio Forte, the reconstruction of an Etruscan farm with grape processing facilities to make wine, the reconstruction of an Etruscan *amphora* and *dolia* workshop, and the start of wine production in the experimental vineyards. Finally, a 1:1 scale reconstruction of an Etruscan transport ship, with the *amphora* cargo crammed into the hold, to be anchored and prepared for visiting at the coastal gate of the networked park conceived by ArcheoVino, in the Orbetello Lagoon.

The design and planting of the experimental vineyard, which accommodated wild vines selected by phenotype and genotype extracted from the populations surveyed in the Albegna and Sanguinaio valleys, were made possible thanks to the availability of a 1,000 square metre plot of land, offered by the Ghiaccio Forte-Aquilaia Farm, in an agreement relationship with the Municipality of Scansano. The vineyard, authorized by the Tuscany Region, was planted in several stages in spring 2015, about 1 kilometre north of Ghiaccio Forte along the slopes of a plateau with a typically Pliocene substrate, consisting of sands, sandstones and polygenic marine conglomerates (Figure 4.7).

The cultivation technique is twofold: on 600 square metres, the Etruscan tree-trained vineyard was set up on an arboreal trellis (two vines married to each of the twenty-four field maples staked at a height of about 1.8–2 metres, arranged in three rows); on the remaining 400 square metres, the Graeco-Roman type vineyard was set up, with four rows of vines trained as saplings on a dead trellis (chestnut stakes), alternating with the Etruscan tree-trained rows. The rows of maple trees and vines in the experimental vineyard should reach a maximum height of three metres, for reasons of practicality in caring for the plants and the braces. The inter-row spacing will allow lateral development of the vine shoots, which will be facilitated by supports formed from swamp reeds, to which the shoots will be secured by tying with willow fibres.

Only at the end of the experimentation will it be possible to think about the eventual selection of plants with a higher yield in terms of fruit quality and quantity, thus arriving at micro-vinification trials designed to produce wines similar to Etruscan or Roman wines, starting from the cultivation techniques and using vines of highly probable ancient origin. The final goal is to perfect genetic research on wild vines in the future, carrying out extended population surveys and expanding our database on local germplasm. At the same time, micro-vinification operations will continue with grapes from the experimental vineyard, under the oenological coordination of the Consorzio di Tutela del Morellino di Scansano.

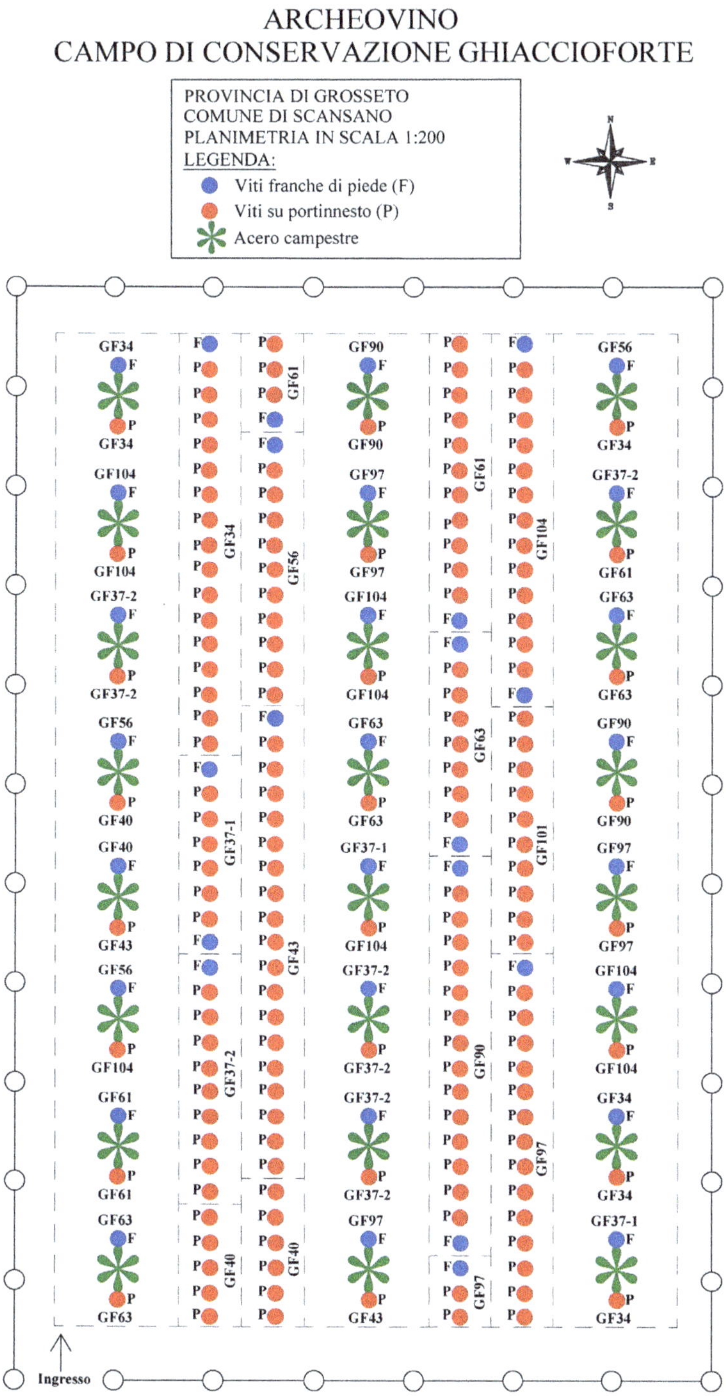

Figure 4.7 Planting plan of the Ghiaccio Forte (Scansano, Italy) experimental vineyard. Drawing by A. Zifferero and V. Zorzi.

The purpose of ArcheoVino is therefore twofold. On the one hand, to stimulate tourism designed around the values and slow-time nature of agriculture, promoting the environmental and cultural features of the Maremma in an integrated way. On the other hand, to add value to local wine production through the experimental production of an 'ancient' wine, suggesting innovative directions for activities that can increase the historical identity of the area's traditional production.

Conclusions

The most significant results that have emerged from the research on wild grapevines in the Albegna Valley consist of the forms of introgression of wild grapevines with the *Sangiovese* and *Ciliegiolo* vines, of Magna-Grecian origin, documented in the germplasm of plants located near Etruscan and Roman open sites around the settlement of Ghiaccio Forte. Considering that the *Ansonica* vine would have arrived in southern Etruria with Euboic navigation around the second half of the eighth century BCE, the question is to establish the time of arrival of both red grape varieties in the territory of Vulci. The local archaeological context, centred mainly on the kilns of transport *amphorae*, distributed between the centres of Marsiliana d'Albegna and Doganella, at a short distance from Ghiaccio Forte, suggests a chronological term prior to the beginning of the production and related trade of Vulci wine, which is placed around the last quarter of the seventh century BCE. It is therefore reasonable to think, for now, also considering the chronology of the Ghiaccio Forte site, of a time frame coinciding with the final years of the seventh century BCE for the acclimatization of *Sangiovese* and *Ciliegiolo* in Etruscan soil.[29]

This text has tried to emphasize how much input from archaeological and historical disciplines is needed to improve the quality of biomolecular data. This could be summarized in a protocol consisting of six rules, which are essential to ensure a satisfactory framework for the phenomena of varietal domestication and circulation of grape varieties in antiquity:

1. Always rely on the archaeological and historical context of viticulture, which is useful for focusing on the stratification of viticultural inputs, best defined as 'stratification of cultivations'. This is usually based on phenomena of domestication of wild grapevines that are highly articulated in the scale of time, with the contribution of cultivars coming from outside the area under consideration, through contacts due mainly to commercial activities or colonization phenomena.

[29] As the earliest transport *amphorae* of Vulcentan production, identifiable with the Py 1/2 type, would seem to indicate: Py 1985. On the transmarine trade in Etruscan wine and the *amphora* production of the Albegna Valley, which was probably initiated in competition with that of Cerveteri and represented by types Py 3A-B, followed by the later types Py 3C and 5, with a chronological spectrum between the early sixth and early fourth centuries BCE, cf. Zifferero 2017a, with previous bibliography and Rojo Muñoz 2021.

2. Rely, if possible, on archaeobotanical finds from the contexts under investigation to document their morphometric characters and the possible presence of aDNA.
3. Research and investigate the possible survival of traditional farming methods, which often originate from very old viticultural techniques.
4. Investigating the relationship between the ancient city and the countryside: the city usually promotes and controls the introduction of new varieties of vines into its political territory (so-called *Demic model* of Attilio Scienza).
5. Investigating the forms and mechanisms of ancient trade, with special attention to ports and emporia in the phenomena of concentration and redistribution of cultivars: these were often entrusted with the function of centres of genetic accumulation (so-called *Cultural model* of Attilio Scienza).
6. To be certain of the archaeological attestation in the area under study of at least one of the elements of the operational chain related to winemaking, starting from the pressing and fermentation of grapes to the manufacture of *dolia* and transport *amphorae*.

CHAPTER 5
WINE PRODUCTION, TRADE AND CONSUMPTION IN THE ROMAN WORLD: THE POTENTIAL OF ORGANIC RESIDUE ANALYSIS

Alessandra Pecci[1]

Abstract

This chapter provides an overview of the methods and sampling process of organic residue analysis as it relates to the production, storage and trade of wine in the ancient Roman Mediterranean. This method is one of the most promising in our ability to discern whether facilities were used to produce wine, olive oil or something else. Particular focus is given to the central and western Mediterranean and the study of various production structures and materials. More than a decade's worth of experience is synthesized to survey a range of studies in Italy, France, the Iberian Peninsula and the Balearic Islands using a range of material evidence, from *dolia* and *amphorae* to plasters and mortars that coated production areas and vats. Importantly, discussion is also included on the analysis of experimental materials in controlled environments to test for the ability and efficacy of organic residue analysis in detecting wine residues.

Introduction

Wine in the Roman world was used as a beverage, sauce, condiment, medicine and offering among other things. It could be consumed cold or hot, transformed into vinegar, and used to preserve food. Regardless of its use, wine, as a liquid substance, was absorbed by the materials with which it came into contact. Organic residue analysis (ORA) can therefore detect the chemical traces of grape juice, wine or vinegar and help to identify wine production installations, as well containers used for the storage and transport of

[1]ERAAUB, Institut de Arqueologia de la Universitat de Barcelona, INSA-UB, Departament d'Història i Arqueologia, Universitat de Barcelona. The paper is part of the I+D+i Projects RACA-Med I-II (HAR2017-84242-P and PID2020-113409GB-I00), financed by the MCIN/AEI/10.13039/501100011033, and part of the activities of the ERAAUB Consolidated Group (2021 SGR 00696), thanks to the support of the Comissionat per a Universitats i Recerca del DIUE de la Generalitat de Catalunya, the INSA-UB, María de Maeztu Unit of Excellence 2021, financed by the Ministerio de Ciencia e Innovación, Programme of the Spanish State Research Agency (CEX2021-001234-M), and the Institut d'Arqueologia de la Universitat de Barcelona.

wine (e.g. *dolia* or *amphorae*) and its consumption in a variety of different pottery vessels such as cups, jugs, strainers or jars.[2]

Since attention was first called to the fact that we can identify wine residues in archaeological materials,[3] different methods have been developed, tested and used. This is accompanied by lengthy debate regarding the possibilities (or not) of the absorption, preservation and identification of wine residues. In particular, as one of the key biomarkers of wine, tartaric acid, which is soluble in water, has long been argued that it could not be preserved in archaeological materials and that when identified, it was the result of contamination. However, research carried out especially in the Mediterranean region has gradually enabled increased confidence in our ability to detect preserved wine residues. We have witnessed the development of methods using first liquid chromatography and later also gas chromatography coupled with mass spectrometry (GC-MS) and the development of suitable extraction procedures. The method used for the analyses discussed in this chapter is one that has been developed at the University of Siena since 2000, and since refined at the University of Barcelona.[4] An alternative method using GC-MS, and which is also commonly used today, is that developed by Garnier and Valamoti and later used by other researchers.[5] These diverse developments are building our understanding of the history of wine, providing data on its ancient production and consumption and potential trajectories from the eastern to western Mediterranean.

What follows is the culmination of studies carried out at the universities of Siena, Calabria and Barcelona over the last fifteen years. This included the study of modern contexts and experiments to develop reference data that allows a better characterization of the residues preserved in archaeological materials, followed by analysis of archaeological materials from a variety of contexts, geographies and time periods.

Methodology

The methodology applied, as regularly occurs in the archaeological sciences, and archaeometry in particular, consists of the application of three phases: 1) sampling; 2) analysis of the samples; and 3) interpretation of the data. During the first phase, attention must be paid not to contaminate samples. While contamination from plastic bags can be recognized due to the presence of abundant phthalates, it is better if samples are stored dry in aluminium foil until they are cleaned and sub-sampled in the laboratory. Similarly, it is better if samples are not washed to minimize possible causes of failure in the

---

[2]Pecci et al. 2013, 2020; Garnier and Pecci 2021.

[3]McGovern et al. 1996, McGovern 2004. Also see the contribution of McGovern in this volume.

[4]Pecci et al. 2013, 2020.

[5]Garnier and Valamoti 2016; Drieu et al. 2020. See the contribution by McGovern in this book for an extended overview and discussion of a range of current ORA techniques.

identification of wine biomarkers (though wine residues are still regularly identified in washed materials). Any washing must be done gently and without additional substances (i.e. acids).

The amount of sample required depends on the methods of extraction and analysis to be performed. It is also crucial to embed suitable extraction methods and analytical and control protocols to minimize any human bias and ensure the rigorous identification of appropriate chemical markers. For example, in the laboratory, any contact by human skin or gloves with the materials must be avoided and aluminium foil must always be used. Before any sample is crushed, they must be washed with acetone, chlorophorm and the other solvents used in the extractions, at which time they can then be pulverized using glass or onyx mortars.

Studies carried out in the past, in which only lipid extraction was carried out, did not identify wine residues, not because wine was not present, but because it was not possible to identify it with the lipid extraction.[6] In those cases, only indirect evidence of the presence of wine could be inferred from the identification of a resinous coating on, for example, *amphorae*. However, we now know that this is not a reliable marker – while wine *amphorae* could be coated with pitch, so too could *amphorae* that contained fish sauces and some that carried plant oils.[7]

To be able to confidently state that a container (e.g. ceramic jar) or feature of a production installation (e.g. vat or *dolium*) was in contact with wine, specific markers must be identified. Tartaric acid has long been considered the most important chemical marker for wine. Although it is also present in other fruits, such as the tamarind, in the western Mediterranean, especially for the Etruscan, Roman, Late Roman and medieval period, it can usually be considered a marker for grapes. In other regions, where fruits particularly rich in tartaric acid can be found and in which wine consumption is not already known, ORA must be integrated within a broader multidisciplinary study which involves, besides the archaeological context, also botanical evidence to confirm the validity of results (though this should also be best practice in all regions).

Other markers useful for the detection of fermented grape juice can also be identified, including succinic, malic, fumaric, maleic and citric acids. In particular, the proportion between malic and tartaric acid has been hypothesized to confirm the presence of wine residues;[8] however, at least with the methods utilized below, this is not a valid methodology, as distribution varies with time and ageing, and this assumption requires further confirmation.[9] As discussed below, a recently enriched vessel displays more concentrated tartaric acid compared to malic acid, while when it is aged the proportions change. With the extraction method used here, at an archaeological level malic acid is always more concentrated than tartaric acid.

---

[6]E.g. Heron and Pollard 1987.
[7]Romanus et al. 2009; Pecci and Cau 2010; Garnier and Pecci 2021.
[8]cf. Drieu et al. 2020.
[9]Pecci et al. 2013, 2020.

It is also impossible at this stage to differentiate between wine, vinegar and other grape derivatives – a critical limitation in our interpretive toolkit, given the common use of both wine and vinegar by Roman cultures. Recently, suggestions have been made that succinic acid is a marker of fermentation, and therefore its absence confirms the presence of unfermented grape juice.[10] Interestingly, succinic acid is often identified even in parts of production installations that were in contact with unfermented grape juice, including vats or press components, probably because fermentation often starts very quickly. Syringic acid has also been suggested to be a marker for red wine.[11] However, it could also derive from contamination, as it is present in cork (often used to seal *amphorae*) and other substances, so attention must be paid in the use of this biomarker, too. The importance is thus clear of integrating analytical data with information recovered from the archaeological context, along with different types of literary evidence as much as possible. Finally, control samples should always be taken and analysed from the sediment of the stratigraphic units where archaeological samples were recovered to check for post-depositional contamination.

Identification of wine residues in experimental and archaeological materials

Analysis of experimental materials

To determine the ability and efficacy of ORA to detect wine residues, various experiments have been implemented. The initial studies described here were carried out between 2002 and 2005 at the University of Siena and continued in the following years.[12] A variety of materials and situations were tested, including ceramic vessels used to store and cook wine and grape juice, and mortar bricks made of lime mixed with volcanic scoriae to simulate lime mortars, which were enriched with wine. The storage of wine in jars lasted six months and cooking involved heating wine ten times in an unglazed pan. A fragment of the ceramic vessels enriched with wine and a control sample were put in aluminium foil and then in an oven at 70°C to simulate an ageing process. Simultaneously, half the experimentally enriched ceramic vessels were buried at the archaeological site of Donoratico in Tuscany and sampled both after six months and seven years.[13] To ensure that contamination was not responsible for the presence of tartaric acid, both a control sample from the oven and a sample of the soil where the sherds were buried were analysed.

Results indicated that the proportions among acids vary depending on the processes of storing, cooking and ageing of wine. The concentration of tartaric acid often decreases, while other acids relatively increase (e.g. malic acid). However, this was not a rule. In the

---

[10]Garnier and Valamoti 2016.
[11]Guasch-Jané et al. 2004.
[12]Pecci 2005; Pecci et al. 2013, 2020.
[13]Partial results were published in Pecci et al. 2013.

case of a cooking pan in which wine was cooked twenty times and afterwards buried for seven years, tartaric acid was detected at over three times the level of malic acid.[14] This provides support to the suggestion that the proportion among acids is not reliable for the identification of grape juice residues and that it cannot be used to differentiate grape juice from the juice of other fruits using this method. This reinforces the notion that it is always important to consider other indicators (e.g. botanical remains and contextual information) to confirm the possibility that wine was produced, stored or consumed using a specific vessel. The analysis of experimental samples and archaeological samples from contexts that are almost certainly related to wine, however, confirms the identification of tartaric acid together with succinic and malic acids, as well as fumaric and citric acids. One example is the analysis of Dressel 2/4 *amphorae* recovered at the Vesuvian site of Oplontis B in Italy.[15]

Analysis of archaeological materials

The following case studies relate mainly to the western Mediterranean and focus on an investigation of the wine lifecycle: from production, through the study of architectural structures, including vats, floors and *dolia*, the storage and transport of wine in *dolia* and *amphorae*, and its consumption in drinking vessels, jugs and pots.[16]

Production installations can be sampled and studied to determine whether they were used for wine or to produce another commodity (most often, olive oil). This crucial question continues to plague the archaeological study of production – particularly the differentiation of wine and olive oil production facilities – and ORA is one of the most promising avenues to provide clarity.[17] In the Greek, Roman and Late Antique periods, production structures often utilized a shallow treading area, vat or floor (in Latin termed *calcatorium*) and presses of a variety of forms to extract liquid from grapes, which was then collected in vats or *dolia*.[18] In the vast majority of cases, only the bottom of a vat, part of a press, or fragments of *dolia defossa* (buried in the ground) are preserved. With such fragmentary preservation it is difficult to be sure that they were related to wine production and chemical analysis is crucial to untangle any hypotheses. It has long been known that ceramic material can be analysed to identify organic residues related to contents and coating,[19] but this is also true for mortars and plasters used to coat the vats and the floors of production installations.[20] Like ceramics, these too are a porous material that absorbs liquid substances to which they come into contact, as Luis Barba

[14]Pecci et al. 2013: Figure 6.
[15]Pecci et al. 2017.
[16]Pecci 2019.
[17]Pecci 2007, 2013; Pecci, Cau and Garnier 2013; Pecci et al. 2013. Also see Dodd and Van Limbergen in this volume.
[18]Brun 2003; Dodd 2022a.
[19]Evershed 1993, 2008.
[20]Pecci 2007, 2013, 2021; Pecci, Cau and Garnier 2013; Pecci et al. 2013.

demonstrated several years ago studying Mexican contexts, and as we have demonstrated also for Mediterranean contexts.[21] This characteristic allows the identification of the substance contained in the vats and absorbed by the bottom of presses constructed using *opus signinum* or *cocciopesto* lining.[22]

A variety of vats, with particular attention paid to their lower levels where liquid pooled and maintained greater contact for longer periods of time, have been studied at different archaeological sites in Italy and Spain. Successful results at the archaeological site of Donoratico (Tuscany, Italy), which identified wine residues in a poorly preserved third century CE vat, led to the development of this research direction.[23] We have since tested vats for residues of wine production with positive results at the Late Roman site of Pla de Ses Figueres (Cabrera, Balearic islands), in a room annexed to the Late Roman basilica of Son Peretó (Mallorca, Balearic islands), and at the so-called 'Villa of Augustus' at Somma Vesuviana on the slopes of Mount Vesuvius in Italy.[24] Recently, this technique has also been applied at several Roman and Late Roman sites in Catalonia, including Badalona (in the Predi de la Estrella) and Can Modolell,[25] and in Italy at the rural Roman industrial site of Marzuolo in Tuscany and at a Roman Republican production facility at Mondragone in northern Campania.[26] As with the experiments and the case studies below, samples were recovered from the bottom of vats and from the soil of nearby stratigraphic units to check for post-depositional contamination.

Dolia are also a common feature of Roman production installations, whether for wine, oil or a range of other commodities, particularly in the central and western Mediterranean, with their use extending through the Late Roman period. They are thus an important target for ORA to discern between the production and storage of wine and other commodities. Analysis of *in situ dolia* with visible coatings was completed at the so-called 'Villa of Augustus' at Somma Vesuviana, covered by the later eruption of Mount Vesuvius in 472 CE, where analysis identified traces of both pitch and wine.[27] Other analyses that indicated chemical signatures for wine have been undertaken at a Roman installation identified in Aspromonte (Calabria, Italy),[28] the Roman villa del Pont del Traball (Barcelona),[29] and at a *cella vinaria* identified in the so-called *thermopolium* of region 5.4.6–8 at Pompeii.[30] On the other hand, the identification of wine residues in

---

[21]Pecci 2005, 2013.

[22]Pecci et al. 2013.

[23]Pecci 2005, 2007; Pecci, Cau and Garnier 2013.

[24]Allevato et al. 2012; Pecci, Cau and Garnier 2013; Pecci et al. 2013.

[25]In collaboration with Clara Forn (Museo de Badalona) and Victor Revilla (University of Barcelona) respectively. The results will be soon available with the publication of the proceedings of the congress 'El vino en la antiguedad', held at the Museu de Badalona in October 2022.

[26]With L. Crimaco and A. Carcaiso, of the Museo Civico Archeologico 'Biagio Greco' of Mondragone, unpublished.

[27]Allevato et al. 2012; Pecci, Cau and Garnier 2013.

[28]Pecci 2016.

[29]Investigated in collaboration with Atics and the Servei d'Arqueologia de Barcelona (Pecci 2018–2019, unpublished reports by Pecci, Inserra, and Cau). These data will also be available with the publication of the proceedings of the congress held in Badalona in October 2022.

[30]Pecci and Mileto, unpublished.

dolia with the remains of fish bones from the so-called *garum* workshop at Pompeii (Region 1.12.8) suggest that they could have been used in wine production or storage and then reused for the production of *garum* (fish sauce), or, alternatively, in the production of a sauce made using fish by-products and wine (i.e. *oenogarum*).[31] Finally, wine residues were identified in *dolia* at the so-called Etruscan 'House of the Amphorae', at the site of La Marsiliana (Manicano, Italy), suggesting it might have been a wine production installation.[32]

Many vats and all the *dolia* analysed that presented positive results for wine production in the above examples were coated with *Pinaceae* resin or more commonly pitch, which was produced by burning *Pinaceae* wood. These products were used to waterproof vats and ceramic jars, and possibly to preserve and give flavour to the wine. They might also be used to prevent erosion caused by the fermentation process.

Sulphur has also been identified in samples of *amphorae* and other ceramic vessels from several archaeological sites, and it has been suggested that it was likely used as a preservative for wine.[33] However, so far, sulphur residues are almost never identified in pottery from production facility contexts – the only known case comes from the *dolia* of the Aspromonte installation.[34] Similarly, no evidence of solid sulphur, possibly used to coat ceramic vessels, has yet been recovered at production installations. The exploration of sulphur in archaeological contexts, its origin and its role within the vinicultural lifecycle thus forms an important avenue of future investigation.

After production and its initial fermentation, wine was stored and transported. While jars and other larger ceramic vessels were regularly used for this in antiquity, in certain regions and periods, containers that were made from perishable materials, such as wooden barrels, were used and could also be analysed using ORA. These are, however, rarely identified archaeologically.

Amphorae, as one of the most common wine storage and transport containers often need to be analysed with ORA to determine their contents. In fact, traditionally, specific types and forms of *amphorae* have been associated with the transport of specific commodities (e.g. wine or olive oil); however, ORA is nuancing our reliance on morphology to determine commodity and is providing new data on wine production and transport in unexpected *amphora* types, thus becoming fundamental in our understanding of wine history.[35] For instance, ORA confirmed the presence of wine in Dressel 2/4 *amphorae* from Oplontis near Pompeii,[36] Dressel 1B *amphorae* recovered in Sa Mesquida (Mallorca) and originating from the Tyrrenian area,[37] and in some *spatheia* (in particular Keay 25.1) recovered from the wrecks of Heliopolis I, Ponte de la Luque B, and the bed of

[31] Pecci et al. 2018.
[32] Pecci and Giorgi, forthcoming.
[33] Garnier et al. 2011; Pecci et al. 2017, 2020; Garnier and Pecci 2021.
[34] Pecci 2016.
[35] cf. Garnier and Pecci 2021; Pecci et al. 2021a, 2021b.
[36] Pecci et al. 2017; Toniolo and Pecci 2020.
[37] Cau et al. 2018.

the river Rhône[38] – although, in the case of the latter, wine was not necessarily the only content of *spatheia*, as previous analyses also identified castor oil in examples from the port of Classe.[39] Wine (or its derivative products) has also been identified in ovoid *amphorae* from *Baetica* and in some early African *amphorae* found at Pompeii, indicating that wine was potentially produced in these regions.[40] While some of these results were unexpected due to a traditional belief that they carried other commodities, the fact that most of the vessels revealed markers for wine (or grape derivatives) suggests that it might have been their original content. Moreover, and perhaps most importantly, the results of ORA also suggests that *amphorae* were commonly reused, even for the same commodity.[41]

The last phase of the wine life cycle is consumption. Different forms of ceramic tableware have also been analysed using ORA, such as jugs of a range of sizes that were used to serve wine, and other types of containers used in the preparation and consumption process. In Tuscany, a small medieval fine ware jug from Grosseto, along with coarse ware jugs from Piombino, provided indicators for wine.[42] Analyses are currently in process on bottles and jugs recently excavated from the *thermopolium* of Region 5 at Pompeii, and on *sigillata* (a type of polished, usually red pottery with impressed designs) found during the Thameslink project near London in the UK. Interestingly, ORA has also identified indicators for wine or grape derivatives in cooking pots, perhaps shedding light on the heating of wine or vinegar for consumption or to make sauces, as suggested by the ancient literature.[43] Examples come from a wide chronological span, from Roman and Late Roman through to medieval.[44] Recently, preliminary experiments on the analysis of glass vessels have been carried out, suggesting that they could be tested and provide valid data, although a substantial amount of future work is required to make this procedure standardized.

Conclusion

The development of ORA, and in particular diverse methods of extraction and analysis, continues to build a more solid and nuanced evidence base for understanding wine production, transport and consumption across Graeco-Roman antiquity. The examples provided in this chapter illuminate the possibilities of ORA and a pressing need to integrate controlled and consistent sampling processes and analysis throughout archaeological research projects, especially during excavation. Our ability to chemically discern indicators for wine, other grape derivative products, and additional non-grape products that might present similar markers will continue to improve and, in doing so, fine tune our interpretation of materials, uses and contexts.

---

[38] Woodworth et al. 2015.
[39] Pecci et al. 2010, 2018.
[40] Pecci et al. 2021a, 2021b.
[41] Pecci et al. 2017; Garnier and Pecci 2021; Pecci et al. 2021a, 2021b.
[42] Valdambrini et al. 2007; Pecci 2009.
[43] E.g. Apicius (*De Re Coquinaria* 4.2.138) mentions sprats or smelts, and pumpkin (*De Re Coquinaria* 3.4.73).
[44] Pecci 2005, 2009; Inserra et al. 2015; Toniolo and Pecci 2020.

CHAPTER 6
A CASE STUDY ON THE IMPORTANCE OF RESIDUE ANALYSIS FOR THE INTERPRETATION OF THE ESTATE ECONOMY: THE VILLA OF SANTA MARINA (ISTRIA, CROATIA)

Corinne Rousse, Nicolas Garnier, Gaetano Benčić and Davor Munda

Abstract

At the time of Augustus, the Istrian peninsula in the northern Adriatic became famous for its oil production, however wine was also produced and perhaps oriented more towards a local market. Indeed, many presses found at villas in this region must have been used to produce wine but are architecturally challenging to discriminate between commodities. Organic residue analysis is thus a critical analytical option to utilize. In this chapter we explore the use of residue analysis at a large first century CE Roman villa built on the bay of Santa Marina (Istria, Croatia). Samples of plaster and *spica* were analysed with the aim of detecting chemical markers for grape wine or olive oil, with signals for malic and tartaric acid present across the entire group suggesting the production of wine. This methodology, combined with more traditional structural analysis, generated a new understanding of the villa's production economy and illuminated vinicultural activities, which were until now only visible indirectly.

Introduction

At the time of Augustus, the Istrian peninsula in the northern Adriatic became famous for its oil production.[1] The multiplication of villas along the coast reflects the rise of a speculative agricultural regime, which combined substantial oil production, mainly intended for export to northern Italy and the Danubian *limes*, but also the production of wine, oriented more to local consumption. Recent archaeobotanical studies underline the clear development of the growing of vines, olive trees and other fruit trees from the

[1] Pliny, *HN* 15.8 ; Martial, *Ep.* 12.63 ; Paus., 10.32.19; Cassiod., *Var.* 12.22–26. See Matijašić 1993, 1998; Bezeczky 1998; Tassaux et al. 2001.

first century BCE, when the colonies of Pula (*Pola*) and Poreč (*Parentium*) were founded.[2] Viticulture is attested through iconography, epigraphy and the presence of large cellars with *dolia* in the *pars rustica* of some villas, in particular those of Brioni island.[3] At the large *figlina* of Loron, in the territory of *Parentium*, small-scale production of wine *amphorae* has recently been identified (Dressel 2/4 *amphorae* and flat-bottomed *amphorae*) alongside thousands of oil *amphorae* of the Dressel 6B type intended for export.[4] In light of this, some of the numerous presses found in the Istrian villas and on their *fundus* must have been used to produce wine for their owners. As the presses are architecturally similar for the production of both commodities, it can be difficult to distinguish wine or oil production, without other archaeological evidence.[5] Biochemical analyses are therefore useful in their attribution.

Oil and wine production at the villa of Santa Marina

Since 2014, research carried out by the Museum of the territory of Poreč (Zavičajni muzej Poreštine – Museo del territorio parentino) and the Centre Camille Jullian (Aix Marseille University, CNRS) has focused on a large Roman villa facing the bay of Santa Marina, about 10 kilometres north of the colony of *Parentium* (Istria, Croatia). The residence was part of an aristocratic estate that also included a large *figlina* at the neighbouring site of Loron, connected to the villa that mainly produced oil *amphorae* (Figure 6.1). By studying the stamps on Dressel 6B *amphorae* produced at Loron, we know that the complex was erected around 10 CE, then became an imperial property at the time of Domitian.[6] The villa, measuring *c.* 100 x 50 metres, was built 400 metres to the north of the *figlina*, on sloping ground, with two superimposed terraces facing the sea. The residential rooms seem to be situated mainly on the lower terrace around a large courtyard. Service buildings are identifiable on the upper terrace, including a set of rooms for oil and probably wine production, corresponding to the *pars rustica* of the villa (Figure 6.2).[7]

---

[2]Kaniewski et al. 2018; Vaschalde et al., forthcoming.

[3]For iconography, see the altar of Volusius Hermes in Poreč; epigraphy: Matijašić and Tassaux 2000: 11–12. The synthesis of archaeological data is provided by Matijašić 1993, 1998, updated by D. Bulić and R. Matijašić in the recent *Vine-growing and Winemaking in the Roman World* conference.

[4]Tassaux et al. 2001. On the local production of flat-bottomed *amphorae* and Dressel 2/4 see Maggi and Marion 2011.

[5]Brun 1993: 523–6.

[6]Tassaux et al. 2001. The stamps mention the names of famous owners, beginning with T. Statilius Taurus Sisenna (consul 16 CE) who built the complex around 10 CE. The estate then became an imperial property from the time of Domitian (81 CE) to at least the Hadrianic era. The high activity of the *figlina* continues during the entire third century CE (Vaschalde et al. 2021) and most probably, as with the villa, was under the control of the emperor. Both sites are completely abandoned during the fourth and fifth centuries CE.

[7]Rousse et al. 2020, 2022.

Figure 6.1 Composite aerial image and plan of the Santa Marina villa and estate in its broader context. Courtesy V. Dumas and N. Basuau, AMU-CNRS-CCJ.

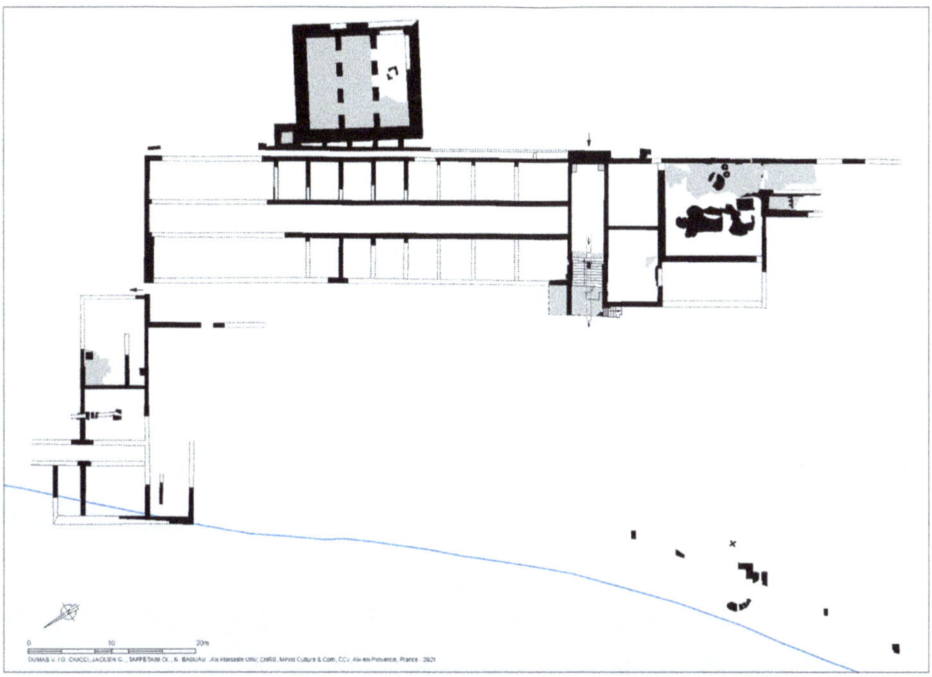

Figure 6.2 Plan of the Santa Marina villa. Courtesy V. Dumas, AMU-CNRS-CCJ.

Olive oil production took place in the main room (11 x 11 metres). The remains of two olive mills of the *mola olearia* type and two large presses have been identified (Figure 6.3). Despite their destruction, numerous elements of the machinery can be identified, such as the rectangular press beds built with a masonry foundation and one monolithic block that was used as a frame for the wooden uprights (*arbores*) onto which the lever was fixed. It is probable that the press beams may have reached a length of 10–11 metres with the use of a winch to power the lever.

On the southern side of the room a rectangular basin over 10 metres long was connected to the presses by a pipe. The pipe is 1.8 metres large and very shallow in depth (0.57 metres) and the tank is sealed with a thick layer of mortar, coated with plaster and *opus spicatum* lining (Figure 6.4). In a later phase, the basin was divided by a small transverse masonry wall which rested on *imbrices* allowing the liquid to be filtered. To the west of the basin, another room located below, which has not been studied yet, could have been used for storage. These structures are very similar to those observed at the villas of Brioni island (Val Catena/Uvala Verige, Monte Collisi/Kolci and Val Madona/Uvala Dobrika),[8] in which the presence of great cellars, that could hold up to 60–80

Figure 6.3 The *pars rustica* at the Santa Marina villa with two presses, millstones to produce olive oil, and the basin. Courtesy E. Botte, AMU-CNRS-CCJ.

[8]Gnirs 1915; Matijašić 1998.

Figure 6.4 The settling basin with *opus spicatum* paving at the Santa Marina villa. Courtesy C. Rousse, AMU-CNRS-CCJ.

dolia, associated with rectangular basins of large dimensions suggest their use for wine production.[9]

To test this hypothesis in the case of the villa of Santa Marina, four samples were taken from the long basin: one from the mortar of the northern wall (under the trace left

[9]Brun 2004b: 50–6. The villa of Val Catena/Uvala Verige had a large cellar that could hold 40 *dolia* and probably a second one containing *c.* 20 *dolia*. These facilities suggest a vinification capacity of 600–900 hectolitres. The villa of Monte Collisi/Kolci had a long cellar of about 100 *dolia* and thus storage capacity of *c.* 1,500 hectolitres (cf. Matijašić 1993: 249 who provides 60 *dolia*). It should be remembered that these facilities were also interpreted as related to the production of oil, which was the region's main export product (Gnirs 1915; Bezeczky 1998: 53). The enormous storage capacities, however, strongly suggest the production of wine (Brun 2004b). See also Matijašić 1993: 249–51.

by the pipe); two from the plaster covering the pavement made in *opus spicatum*; and one directly from the pavement (*spica*).

Biochemical analysis

Methods

To identify the organic markers embedded in the porous inner surface of the basin, we analysed the four samples (plasters and *spica*) according to a well-established methodology adapted for the research of grape and wine markers.[10] This method was developed by optimizing the extraction of grape markers from model ceramic sherds permeated by different grape products: white grape, dark grape, fresh pressed juice, the same juice after fermentation, white and red wines obtained from different fermentation and ageing processes (e.g. natural fermentation, port wines, *vin jaune*, naturally oxidized wine over several years). The detection of grape markers, especially malic and tartaric acids, was quasi quantitative with impregnations of diluted pressed juices. Blank samples of non-permeated ceramics gave no aldaric acids, and thus no false positives. The implementation of gas chromatography-mass spectrometry (GC-MS) allows the identification of grape and fermentation markers even if other organic materials are present. High-resolution mass spectrometry using an Exactive mass spectrometer, with an Orbitrap-type analyser, greatly enhanced the sensitivity helping to solidify the results.

The side of the plaster is washed under water to eliminate sediment residues. After air drying, the surface is cleaned with a Dremel rotary polisher and the first layer discarded. Then the plaster is drilled to a depth of 2–4 millimetres with the same Dremel tool. An aliquot of cholestane solution (internal standard, 0.1 mg/mL) is added and the powder obtained is extracted a first time according to the classic method using a mixture of dichloromethane/methanol 1:1 v/v (2 x 3 mL). The combined extracts are centrifugated and the organic layer filtered over silica gel (230–400 mesh). The filtered extract is evaporated to dryness under a slight nitrogen flux. Pyridine (5 μL) and BSTFA (50 μL) are added for the derivatization (30 minutes, 80 °C) and then the mixture is evaporated to dryness on a nitrogen flux and dissolved in cyclohexane (100 μL). An aliquot (1 μL) is injected in the GC-MS using a high-resolution GC column (Phenomenex Zebron-5MSi 20 m x 0.18 mm i.d. x 0.18 thickness phasis) and an Orbitrap high-resolution mass spectrometer (EI source at 230°C, 70 eV, mass range from 50 to 650 amu with a 60,000 resolution).

The extracted plaster powder is then dried under a vacuum. An aliquot of cholestane solution (internal standard) is added, followed by cyclohexane as solvent and co-solvent (1 mL, aiding the transbutylation reaction to proceed, an innovative reaction combining acid-catalysed hydrolysis and esterification), a solution of boron trifluoride in ether

[10] Garnier and Valamoti 2016.

(200 μL), and butan-2-ol (400 μL). The vial is placed under a microwave at 100°C for one minute, then at 150°C for two minutes. After neutralization of the chemical mixture with a sodium carbonate solution and extraction with dichloromethane (2 x 2 mL), the combined organic layers are washed with water (2 mL), dried magnesium sulphate, and filtered over nanosilica gel. After evaporation to complete dryness, the solid residue is derivatized as before with pyridine/BSTFA, evaporated to dryness and redissolved in cyclohexane. 1 μL is injected for GC-MS analysis.

Results

The results of these biochemical analyses confirmed that the long basin of the villa of Santa Marina contained grape juice, with malic and tartaric acid present in all the samples. Syringic acid, coming from malvidin and its derivatives, an anthocyanin suggested to be specific to dark-coloured berries and grapes, was also detected. Two samples from the bottom of the basin revealed very clear traces of grape and another fruit rich in malic acid (mainly *Rosaceae* such as apple, pear or quince, but also cherry or hawthorn). They also contained traces of acids from alcoholic fermentation (malonic, maleic, succinic, fumaric and pyruvic acids) indicating a fermented drink, in this case probably red wine. Only the sample from the north wall of the basin indicated traces of a strong permeation with fatty acids, suggesting that it was in contact with vegetable oil. During the survey, traces caused by oleic acid were also noticed on the mortar.

Discussion

As Brun pointed out for the Brioni villas, producing and having one's own wine was a matter of prestige for owners.[11] We should not be surprised that a villa like that at Santa Marina produced both oil and wine since its foundation – the question is to understand *where* the wine presses and the cellars are. It is certain that the large presses discovered next to the basin were used to produce oil in the later phases of the villa's activity. The presence of millstones is a strong indication of this, as well as the traces of vegetable fatty acids on the mortar. But were the presses first used to produce the wine that was detected on the structure of the basin? That the tank had to be subdivided later to facilitate the decantation of oil in the separation process could be a sign of this substitution.

The same framework can be observed in a large tank belonging to the *pars rustica* of the nearby villa of Stanzia Blek (a long shallow basin, divided by a low wall in a second phase, already interpreted as a possible structure for wine production)[12] and in the basins of the complex of Barbariga, in the territory of Pola, clearly intended for industrial-scale oil

[11]Brun 2004b.
[12]Benčić 2006; Carre et al. 2011: 136–9 with Figure 14.

production.[13] On the other hand, at Santa Marina the entire *pars rustica* has not yet been studied and it is possible that vinicultural installations are located beyond the excavated area. In the later stages of the villa, a block that can be interpreted as a small anchor stone more suitable for a wine press was reused in a wall. Fragments of *dolia* indicating the existence of storage facilities were also found in the rooms surrounding the presses. By comparing the plan of the installations of the Brioni villas, which are very similar to Santa Marina, it is possible to suggest the presence of a cellar under the basin, like at Val Catena.

Conclusion

The chemical analyses performed in a basin nearby presses and outlined in this case study allow us to identify an essential part of the economy of the estate, for which only indirect data was available until now. This facilitates improved discussion regarding the status of the villa, from its foundation within a prestigious estate to its last phases of occupation. The results also highlight the importance of wine in Istrian villas alongside speculative olive oil production, which is not always easy to interpret based solely on archaeological remains. These results will be useful in defining future research at this villa and the other sites in the territory of the colony of *Parentium*.

[13]Matijašić 1998: 188–92.

PART II
ARCHAEOBOTANY AND PALYNOLOGY

CHAPTER 7
ARCHAEOBOTANY IN THE ARCHAEOLOGY OF WINE: CURRENT APPROACHES AND FUTURE POSSIBILITIES

Patrizia Basso and Diana Dobreva, with a contribution by Gianni de Zuccato[1]

Abstract

Interactions between archaeology and archaeobotany, chemistry and archaeometry have in recent years forged promising pathways to illuminate some of the remaining unresolved grey areas in the study of Roman viticulture and wine production. This chapter will focus on key topics for the study of vine-growing and winemaking: in particular, macro remains, ancient DNA, charcoal, pollen and phytoliths. We examine for each how appropriate methodologies or field practices can significantly advance our historical knowledge. This chapter is enhanced by close collaboration between the authors and archaeobotanists, as well as experts in chemistry and archaeometry, illuminating how a greater degree of communication and collaboration between archaeologists and scientists can improve the historical and archaeological reconstruction of ancient viticulture and winemaking. The discussion is restricted mainly to the western Mediterranean, with reference to northern Italy and the Valpolicella area, where a multidisciplinary project has since 2020 produced a wealth of data that is enriching our understanding of viticulture and wine production in Graeco-Roman antiquity.

[1]We would like to thank all the colleagues involved in the project *In Veronensium mensa: Food and Wine in Ancient Verona*, in particular Diana Bellin and Giada Bolognesi for sharing preliminary data on ancient DNA of grape seed analyses and Marco Marchesini for helpful feedback on an early draft. Many thanks also to Nicolas Garnier who performed the organic residue analyses and to Claudio Capelli for conducting the petrographic analyses on the pottery samples. Thanks to Alberto Manicardi and the SAP s.r.l. for the use of graphic documentation of the Negrar site. Our appreciation goes also to Gianfranco Valle for informing us about an ongoing project on some recent core samples on the city of Verona. Finally, we would like to thank the editors Emlyn Dodd and Dimitri Van Limbergen, the anonymous reviewers for their insightful comments and guidance and Max Norman for assistance with editing this article. The authors of the following sub-chapters include: introduction by PB and DD; macro remains, ancient DNA, wood and charcoal analysis by PB; pollen and phytolith analysis by DD; considerations by PB; a case study by GdZ and PB; vine growing analyses by PB; new approaches on the study of winemaking and trade and conclusion by DD.

Introduction

The continuous and constant interaction between archaeology, archaeobotany, molecular biology, chemical residue analysis and archaeometry has in recent years promised to illuminate some unresolved grey areas in the economy of ancient wine. In this chapter, we will focus in particular on the contribution of archaeobotany to the study of vine-growing and winemaking, addressing five key methodologies: analyses of macro remains, ancient DNA of grape seeds, charcoal, pollen and phytoliths. Since we are not archaeobotanists, but archaeologists in dialogue with archaeobotanists, we will not dwell on methods of sampling and study in the laboratory.[2] Instead, we try to reflect on the answers that an archaeologist would obtain from these analytical approaches in relation to the three phases of the wine economy: vine-growing, winemaking and wine marketing. These are in fact distinct activities both in terms of the spaces/tools used and the personnel involved, and therefore require different field research methods and analyses.

This chapter concentrates largely on the Roman period, although archaeobotanical analyses have so far been widely applied to sites from pre- and protohistoric periods, with a particular focus on the chronology and geographical playgrounds of the domestication of the grapevine for winemaking.[3] There is less research of this kind, however, for the Roman period, when wine processing had been firmly established for both domestic and market use, and was traded on a vast scale.[4] Without any presumption of being exhaustive, the following discussion touches on a few particularly significant studies from the Mediterranean area, with particular reference to the territory of Verona, where since 2020 a multidisciplinary project has produced data that has enriched our understanding of ancient vine-growing and winemaking in an area still famous for its wines.[5]

---

[2]Lodwick and Rowan 2022 provide a recent overview analysing the interaction between archaeobotany and classical archaeology, including some more general comments on sampling, preservation and methodology.
[3]For a recent summary of the current understanding of protohistoric wine in Italy, with an extensive collection of archaeobotanical case studies, see Acconcia and Piergrossi 2021. On the domestication of the grapevine, addressed through morphometric analyses, see recently Pagnoux et al. 2021 (Greece); Moricca et al. 2021 (the Phoenicians and western Mediterranean); similar issues are also addressed through genetic studies in Magris et al. 2021; Mercati et al. 2021; for Italy, see also Marchesini, Marvelli and Rizzoli 2021.
[4]For an overview of archaeological evidence for wine production in Roman Italy, see Dodd 2022a.
[5]The project *In Veronensium mensa: Food and Wine in Ancient Verona*, is financed by the Cariverona Foundation and is conducted by the University of Verona's Department of Culture and Civilisation, with collaboration from the Department of Biotechnology, as well as the Soprintendenza Archeologia, Belle Arti e Paesaggio per le province di Verona, Rovigo e Vicenza, local museums, University of York, and the Accademia di Agricoltura, Scienze e Lettere di Verona. The project aims to identify the historical roots of the production and consumption of both wine and food in the Verona area from prehistoric times to the Middle Ages, using a diachronic and interdisciplinary approach. For initial results, see Basso et al., forthcoming.

The role of archaeobotanical analysis in the study of ancient wine

Archaeobotany is a comparatively recent discipline and its areas of investigation have expanded considerably in recent decades to include phytolith and molecular analyses, in addition to the more traditional pollen, charcoal and macro investigations. Genetic sequencing has also been used to identify plants and study their evolution. For all these approaches, the importance of field sampling protocols must be emphasized, as the results of the analyses are only valid if the samples are well excavated, well dated, well preserved and uncontaminated after their deposition.[6]

Macro remains

Archaeobotanical study of the macro-scale remains left behind by spontaneous or cultivated plants (seeds, fruits and floral annexes) helps us to reconstruct the ancient environment, as carpological finds often make it possible, for example, to identify a plant's species. Carpological analyses thus include pips and even in some cases grape skins and berries found during excavations.[7] The sites, environments or stratigraphic units where these elements are best preserved — and can therefore offer more significant data — are those characterized by waterlogging, in particular wells. This was the case, for example, at the two farms of Guasquinoy (France), in which abundant crushed residues including pips, grape berries, fragments of grape skins, and pedicels were recovered.[8] At these sites, the residues of winemaking prove that grapes were processed to produce wine, but at others they may have been consumed as food. For example, the macro remains of grapes frequently found in tombs indicate bunches placed as offerings of food to the deceased.[9]

When grape pips are well preserved a morphometric study can be applied. Considering that the shape of grape seeds seems to be controlled at a genetic level, while soil characteristics, climatic and ecological parameters seem to influence only their size, this analysis may in fact distinguish the wild grapevine *(Vitis vinifera* L. spp. *sylvestris)* from the cultivated vine (*Vitis vinifera* L. spp. *vinifera*).[10] Grape seeds from wild plants are often considered to be smaller, shorter, rounder and more heart-shaped with a 'slightly pronounced beak, flattened or slightly prominent in the ventral part, with two narrow and deep furrows separated by a longitudinal bridge and with an evident depression on

[6]For methodological issues in archaeobotany, see Lodwick and Rowan 2022.

[7]For an example of the discovery of grapes from the Roman period in Italy, see, in addition to the Vesuvian area, the site of the villa in the locality of Mercatello in Castello di Serravalle (Bologna), which also yielded pips and vine wood (Badiali 2009: 90, Figure 7).

[8]Figueiral et al. 2010; Bouby at al. 2013.

[9]For an analysis of the presence of *Vitis vinifera* remains in northern Italian tombs, see Rottoli and Castiglioni 2011: 502.

[10]On the shape and size of grape seeds, see Terral 2002.

the back, while those from cultivated plants are larger, long, slender, oval or pyriform with a long and prominent beak' (Figure 7.1).[11] Since the 1990s, morphometric analysis has become increasingly refined, relying on increasingly complex biometric parameters and computerized visual analysis extending to statistical and multivariate investigations[12] in some cases even hypothesizing the grape varieties of the grape seeds.[13] However, we should be very cautious in interpreting grape use based only on these analyses. Indeed, we cannot be confident that if excavated grape seeds belong to the wild vine then grape consumption was related to the harvesting of wild fruits, nor if they appear to be of a cultivated type that wine making was occurring.[14]

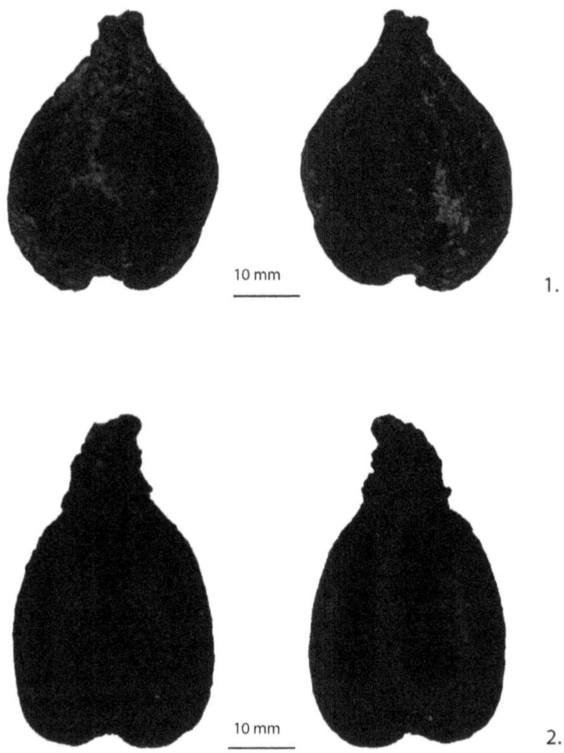

Figure 7.1 Examples of grape seeds classified as wild, *Vitis vinifera* L. spp. *sylvestris* (1) and as domestic, *Vitis vinifera* L. spp. *sativa* (2) found at Negrar in Valpolicella, Italy. Photographs by D. Bellin and G. Bolognesi, FaW Project, University of Verona.

[11]Marvelli 1999: 36.
[12]From the analyses of Mangafa and Kotsakis 1996 or Perret 1997 to more recent studies (e.g. Terral et al. 2010; Bouby et al. 2013; Orrù et al. 2013). Also see the contributions of Marchesini et al. and Carmenati et al. in this volume.
[13]Cf. Terral et al. 2010.
[14]Grasso and Fiorentino 2012: 688.

Ancient DNA

Genetic analysis provides more robust data to distinguish wild and cultivated grapes.[15] Thanks to comparisons with the DNA of modern grape varieties, such analyses make it possible to identify cultivars from sites when found in good condition, carefully sampled and analysed in controlled environments. Over the last decade, DNA research has advanced considerably due to the development of new extraction methods and sequencing libraries,[16] opening up new horizons for studying not only past landscapes and their evolution, but also the history of viticulture, spread of grapevines, and the allochthonous or autochthonous origins of grape varieties. Insights into the origins of modern grape varieties and historical cultivation processes have come from projects in Tuscany[17] and Lucania.[18] Most important, though, is a French project that was able to identify a sample dated to 1050–1200 CE as 'Savagnin', a grape variety still cultivated in France for vinification, demonstrating the continuous use of a grape variety for at least 900 years.[19] In some cases, such analyses have also shed light on genetic changes within grape varieties. A study carried out at the Etruscan-Roman site of Cetamura, in the Chianti region of Italy, combined genetic and morphometric analyses on waterlogged grape seed remains to highlight the presence of different cultivars in the area, as well as an overall increase in grape seed size over time.[20]

Wood and charcoal analysis

Study of the anatomical structure of wood and charcoal recovered during excavations can confirm the presence of grapevines in the area, given that its wood was not usually used for construction.[21] Wood and anthracological analyses can also provide insights into vine-growing strategies.[22] Indeed, by finding grapevine wood/carbon along with that of other shrubs or trees (e.g. the elm in particular), we can discern the cultivation method known as 'maritata' or 'della piantata'. There are numerous attestations in Roman literary sources on the relationship between the vine and other planted trees, such as the elm (Virgil, *Bucolic* 2.70; Pliny, *HN* 14.3.12), but also the poplar, especially in Campania (Pliny, *HN* 14.3.10) or the *rumpontinus* or loppio in the Transpadane area (Pliny, *HN*

---

[15]For an integrated analysis of morphometrics and genetics, cf. Bacilieri et al. 2017.
[16]Cf. recent studies by Wales et al. 2015.
[17]Cf. Ciacci, Rendini, Zifferero 2012 and the contribution of Zifferero in this volume.
[18]The Basivin Sud project maps the region's ancient wine-growing heritage using genetic analysis of archaeobotanical material: Alba et al. 2015; Del Lungo 2017.
[19]Madrigal et al. 2019.
[20]Aversano et al. 2017. The increased number of pips around 200 BCE was explained not only by the selection and introduction of new varieties, but also by changes in vineyard management practices and in particular by the introduction of pruning, which allegedly created more favourable physiological conditions for vine-growing. The dating of the grape seeds and the proposed interpretation have been corrected by de Grummond 2018, who explain that the pruning hook was known to both the Etruscans and the Romans from a very early date.
[21]Cf. Grasso and Fiorentino 2012: 688.
[22]Anthracology is the study of charcoal macroremains.

14.3.12).[23] It is evident, however, that one cannot simply equate the presence of wood from certain shrubs with viticulture, because plants of these species may have characterized an ancient landscape independently of vine cultivation.

However, it should be noted that grapevine wood/charcoal might also evidence fruit used as food and not for wine production – Pliny (*HN* 14.4.40-42) himself mentions several grapes for eating (*escariae appellatae*). Moreover, these analyses rarely succeed in distinguishing the remains of *Vitis sylvestris* or *Vitis vinifera*, as the wood in excavations is mostly preserved in carbonized form. Of particular interest, therefore, are some cases where the structure of vine wood has been preserved, in vine shoots or stumps of a size that can be morphologically and anatomically identified as attributable to *Vitis vinifera*. The French site of Lattes provides key evidence in this regard, where both the harbour's wetland area and various wells have yielded large pieces of wood dating from Bronze Age to Roman era.[24] Notable among these discoveries is the rare testimony of a grapevine branch several decimetres thick, which shows traces of pruning carried out with a serpentine cutting tool, shedding light on how the vines were cultivated and worked.[25]

Pollen analysis

Pollen can attest to the presence of grapevines at a specific site or in a general area, but some characteristics of their distribution must be considered.[26] On the one hand, the grapevine is a plant that produces a limited amount of pollen, and even in regions where viticulture is widespread, vine pollen is usually not produced in large quantities.[27] On the other hand, certain pollen can fly far from the plants that produce them, and their presence in archaeological strata must be analysed with caution. Furthermore, while palaeoenvironmental analysis can confirm vine exploitation by detecting the presence of grapevine pollen, it does not advance hypotheses around the production of wine, unless supplemented by other pieces of evidence.

This limitation of pollen analysis can be overcome by combining it with other types of archaeobotanical analyses, as in the case of the Roman villa at La Lesse in Sauvian, located in the lower Orb valley in Narbonne Gaul, where pollen analyses revealed how vineyards were planted alongside an orchard, with the latter also confirmed by the thousands of macro remains found in two wells.[28] Pollen grains, charcoals and grape seeds were also found in the villa of Domagnano (Republic of San Marino) where wine making was evidenced structurally by a tank and *torcular* base. In northern Italy, Badia

[23]On some Emilian cases of the discovery of grapevine wood together with elm or maple, which led to the hypothesis of a 'piantata', cf. Marvelli 1999: 46–7; for a discussion of the system as a whole, see Van Limbergen 2020b.

[24]Cf. Chabal 2005: 224. In addition to Lattes, other examples of vine wood found in French wells are also cited, such as at Candillargues or at the Ateliers Saintes site in Charente-Maritime.

[25]Chabal 2005: 225–7.

[26]See also the contribution of Brown, Meadows and Turner in this volume.

[27]Marvelli 1999: 37.

[28]Bouby et al. 2013: 5.

Polesine, nearby Rovigo revealed high percentages of pollen in a well dated to the Late Roman period, together with significant number of grape seeds and pedicels.[29]

Phytolith analysis

Although a more recent addition, the study of phytoliths has fast become increasingly important in this field.[30] Phytoliths are tiny mineral particles (silica and calcium oxalate) formed during plant decomposition and fossilized in rock. Researchers use them to answer both traditional and novel archaeobotanical questions in archaeological and palaeoenvironmental studies. As phytoliths reflect the original cell morphology of the plant, they are used to identify plant taxa at sites for purposes of diet and nutrition, agricultural practices and building materials.[31] Indeed, by using multivariate and morphometric analyses of single-cell forms, phytolith analysts can identify different crops and varieties. Thanks to the creation of numerous reference collections, scientists have made considerable progress in identifying plants even in hostile environments. New techniques are continuously being developed, such as the extraction of phytoliths from ceramics or their dating by means of thermoluminescence.[32]

Palynological and phytolith analyses can also help to reveal plants that are poorly preserved in the sediments of an archaeological excavation.[33] This invisibility may depend on climatic conditions, the characteristics of certain categories of plants (tuberous plants, for example, are usually poorly preserved because they are too soft), or the fact that certain plants produce little pollen (as is the case for grapevines). Alternatively, the absence of this material in archaeological sedimentary deposits could be a result of specific post-depositional processes. For example, at the ritual context of the ash altar to Zeus on Mount Lykaion in Greece, the absence was explained by the selection of woody plant tissues containing little silica as fuel for fires practised on the altar surface.[34]

Phytoliths are easily recovered from archaeological deposits. A small amount of sediment (*c.* 2 to 30 grams) is collected with a clean tool and placed in a sealable plastic bag. In the laboratory, researchers extract the phytoliths, which are then washed, mounted on slides and examined under a microscope. Quantifications can provide relative percentages or absolute numbers per gram of sediment, depending on the research methodology. Using multivariate and morphometric analyses of single-cell forms, phytolith analysts are able to identify different crops and varieties. Studies have, for example, identified the grapevine phytoliths (*Vitis vinifera* L.) of the Agiorgitiko variety, a black berry vine indigenous to Nemea in Greece, and of the olive tree (*Olea europaea* L.) of the Megareitiki and Kothreiki varieties, widespread mainly in the Peloponnese.[35] The

[29]Further details in Marchesini et al. in this volume.
[30]Cf. Piperno 1988; Rapp and Mulholland 1992; Pearsall and Piperno 1993.
[31]Rosen 2008.
[32]Rowlett and Pearsall 1993; Piperno 2006b.
[33]Pearsall and Piperno 1993; Rosen 1999.
[34]Mentzer et al. 2017.
[35]Tyree 1994.

aim of this study was to show whether the two plant species produced different phytoliths that can be distinguished even under an optical microscope. The results, however, are not clear enough to argue that we now adequately understand the characteristics of the *vitis* phytoliths to easily identify them in archaeological records and use them for deep-insight analysis. Analysis of phytoliths from the residues of storage vessels, as in the case of the former study, makes it possible to identify the products and distinguish the types of vessels used to store and transport certain goods. Phytolith analysis can thus offer an advantage over, or at least complement, chemical analysis: samples require minimal specialist handling by the excavator and do not need expensive equipment to be analysed.

Considerations

An analysis of methods of vine-growing, winemaking and the sale of surplus production is useful to deepen our understanding of the wine economy. These issues, however, must be considered both separately and correlated in a synthetic framework. In fact, it is evident that the usefulness of our sources and methods – from more traditional to innovative – depends on the topic in question, and they therefore need to be applied dependent on context and question. Archaeobotany together with DNA analysis can help us understand whether grapevines were cultivated at a site, whether they were used to produce wine, and what types of grape varieties were grown and wines produced. Other analyses, such as organic residue analysis and those discussed elsewhere in this volume, can be used in tandem with these to better refine our knowledge of wine production.

It is evident, however, that these investigations must be combined with the study of literary, epigraphic and iconographic evidence, and with material remains, from the traces of vineyards and their irrigation systems to the spaces and tools for processing and storing wine, and the containers for its use and commercial transport. In this regard, what follows is a case study recently explored as part of the previously mentioned interdisciplinary project in the region of Verona, Italy.

A case study in Valpolicella (Verona, Italy)

A Late Antique villa located at Corteselle in Negrar of the Valpolicella region (Verona) has recently been subject to extensive archaeological investigations by the Soprintendenza Archeologia, Belle Arti e Paesaggio per le province di Verona, Rovigo e Vicenza, in collaboration with the University of Verona.[36] The complex included a large production

[36]The excavation is still in progress and many results are thus unpublished, apart from a few studies on the investigations conducted here in the last century or on the better-known mosaics: cf. De Stefani 1887; Campanile 1922; Tosi 1983–4; Rinaldi 2003.

sector, where several pieces of archaeological evidence testify winemaking activities. A room with a *cocciopesto* floor, thought to be a *calcatorium* used for treading grapes with the feet, linked to a small *lacus* or collection vat, of which the stone base remains with recesses for the insertion of vertical slabs. In the building's disuse phases, three large fragments of counterweights for a press were found.

The site is located in Valpolicella, an area of approximately 240 square kilometres to the northwest of Verona (Figure 7.2). The region is well suited to vine-growing, with suitable soils for grape cultivation, cretaceous limestone and extensive basalt layers in the hilly areas, and, at the bottom of the valley, soil composed of the sediments of the Adige River and the three streams that run through the territory. Valpolicella also has an ideal climate for the grapevine – mild and not overly rainy, with an annual average rainfall of 850–1000 millimetres, due to the protection offered by the Lessini mountain range to the north, the proximity of Lake Garda, and the southern exposure of the hilly and valley floor soils. Literary sources from the second century BCE attest to vine-growing in the area. Latin authors mention Rhaetic wine, including Cato (in the precepts *Ad Marcum filium*: cf. Serv., *In Georg.* 2.95), Virgil (*G.* 2.95-96) who ranks it second after the prized Falernian wine, and Suetonius (*Aug.* 77) who recounts how the emperor Augustus greatly appreciated it. Roman wine production in the Veronese territory is confirmed by Pliny (*HN* 14.16, 14.67), who celebrates the Rhaetic grapes and wines produced in the countryside around the city, and by Martial (*Ep.* 14.100.1-2) with mention of the Rhaetic wines of the land of the learned Catullus. Strabo (4.6.8) locates its production in the hilly area around Verona, stating that it was a wine produced at the foot of the Rhaetian Alps above Verona.

The production of wine in this area is also attested archaeologically in the villa of S. Pietro in Cariano–Ambrosan (first to sixth century CE), where various rooms presumably dedicated to the fermentation of must were brought to light,[37] and nearby Mattonara (first to fifth century CE), where a *dolium* was found, buried in an earthenware floor beneath limestone slabs, hypothesized to serve as a *lacus* for collecting the must inside a *calcatorium* or *torcularium*.[38] Furthermore, dozens of counterweights from lever and screw presses found in the area indicate pressing for either wine or olive oil production.[39]

Vine-growing analyses

The various types of archaeobotanical analyses presented above (pollen, macro remains and charcoal) were all applied in the study of this site, including *c.* fifteen samples of pollen grains and *c.* ten samples of wood/charcoal. A range of stratigraphic contexts was

[37]On this villa, see Maria Stella Busana's presentation during the Vine-growing and Winemaking in the Roman World conference (Rome, 27–9 October 2021), and Busana 2002: 344–50; 2003: 119, 125; forthcoming.

[38]Busana 2002: 176, 351–2; Busana 2003: 117–19. Use related to the fermentation of wine is excluded, both because it is isolated (when in the *cellae vinariae* of the villas there are generally from a few units to several dozen) and because it is totally underground.

[39]Unfortunately, these are very difficult to date when found out of context, and they could well be from medieval or early modern periods (Liverani 1988; Busana 2003: 120).

1.

2.

Figure 7.2 Topographical position and plan of the 'Villa dei mosaici' at Negrar in Valpolicella, Italy. Red dots in (2) indicate sampling areas for pollen analyses, and blue dots the areas for macro remains sampling. Courtesy OpenStreetMap and SAP s.r.l.

selected to analyse the entire chronological development of the area, from its earliest phases to Late Antiquity and the early Middle Ages (Figure 7.2).[40]

Pollen analyses were conducted on soil samples taken with a syringe vertically every 5 centimetres in a particularly complete stratigraphic section, brought to light in the porticoed courtyard that was the heart of the building (Figure 7.2). These analyses allowed the vegetational framework of the area to be reconstructed, revealing a total of around 100 taxa. However, few pieces of grapevine pollen were recovered, proving that such results must be analysed with caution since vine pollen, as mentioned previously, is produced in relatively low quantities. Analysis of charcoal and macro remains instead attested to the extensive presence of grapevines in the area, identifying vine wood charcoal, as well as grape seeds in all samples related to the Late Antique phases of the villa. Of great interest among the anthracological data is the discovery among the wood used as fuel in the *praefurnium* of the thermal sector of elm wood charcoal, which could indicate the use of elm as a grapevine rootstock and thus attest to its cultivation on the site in the form of the *vite maritata* (tree-wedded vine). The most significant find was, however, the grape seeds, with more than sixty found. Some particularly well-preserved examples were found in the latrine drainage of the thermal sector, where the damp environment aided preservation. A morphometric study of these pips identified both *Vitis sylvestris* and *vinifera*,[41] but the context of discovery in this case suggests they were remains of table grapes. In contrast, grape seeds found in other layers of the villa, especially from the area of the courtyard and rural sector could further evidence local vine-growing and wine processing that was previously hypothesized from archaeological data.

DNA analysis of these grape seeds is still in progress, undertaken by colleagues from the University of Verona Department of Biotechnology. It will be interesting to compare the results of these analyses with the text of a letter by Cassiodorus (*var.* 12.4.2-5), in which he asks for the cellars of King Theodoric to be supplied with a wine called *Acinaticium*. The processing techniques mentioned by Cassiodorus (drying until December for grapes harvested in the autumn; preserving the bunches in special containers and then leaving the grapes to harden so that the 'insulting humours' ooze out and the product gently softens) recall modern techniques for making Recioto, one of the most famous wines of present-day Valpolicella. Similarly, Cassiodorus' description of *Acinaticium* –an extraordinary product, ineffably sweet and mellow and so red in colour as to seem squeezed from purple dye – recalls colour, sweetness and mellowness of this modern wine.

The case of the villa at Negrar demonstrates how diverse sources of multidisciplinary data, with their various strengths and weaknesses, can bring different pieces of information to the table. The continuous interaction between traditional sources and more recent scientific approaches could confirm how archaeobotanical analysis is

40 Marchesini et al., forthcoming.

41 Cf. Bellin and Bolognesi, forthcoming.

particularly useful in the reconstruction of ancient landscapes and in attesting to not just to the cultivation of grapevines in a given area, but also the type grown.

New approaches to the study of winemaking and trade

While archaeobotanical analyses generate information about the presence and cultivation of grapevines, they tell us less about the production and trade of wine. To address these issues, a combination of methods is required that integrates data from the archaeobotanical analyses with that of chemical residue analysis. Regarding the production phase, the ongoing chemical analyses of working surfaces of the *calcatorium/lacus* at the villa and of the counterweights will provide crucial evidence regarding the functional interpretation of these spaces and tools.

The traditional type-chronological approach to the study of the wine trade in this area was based on the containers used for storing and transporting wine. Above all, this involved the analysis of a group of *amphorae* morphologically identifiable with the Dressel 6A type. It is now generally accepted that the Dressel 6A includes the wine *amphorae* produced from the early Imperial age between Cisalpine Gaul and *Picenum.*[42] Among these is a homogeneous nucleus of containers produced in the Emilia-Venetian area. These are widespread in the eastern sector of the Po Valley and distinguished by stamps recalling the *gentes* of the *Ebidii/Ebidieni*, *Hostilii*, *Gavii* and *Valerii*, very often associated with various names of the enslaved *officinatores.*[43] The distribution patterns of these *amphorae* point to Verona as the epicentre of the diffusion of some of these series, and those marked with the *gens Valeria* (Figure 7.3) were selected on which to apply innovative methods to the study of wine consumption and trade.[44] But who were these *Valerii*? The lapidary epigraphy evidences a picture too large for us to think of a single family. The *gens Valeria* is widely attested in Cisalpine Gaul and, indeed, in all the territory of the Empire,[45] but the numerous testimonies in Verona are striking.[46] Furthermore, the *gens Valeria* is widely attested in today's Valpolicella (*CIL V*, 3932, 3939, 3973-78, 3980) and the Lake Garda area, where the monumental villa of Sirmio

---

42 For a recent summary of all the issues presented here, see Carre, Pesavento Mattioli 2003; Carre, Pesavento Mattioli, and Monsieur 2014; Van Limbergen 2018.

43 The containers in question are characterized by an oblique banded rim, small dimensions, and a pinkish, soft fabric with tiny calcite inclusions.

44 Thanks to a recent synthesis, fifteen specimens of Dressel 6A *amphorae* bearing stamps of the *gens Valeria* are known so far. Some of these refer to *Aulius Valerius*, probably the *dominus*, mentioned both alone and in association with a *servus*, i.e. *Felix* and *Philom(---)*. Moreover, a third member of the family, *Quintus*, is documented in association only with *Pilom(usus?)*, probably a different *officinator* than the former *Philom(---)* (Pesavento Mattioli and Mongardi 2018: 322–5, Table 1).

45 *OPEL IV*, 2002: 142–6.

46 Two examples are particularly significant. One testifies to the involvement of a member of the *gens Valeria* (the *quattuorvir P. Valerius*) in the construction of the gates and city walls when the city was founded in the bend of the Adige River in the mid-first century BCE (Buchi 1987: Figures 9 and 11). The other was found in the loggia of the *summa cavea* of the theatre and could indicate the reserved seating for the family or some act of euergetism related to the building (Bolla 2008: 79–85).

Figure 7.3 Dressel 6A *amphorae* found in Verona, Italy, bearing the stamps Q.VALERI PILOM (1) and VALERI (2). Photograph by S. Zago, FaW Project, University of Verona.

was once attributed to the family of the Veronese poet Catullus, a member of the Veronese *Valerii*.[47] The widespread diffusion of the *Valerii* name leads us to imagine a powerful family that had an important role probably not only in local production activity but also in commerce.[48]

To obtain further data on the family and its role in the Veronese economy, we used multidisciplinary methods to analyse two *amphorae* bearing the stamp Q.VALERI PILOM. These *amphorae* were found during largescale rescue excavations in the area of the ex-Campo Fiera–ex-Cappuccini Convent, located on the banks of the Adige River, east of the presumed course of the southern section of the Via Claudia Augusta, nearby a large funerary area in use from the first century CE onwards.[49] At the time of discovery, the banks contained several hundred *amphorae*, placed upside down as part of a drainage system with the intention of drying soil that was in a low-lying area subject to variations in groundwater levels. The good state of preservation of the *amphorae* offered the chance to conduct organic residue analysis to detect the residues of commodities originally stored in the jars and now soaked into the porous ceramic surface using a methodology involving two extraction phases.[50] The results show that both *amphorae* contained wine and were waterproofed with conifer pitch (Figure 7.4). The second extraction, in particular, showed concentrated levels of tartaric acid (cTar 122.6 µg/g) and abundant malic acid – both key indicators for the presence of wine or grape derivatives. The combination of the presence of syringic acid that suggests black grapes and markers of alcoholic fermentation hints at a red wine.[51] This not only scientifically confirms the hypothesis that Dressel 6A vessels contained wine, but also proves that *amphorae* bearing the stamp of the *gens Valeria* were used to trade red wine. As for the pitch, the toponym *Pésina* (widespread in the Caprino Veronese area) could have derived from *pix (*picina)*, suggesting the possible existence of an ancient pitch market in the foothills of Monte Baldo.[52]

Archaeometry was then used to understand where these *amphorae* were produced. Thin-section analysis of the vessel fabric shows that they belong to the same petrographic group, which is the first important sign of a common origin. Both fabrics are characterized by clays of marine origin with inclusions of sandstone and granite or metagranite (cf. Figure 7.3). The characteristics of the fabric are consistent with a large area of the Po plain and reflect the strong homogeneity of Adriatic clay beds and make pinpointing certain places of production challenging. Nevertheless, since there are few sources of

---

[47]Wiseman 1987: 311–70. The epigraphic sources also refer to the presence on Lake Garda of members of the Senatorial order (in the case of the *Valerii Nasones*), who must have invested large amounts of capital in the agriculture of this rich territory (cf. Tassaux 2005: 150–2).

[48]An inscription found in Peschiera, dating to the second half of the second century CE, confirms the latter, certifying the presence of a *Valerius* among the members of the *collegia* di *n(autae)* or *n(avicularii)* (see *CIL V*, D 4017; Bertolazzi and Guidorizzi 2017: 335–7).

[49]Cavalieri Manasse 1998.

[50]See contributions by McGovern and Pecci in this volume.

[51]See Garnier, forthcoming for further data on this aspect.

[52]Franzoni 1990: 61.

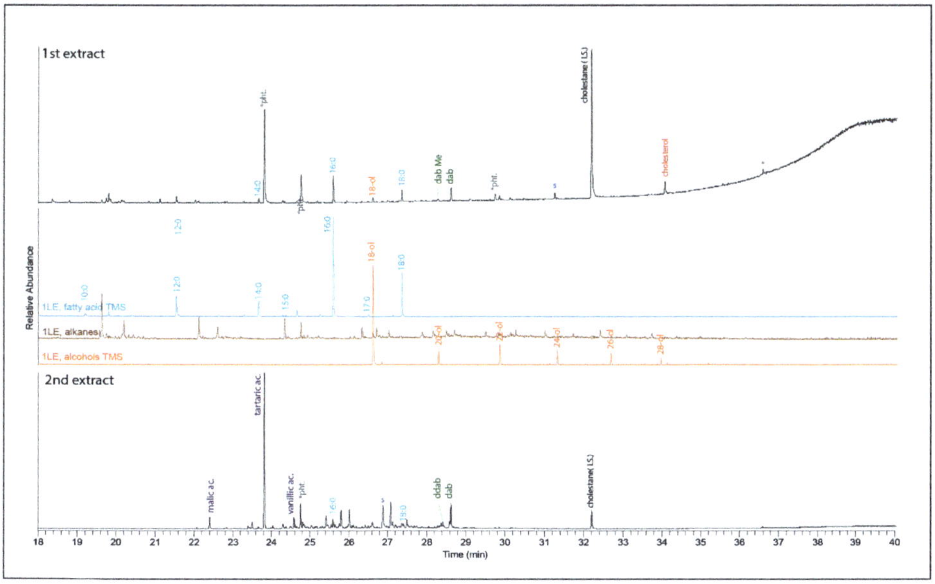

1.

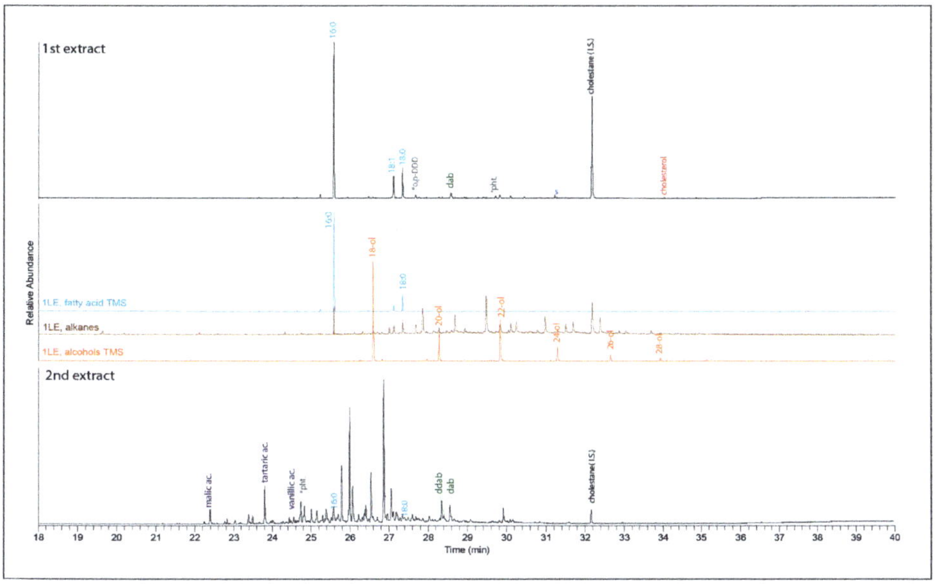

2.

Figure 7.4 Chromatograms of the first and second lipid extracts obtained from the inner wall of two Dressel 6A *amphorae* bearing the stamps Q.VALERI PILOM (1) and VALERI (2). Graphs by N. Garnier, SAS Laboratoire, FaW Project, University of Verona.

granite in the region and in Italy as a whole, the inclusion of granite here limits the possible territory from which the clay might have originated. While research is still at an early stage, recent core samples from the city have shown that the Adige River transported granite pebbles, so we cannot exclude the hypothesis that the clay basin which furnished the materials for these *amphorae* may have been located quite close to the river.

We thus have relatively direct evidence for a family of businesspeople – the *Valerii* – who took advantage of the natural resources of the Veronan countryside, becoming notable in local city society and significantly enriching themselves. The new pieces of information that emerge from these investigations (wine in the *amphorae*, conifer pitch, numerous *Valerii* inscriptions in Valpolicella) bring to light new suggestions, which we hope can be confirmed as research progresses, helping us to reconstruct the social and economic picture of Roman Valpolicella on the basis of more recent archaeological and scientific techniques.

Conclusion

This chapter highlights the importance of combining traditional methods to analyse containers (such as typology, chronology, epigraphic and iconographic investigations) with more innovative multidisciplinary approaches, including organic residue and archaeometric analyses to determine the origin of the raw materials from which the vessels used to store and trade wine were made. A range of multidisciplinary techniques within the broad field of archaeobotany simultaneously show their essential utility to forge a new understanding of viticulture and winemaking in antiquity. A greater degree of collaboration between archaeologists and various archaeobotanical specialists is needed. The three phases of the wine economy (vine-growing, winemaking and wine marketing) should be considered both separately and correlated in a synthetic framework that joins the traditional (stratigraphic and typo-chronological) to more innovative analytical approaches (e.g. biomolecular archaeology, pollen, charcoal, organic residue analysis). Future studies will undoubtedly provide more and better data, thereby advancing our understanding of the production of one of antiquity's most important commodities.

CHAPTER 8
APPLICABILITY AND USE OF ARCHAEOBOTANY FOR THE STUDY OF VINE CULTIVATION AND WINEMAKING IN THE ROMAN PERIOD

Marco Marchesini, Silvia Marvelli,
Anna Chiara Muscogiuri and Elisabetta Rizzoli

Abstract

This contribution presents a synthesis of the archaeobotanical studies of vine remains (pollen grains, woods, charcoals, seeds, fruits and other botanical material) in northern Italy during the Roman period. Carpological remains are among the most important pieces of evidence because they often make it possible to distinguish between cultivated and wild grapevines, while the presence of grape pomace is highly suggestive of the presence of winemaking processes. Remains of grape pomace are documented at numerous sites across northern Italy and the high presence of pollen grains and xylo-anthracological as well as carpological remains testifies to the grapevine's widespread presence in the Po Valley and neighbouring hilly areas.

Introduction

In Italy, the grapevine (*Vitis vinifera* L.) is present with two subspecies: *Vitis vinifera* L. ssp. *sylvestris* (c.c. Gmelin) Hegi (wild grapevine) and *Vitis vinifera* L. ssp. *Vinifera* (cultivated grapevine). As a synonym for the cultivated form, the name *Vitis vinifera* L. ssp. *sativa* Hegi is also commonly used.[1] Despite abundant archaeological, archaeobotanical, historical and genetic data, the origins, historical biogeography, identity of ancient grapevine cultivars, and mechanisms of domestication are still debated. This chapter presents a synthesis of the archaeobotanical data on grapevine finds in northern Italy during the Roman period. It focuses on the use of these studies to document the cultivation and transformation of grapes into wine through winemaking processes.

The cultivation and domestication of the grapevine are now widely accepted to have occurred between the seventh and the fourth millennium BCE in the hygrophilous

[1] Webb 1968; Pignatti 2017–19.

forests located between the southern coast of the Caspian Sea and the eastern coast of the Black Sea.[2] Cultivated forms would have been spread via human agency through the Near East, Middle East and Central Europe (see Chapter 2). In northern Italy, archaeobotanical data shows that during the Neolithic and Early Bronze Age a range of knowledge concerning the harvesting and use of grapevine fruit had already been acquired or developed. During the Middle and Late Bronze Age, the spread of grapevine cultivation appears to have intensified, although it is not clear whether this was due to technologies directly or partly acquired from earlier contacts with the Mycenaean world or with the Greek world at a later stage. Based on current data, it is more plausible to think that an evolution of spontaneous Italian vines from indigenous populations was accompanied by the introduction of allochthonous vines from areas where viticulture had already been established for some time (e.g. the eastern Mediterranean basin, where viticulture was already practiced with a relatively high level of technology). The period critical to the spread of grapevine cultivation lies between the Bronze and Iron Ages, after which viticulture became an established activity.[3] Such conclusions are further supported by chemical analyses carried out on ceramic vessels found at some Middle Bronze Age sites, in which the presence of tartaric acid, a fundamental component of wine, is documented.[4]

Materials and methods

In this study, pollen grains, xylo-anthracological and carpological grapevine finds from various Roman sites in northern Italy were considered to investigate how archaeobotanical evidence can contribute to our understanding of the transformation of grapes into wine through vinicultural processes.

The different types of grapevine finds in the archaeobotanical field play an invaluable role in archaeological research but, from a methodological point of view, the identification of cultivated vines is somewhat problematic and discrimination between *Vitis vinifera* L. ssp. *sylvestris* and *Vitis vinifera* L. ssp. *vinifera* is not always possible. Understanding when grapes were used to produce wine is even more complex, since this can only be inferred based on the type of assemblage of material found (e.g. grape pomace) and the archaeological context in which it was found (e.g. wine presses, production facilities or *amphorae*). The congenital characteristics of the vine – a plant that produces relatively limited quantities of pollen – also must be taken into consideration (see Figure 8.1 for an example of grapevine pollen). Subspecies identification through grapevine pollen grain

---

[2]For more discussion on the origins of cultivation, domestication and wine production, see the contributions by Dodd and Van Limbergen and McGovern in this volume as well as McGovern and Rudolph 1996; Zohary and Hopf 2000.
[3]Marchesini et al. 2021.
[4]Pecci et al. 2020; Borgna et al. 2022.

is also somewhat problematic and this field is still developing.[5] It is only possible to reach an indirect discrimination between wild and cultivated grapevine when pollen quantities are sufficiently recovered from suitable archaeological contexts.[6] Finally, all these aspects must be correlated with the state of preservation of the materials recovered.

On the one hand, as far as trunks and shoots are concerned, it is possible to morphologically distinguish between cultivated and wild grapevines; however, the limited size of xylo-anthracological finds makes it difficult to distinguish between the two subspecies and their ability to contribute to the identification of *vinifera* subspecies is only possible in certain cases. Microscopic analysis of the wood structure, on the other hand, does not enable a clear distinction between the two types. Like pollen analysis, therefore, the presence of a quantitatively significant xylo-anthracological assemblage and a correlated archaeological context is fundamental.

Grape seeds (also called pips) are the most common proxy among macro remains and the only type that, thanks to their morpho-biometric characteristics, can partially offer a distinction between wild and cultivated subspecies. To assess these parameters, it is necessary to have a sufficiently large number of grape seeds to accurately observe variations and conduct appropriate statistical research. Studies are based on the measurement of dimensions (length, width and thickness) of the seed, length of the beak and the chalaza. Equally diagnostic are the indices and parameters of morphology, such as the germination area and the fossettes pattern.[7] Regarding size ratios, specifically the ratio of width to maximum height, grape seeds can be discriminated on the basis of the intervals established by Stummer's Index, according to which a result between 0.44 and

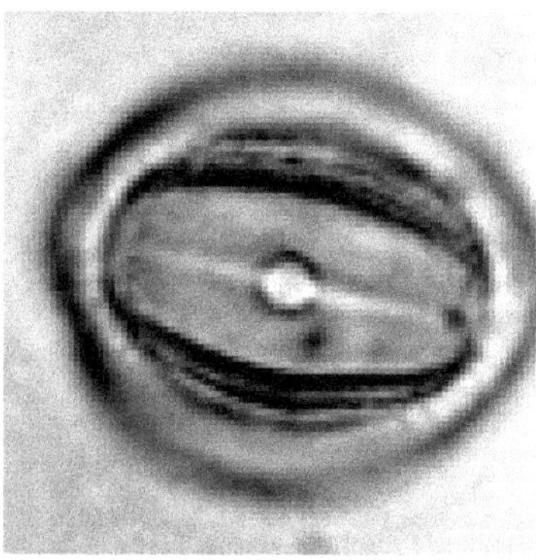

Figure 8.1 One 23 μm grapevine pollen grain. Courtesy Centro Agricoltura Ambiente Giorgio Nicoli.

[5]See contributions by Brown et al. and Van Limbergen and Gurdebeke in this volume.
[6]Marchesini et al. 2021.
[7]The distinctive ventral infolds of the grape, also known as seed folds or rumination in-growth.

0.53 would indicate a wild form, whereas a value between 0.76 and 0.83 would indicate that it is cultivated.[8] Nowadays this criterion is losing some of its original and established reliability and attempts are being made to identify new, more rigorous parameters.[9]

Results

One of the most widespread mixed forms of grapevine cultivation in northern Italy in Republican and early Imperial times was *arbustum gallicum* (Figure 8.2).[10] According to this method the grapevine was 'married' to elm, maple and horn-beam trees – a practice also attested in the Roman literary sources.[11] From an archaeobotanical perspective, the earliest evidence of *arbustum gallicum* was found in the nineteenth century during drilling work for the construction of several artesian wells. Elm stumps married to vines

Figure 8.2 Grapevine training using the system of *Arbustum gallicum*. Courtesy Centro Agricoltura Ambiente Giorgio Nicoli.

[8]Stummer 1911.
[9]cf. Mangafa and Kotsakis 1996; Perret 1997; Castelletti et al. 1998; Milanesi et al. 2011; Terral et al. 2010.
[10]Sereni 1964; Forni 1996; Van Limbergen et al. 2017; Van Limbergen 2020b.
[11]E.g., Columella, *Rust.* 3.2.17-24, 5.6.5, 5.6.24, 5.7.1; Pliny, *HN* 14.3.34, 17.35.199-203, 14.4.31, 14.4.22, 17.15.78; Varro, *Rust.* 1.7.2; Virgil, *G.* 2.217-21, 2.277-78.

were recovered between 10–15 metres deep during the excavation of a well near the parish church of S. Biagio del Carmine and in the courtyard of a private house in Via del Muro in Modena.[12] Numerous archaeobotanical studies relating to the xylological finds from the storage pits in Cognento (Modena),[13] Rubiera (Reggio Emilia)[14] and Badia Polesine (Rovigo)[15] indicated the presence of grapevine, maple and elm branches in the levels dated archaeologically to the late sixth and early seventh centuries CE. This supports a hypothesis of vine cultivation using *arbustum gallicum* nearby the pits. Moreover, the presence of charred remains of grapevines, maple and elm trees indicates the use of *arbustum gallicum* as a source of firewood.

In the Roman villa of Casteldebole (Bologna), from a pit datable to the second to third centuries CE, the remains of grape pomace consisting of a total 7,339 grape seeds, skins, berries and pedicels were recovered, documenting winemaking processes in some tanks found in the villa's production area.[16] In a well in Via Don Dossetti at Sant'Agata Bolognese (Bologna) in levels dated to the first century CE, numerous grape seeds and some perfectly preserved charred grapes were found (Figure 8.3). The presence of a press (*torcular*) (Figure 8.4) and numerous *dolia* suggest that one of the settlement's main agricultural activities was related to wine production.[17] Grape seeds and charred grapes were recovered at the Imperial-era villa of Mercatello in Castello di Serravalle (Bologna). The large quantity of grape seeds found near a hearth here could be interpreted as a

Figure 8.3 Charred grape (14 millimetres) found in a well at Via Don Dossetti in Sant'Agata Bolognese, Italy. Courtesy Centro Agricoltura Ambiente Giorgio Nicoli.

[12]Mazzetti 1892; Malavolti 1948; Forni 1996.
[13]Accorsi et al. 1998; Marchesini and Forlani 2002.
[14]Marchesini et al. 1998.
[15]Malaguti et al. 2011.
[16]Bandini Mazzanti et al. 1995; Marchesini 1998.
[17]Trocchi et al. 2014.

Figure 8.4 Foundation remains and reconstruction of the *torcular* found at Via Don Dossetti in Sant'Agata Bolognese, Italy. Courtesy Centro Agricoltura Ambiente Giorgio Nicoli.

secondary use of the grape pomace after pressing, dried and burnt as fuel, according to similar practices attested traditionally in many viticultural regions in later centuries.[18] Pollen grains, charcoal and grape seeds were also found at the villa of Domagnano in the Republic of San Marino. Here, the transformation of grapes into wine is evidenced by a tank and a press base.[19]

The spread of Roman viticulture throughout the Po Valley is supported by archaeobotanical material, especially grape seeds, at various sites in the Veneto region. At the villa of Marina di Lugugnana in Portogruaro (Venezia), dated to the end of the first century BCE and the beginning of the first century CE, grape seeds in sufficient quantities for the production of one hectolitre of wine were recovered inside a tank and press rooms.[20] In the filling of a well at Badia Polesine (Rovigo), from the late Imperial age, several hundred grape seeds, pedicels, fragments of vine shoots and pollen were found (Figure 8.5), which, in addition to testifying to the cultivation of vines, could also suggest the production of wine.[21] In the excavation of the Roman villa at Negrar (Verona), a region still dedicated to viniculture today, grape seeds and charred trunks (Figure 8.6)

[18]Badiali 2012.
[19]Mercuri et al. 2009.
[20]Di Filippo Balestrazzi 2004.
[21]Malaguti et al. 2011.

Figure 8.5 Grape seeds found in the well at Badia Polesine, Italy. Courtesy Centro Agricoltura Ambiente Giorgio Nicoli.

were found supported by quantities of pollen high enough to attest productive continuity from the Roman era onwards.[22] Excavations conducted at the beginning of the 1960s in the Venetian lagoon on the island of Torcello (Venezia) brought to light numerous macro remains including a large quantity of grape seeds of different types, pedicels and shoots in levels dated to Late Antiquity.[23] The discovery of bundles of vine shoots arranged horizontally to protect the soil from sea erosion is particularly noteworthy.[24]

Intensive vine cultivation is also documented in the Trentino region, as demonstrated by the discovery at the bottom of a well dating to the third to fourth centuries CE in Nago (Trento), of more than 9,000 grape seeds mixed with potential fragments of grape pomace and shoots.[25] Inside rooms interpreted as a kitchen or storage space of the first/second to fifth/seventh centuries CE at an agricultural building in Mezzacorona (Trento), more than 200 grape seeds and charred wooden remains of pruned vines were recovered, suggesting cultivation of grapevines nearby and wine production.[26]

At Genoa, in the Ligurian region, a grape seed and pedicel were found in a cabinet containing votive offerings from levels dating to the late second to early first century

---

[22]See Basso and Dobreva in this volume.
[23]Marcello 1965.
[24]Manzoni 1965.
[25]Ciurletti 1996.
[26]Castiglioni and Rottoli 1994.

Figure 8.6 Charred grapevine shoot fragments (*right*) and light microscope section of *Vitis vinifera* (viewed at 25x, *left*) from the villa in Negrar, Italy. Courtesy Centro Agricoltura Ambiente Giorgio Nicoli.

BCE.[27] More exceptionally, however, was the presence of whole bunches of charred grapes inside an *amphora*, found in destruction layers. In total, there were between 24–28 bunches, with each consisting of 150–200 grapes. The original volume occupied by the grapes was about six litres. Also in Liguria, material from a shipwreck in the Dianese Gulf allowed for analysis of resins inside containers (*dolium* and *doliolum*), which revealed clear traces of *Vitis* pollen (18.4 per cent), providing strong indications for their contents to be wine.[28]

Significant percentages of vine pollen are present in archaeological levels at the Roman city of Aquileia (Udine), particularly in the layers of the harbour and western forum. This, combined with the presence of numerous grape seeds, suggests the cultivation of vines and probable transformation of grapes into wine.[29]

Conclusion

In Italy during the Roman period, grapevine remains have currently been recovered from around seventy sites, at least forty-five of which are in northern Italy. In the Po

---

[27]Castelletti et al. 1996.
[28]Arobba et al. 1999.
[29]Cottica et al. 2017.

Valley during this period there was a progressive colonization of the territory, which is particularly evident in the region around Emilia Romagna where vines were found at over twenty-five sites. The most widespread method of cultivation was the *arbustum gallicum* supported traditionally by Roman agricultural literature but now also by numerous archaeobotanical findings.[30]

Studies that have synthesized material across multiple sites indicate vine pollen at significant percentages in several samples from nine sites on the Bologna plain and from structures of different types (villas, rural buildings, anthropic levels, necropoleis) ranging from the first century BCE to the fifth century CE.[31] Such widespread presence of grape pollen cannot only be due to the use of grapes as fruit but must be linked to some scale of vinicultural industry and wine production – a theory supported by the discovery of grape pomace at the *villa rustica* at Casteldebole, in tanks used for winemaking at the villa of Marina in Portogruaro, and the well of Nago in Trentino. It is further confirmed by the discovery of numerous structures and facilities directly related to wine production (e.g. presses, tanks/vats and cellars) and by reference to viticulture in this region in classical literature. Within this broad multidisciplinary scope, however, the contribution of archaeobotany to research into winemaking is clearly fundamental and, in particular, the continuous presence of grapevine pollen at almost all the sites investigated provides an indication of just how widespread cultivation was in Roman northern Italy.[32]

[30] Bosi et al. 2020.

[31] Marchesini 1998; Marchesini and Marvelli 2010, 2017.

[32] Lodwick and Rowan 2022.

CHAPTER 9
GRAPES UNDER THE LENS: A METHODOLOGICAL APPROACH TO THE STUDY OF A SEED ASSEMBLAGE FROM VILLAMAGNA (URBISAGLIA, ITALY)

Riccardo Carmenati, Francesco Breglia, Girolamo Fiorentino and Roberto Perna

Abstract

Villamagna, a Roman villa in the countryside of Urbisaglia, in the Marche region of Central Italy, housed important *gentes* of the nearby colony of *Pollentia-Urbs Salvia*, such as the *Herenni*, some of whom were active in the wine trade. A rich carpological assemblage, dating between Late Antiquity and the early Middle Ages, was collected from the site. *Vitis vinifera* was the most recurring *taxon*, highlighting the importance of grape exploitation in this production context. A morphometric approach was taken on 163 grape pips from two contexts to shed light on morphological variability through Principal Component Analysis (PCA). Biometric analyses were also applied to better understand archaeobotanical significance and, for most of the analysed pips, the indices fell within the wild subspecies range. This preliminary data allows us to formulate more specific questions about viticulture in this period, especially concerning wild and domestic grapevine exploitation and its historical significance.

Introduction

The study of historical, ceramic and epigraphic documentation related to Roman viticulture has shed light on the economic and cultural importance of wine and the vine. However, many questions remain concerning the spread of grapevine cultivation and harvesting systems, recognition of ancient cultivars, and the selection of cultivars to territorial scale. Archaeobotany, although still young as a discipline, is well equipped to respond to these questions. Villas and farmsteads provide recognizable and articulated rural production centres in the Roman world and were often located in the heart of the landscape. They are of fundamental importance for investigating these questions. One such site forms the object of the present case study. The Roman villa of Villamagna in Urbisaglia was an important agricultural centre and has yielded ceramic and epigraphic remains suggesting connections to the wine trade. During the most recent excavation campaigns, thorough sampling to collect archaeobotanical remains was conducted. The site and our methodology provide a case study, which, if properly implemented on a larger scale, could begin to answer broad questions regarding our understanding of Roman viticulture.

The archaeological context

Located near Urbisaglia in the central Marche (Figure 9.1), the structures at Villamagna lay within the territory of the colony of *Pollentia-Urbs Salvia*, 1.5 kilometres to the northeast. The villa itself was inhabited for over six centuries, but the broader area continued to be populated until the fifteenth century. Although the perimeter has been almost completely identified, some parts of the villa – which covers more than 7,000 square metres – remain to be investigated.[1]

In the Roman era, from the first century BCE to the fifth century CE, the villa was inhabited by notable families from the nearby colony. Numerous stamped bricks with inscriptions attest to the presence of members of the *gens Herennia* and it is possible that a family with political influence from the town decided to base their economic power, production and commercial activities in the villa. There is an analogy between the *bullae*

Figure 9.1 Aerial image of the Villamagna plateau and archaeological area. Photograph by authors.

[1]Paci and Perna 2016.

of the *Herenni* of *Urbs Salvia* and those of the *Marci Herenni Picens*, renowned wine merchants and producers of *amphorae*.[2] The similarities may reveal the involvement of the *Herenni* of *Pollentia-Urbs Salvia* in the production and commerce of wine, as well as links, perhaps including family ties, with the *Marci Herenni*.

The structure at Villamagna is delimited by a perimeter wall and is divided into sections: a *pars urbana*, the main residential area, and *pars rustica*, the working or production areas. The former is made up of a day area with rooms for receiving guests (two of which have apses), a courtyard and a *portico atrium* with pillars. To the southwest of this area is a large open space, perhaps an orchard or vegetable garden. Of the *pars rustica*, a large space for production and storage that is divided into three naves by pillars has been investigated (Figure 9.2a). In the southern part, there is a complex of underground rooms, among which is a vaulted cellar and a square room; to the west is a space in which the paving in *opus spicatum* suggests the presence of presses; and in the northern part, a cistern and a well.

Inside the excavated production and storage space of the *pars rustica*, a basin is built of *opus spicatum* on the south side and a series of four more adjacent basins, lined with *opus signinum*, are located on the east side. Both structures have a large *dolium defossum* on one of their short sides. Such an association suggests a sequential process that entailed the joint use of the basins and *dolia*. The north side of the space is characterized by at least two rows of *dolia*, while in the west corner and central area are large pits filled with dumped building materials, dated between the fifth and sixth centuries CE. Near the central pit, excavations located a structure for combustion, perhaps a kiln from the mid-fifth century CE. Before its abandonment, the area was used for funerary purposes with nine burials found.

Materials

During the 2018 excavation season, 2,885 carpological finds were recovered, of which the majority belong to *Vitis vinifera* (2,248 or 78 per cent). All the pips were charred and found in several contexts.

Found next to the *dolium defossum* near the *opus spicatum* basin – attributable, however, to a phase of structural abandonment – was a layer (SU 176) that yielded 1,823 grape pips, many of which were fragmented (Figure 9.2c). Additionally, at the bottom of the largest of the four linear basins lay a two-handled vessel with a spout, in which was a sediment (SU 145) containing more grape pips, together with the flesh and seeds of other fruits (Pomoideae, *Ficus carica* and other species not yet identified). Also dated to the abandonment phase is a small layer (SU 4012) that yielded 223 mostly intact grape pips (Figure 9.2b). Contemporary with its use as a burial area, the filling of Tomb 8 (SU

[2]Marengo 2002.

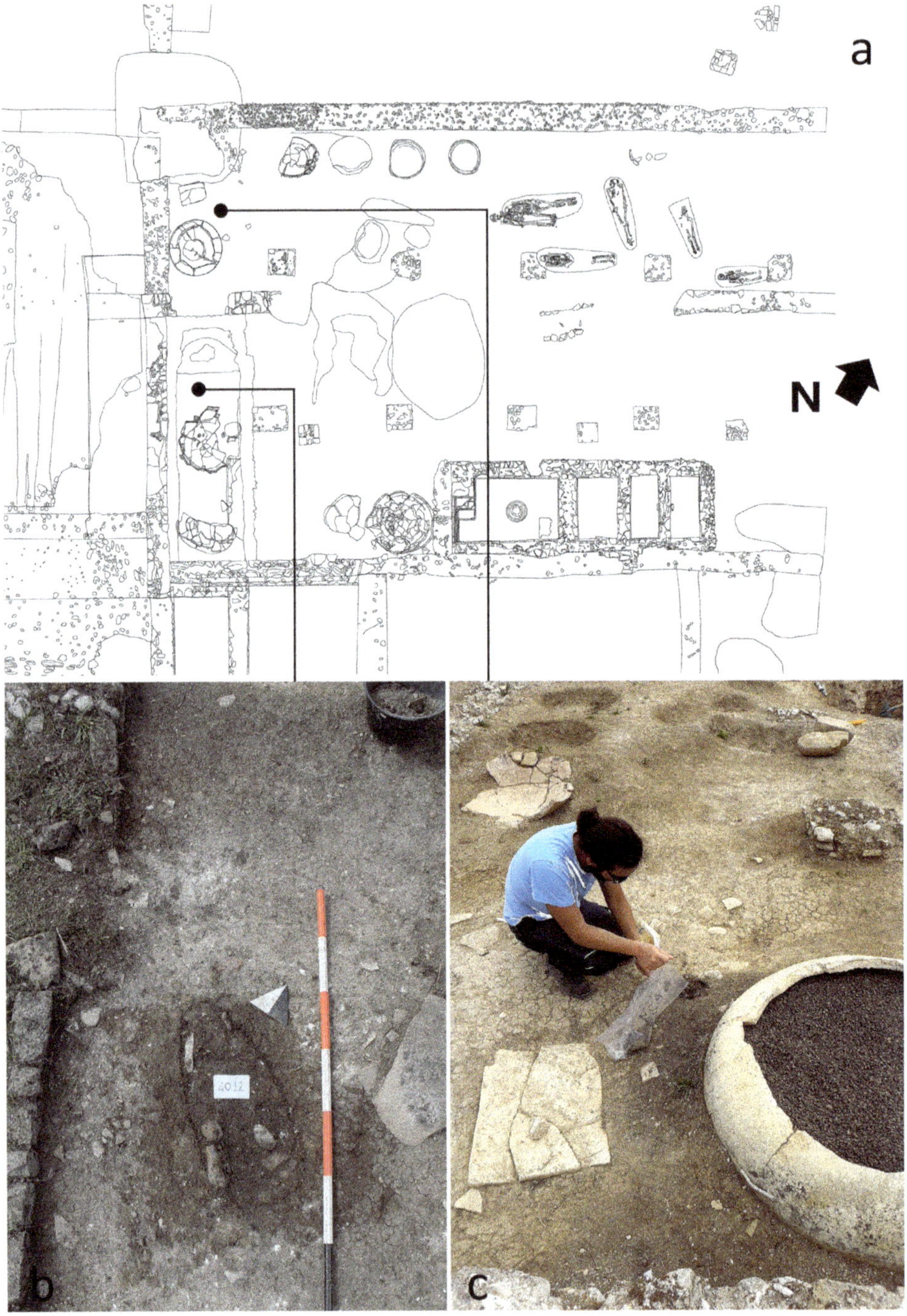

Figure 9.2 (a) plan of the Villamagna *pars rustica* with sampling points indicated by the black dots; (b) SU 4012; (c) sampling SU 176 during excavation. Photographs and plan by authors.

4001) delivered fifty intact grape pips, perhaps linked to ritual offerings of food or purely by chance.[3] Finally, one of the preparatory levels (SU 4163) of the combustion structure was composed of over 2,000 mostly intact grape pips, along with olive stones and fragments of walnut. Excavations of the *pars urbana*, recovered grape pips in sediments datable to the destruction phase (SU 1034) and to the reuse phase (SU 1031). From the latter came a charred grape, almost intact, with at least two pips inside.

A sub-sample of ninety-seven pips from SU 176 and another of sixty-six from SU 4012 were taken for biometric and morpho-geometric analyses (Figure 9.3a). The two stratigraphic

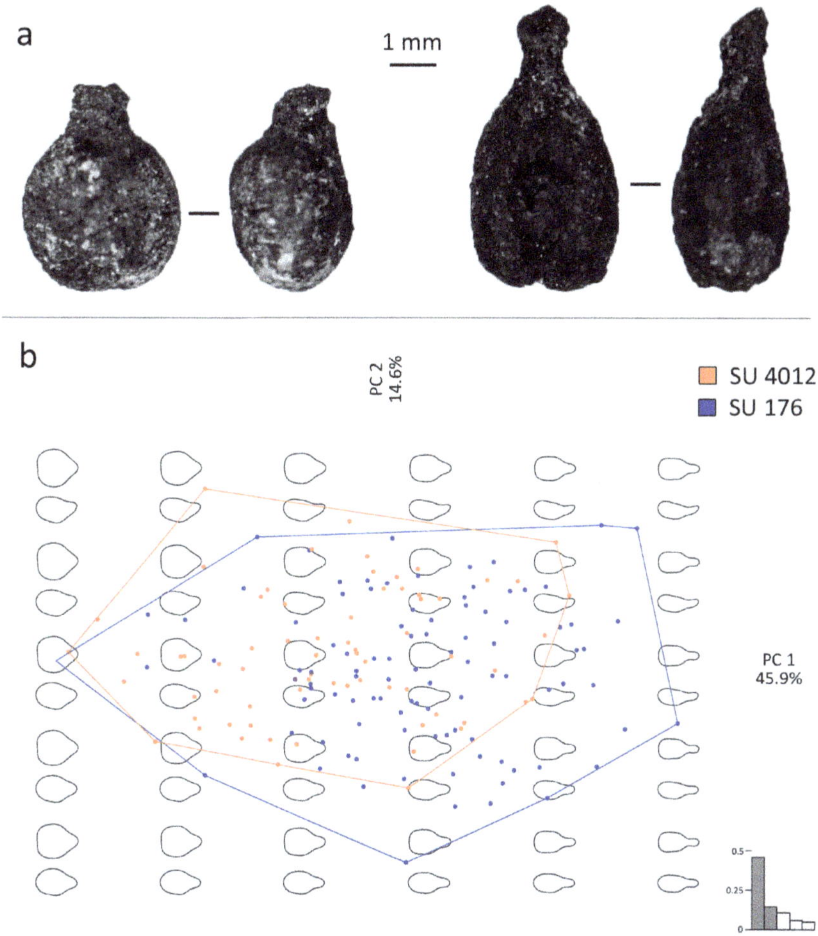

Figure 9.3 (a) example of the sample's inner variability which illustrates wild (*left*) and more domesticated (*right*) morphotypes; (b) PCA plot for SU 176 and SU 4012 grape pips. Photographs and diagram by authors.

[3]Carmenati, Fiorentino and Perna 2019.

Table 9.1 Radiocarbon dating performed on grape pips of SU 176 and SU 4012

Sample ID	*Stratigraphic Unit*	*Radiocarbon Age (BP)*	*δ13C (‰)*	*Calibrated date (2σ)*
LTL21925	176	1440 ± 45	-25.9 ± 0.5	555 CE / 661 CE (95.4%)
LTL21921	4012	1609 ± 45	-28.1 ± 0.4	375 CE / 569 CE (95.4%)

units were dated by direct radiocarbon dating on grape pips, determining a chronological range between Late Antiquity and the beginning of the Middle Ages (Table 1). The samples were dated at CEDAD (Centro di Datazione e Diagnostica of the University of Salento), using high-resolution mass spectrometry (AMS) and calibrated to calendar age using the OxCal software Ver. 3.10 based on INTCAL20 atmospheric data.

Methods

Morphometric and statistical analyses

Morphometric analyses, based on outline acquisition using the Elliptic Fourier Transform method, were performed on all 163 selected grape pips, following the method developed in recent decades by researchers at the Institut des Sciences de l'Évolution de Montpellier (ISEM).[4] Each seed was photographed in dorsal and lateral view, and the images were subsequently semi-automatically processed as to obtain a black shape on a white background. The files were then imported into the R software environment, where they were processed using Momocs and MASS statistical packages for morphometrics analyses.[5] The coordinates of 360 equidistant points were automatically extracted from each outline by the software, then centred and aligned. Following a consolidated practice for this type of analysis, six harmonics were considered for the description of the profile.[6] In this way, forty-eight coefficients per seed (twenty-four for the dorsal view and twenty-four for the lateral view) were obtained and used as quantitative variables describing the geometry of the pip. These coefficients are also used in PCA, which makes it possible to visualize the internal variability of the sample and to compare assemblages from distinct contexts, chronology and sites. In this case, the two stratigraphic units were compared.

Biometric analyses

The same sample of grape pips from SU 176 also underwent biometric analysis to understand whether they were wild or domesticated morphotypes. Specifically, we

[4]Terral et al. 2010; Pagnoux et al. 2015, 2021; Bonhomme et al. 2020; Bouby et al. 2021; Roushannafas, Bogaard and Charles 2022.

[5]Venable and Ripley 2002; Bonhomme et al. 2014.

[6]Terral et al. 2010.

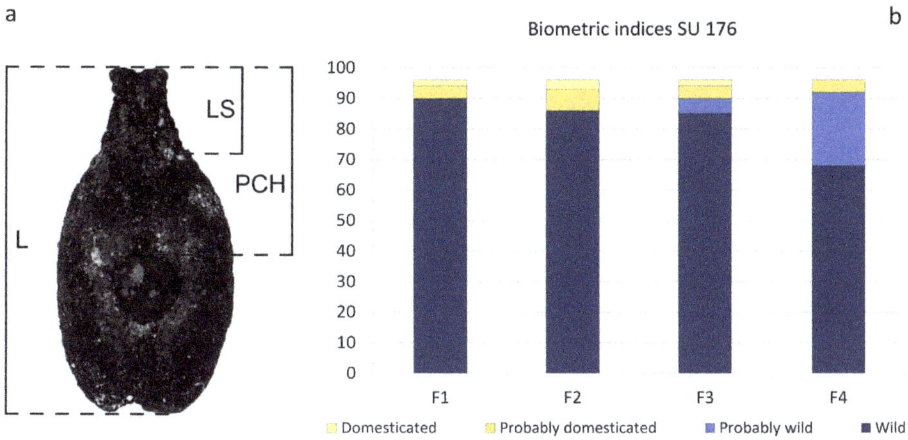

Figure 9.4 (a) anatomical measurements of the grape pip to obtain biometric indices developed by Mangafa and Kotsakis (L: total length; LS: length of stalk; PCH: placement of chalaza); (b) biometric analysis results of grape pips from SU 176. Photograph and diagram by authors.

applied a method that has been shown to be particularly effective at distinguishing wild from domesticated charred grape pips.[7] This is based on the relationship between the measurements of certain anatomical characteristics of the seeds, all along the same axis (Figure 9.4a), thereby minimizing the deformation effects resulting from charring, which acts unevenly on various axes. The measurements make it possible to obtain biometric indices which, when inserted in four mathematical formulae, provide a result that falls within one of the four ranges envisaged by the authors: wild, probably wild, probably domesticated, and domesticated.[8]

Results

The PCA performed on the morphology of the grape pips from SU 4012 and SU 176 highlighted the broadly homogeneous character of the sample: the PC1-PC2 biplot (60.2 per cent total variance) shows the overlap of the two clusters (Figure 9.3b). There is, however, a slight difference between the two assemblages. The morphological characteristics of the grape pips from SU 4012 are shifted slightly towards wild morphotypes, with rounded forms and short stubby stalks (Figure 9.3b, top left). A larger number of seeds from SU 176 show a morphology more alike domesticated types, with an elongated shape and long stalks (Figure 9.3b, bottom right). Both clusters show unimodal distribution of the respective populations, with peaks corresponding to morphotypes with intermediate characteristics.

---

[7]Developed by Mangafa and Kotsakis 1996.
[8]Mangafa and Kotsakis 1996.

In order to further investigate this substantial similarity, biometric analyses were performed on the pips from SU 176 (i.e. those with the more domesticated morphotype). The results show that almost all the pips – from sixty-eight to ninety, depending on the formula used – fall within the wild range, while zero to twenty-four grape pips fall within the probably wild range (Figure 9.4b). A small number of grape pips – from four to seven – fall within the probably domesticated range, while a negligible quantity – from zero to three – fall within the domesticated range.

Discussion

Considering the chronology and the characteristics of the contexts from which the analysed grape pips were recovered, the results above may appear rather surprising. Indeed, in Late Antiquity viticulture was widely practised in Italy, and there is significant archaeological evidence of its economic importance at the site of Villamagna. The almost exclusive presence of wild morphotypes among the grape pips analysed thus raises a series of questions. Were local wild grape varieties used? Or is this evidence of a particular cultivar with seeds that do not bear the typical morphological characteristics of current domesticated varieties?

While a high percentage of the grape pips analysed are attributable to wild morphotypes (72 to 94 per cent), it is unlikely that the economic activities performed at this site were reliant upon intensive wild grape gathering, especially considering that the villa was almost certainly set up as a rural production centre. The possibility that the morphotypes illustrated were composed of feral forms (domesticated types that have been 're-wilded') should not be dismissed. The two stratigraphic units from which the analysed specimens were sampled are datable from the late fourth to the mid-seventh centuries CE. Throughout peninsular Italy, this period was characterized by pronounced geo-political instability, as well as general economic and demographic decline.[9] These problems were compounded by the violence associated with the Gothic War and its aftermath and by the so-called Plague of Justinian in the mid-sixth century.[10] Given this socio-political and economic context, perhaps the management of agricultural activities was neglected and sporadic, causing the vineyards to revert to a natural or feral state (see above) in a relatively brief timespan. This may have caused a reduction of fruit productivity and quality in the absence of human care.[11] That the specific contexts in which the grape pips were sampled are interpreted as the result of abandonment – or disuse – of the villa's production structures might support this argument.

It should also be considered that, although viticulture was widely practiced in the Roman period, previous studies suggest that selective pressure on crops in antiquity was

---

[9]Barzanò 2018.
[10]Fabrini 2000; Ravegnani 2004; Perna 2006.
[11]cf. Grassi and De Lorenzis 2021.

lower than that visible in current domesticated varieties.[12] This is perhaps due to the constant recourse in ancient cultivation to plants arising from sexual reproduction, while modern grape cultivation is mainly based on vegetative propagation.[13] This might lead to the combined presence of wild, intermediate and domesticated morphotypes subject to moderate selection. We must also consider the possibility that the vines cultivated around Villamagna were of a cultivar characterized by small berries, whose seeds may resemble the characteristic wild morphotype despite being of a domesticated species,[14] as is indeed the case with some modern vines.

The questions raised by the preliminary results of this study cannot yet be answered with certainty. They do, however, indicate a crucial research pathway dependent upon high-quality archaeobotanical sampling and study regimes that can ultimately lead towards a better understanding of viticulture in the Roman period and antiquity more broadly. The potential and reliability of morphometric and biometric studies is constantly enhanced by increasing numbers of samples and sites. We must continue to extend this method of enquiry, both temporally and geographically, to archaeologically identify distinct cultivars and regions of production in the sphere of Roman viticulture.

[12]Bouby et al. 2013.
[13]Bouby et al. 2013.
[14]Bonhomme et al. 2020.

CHAPTER 10
ROMAN VITICULTURE FROM PALYNOLOGY: A REVIEW AND NEW DATA FROM THE BRITISH ISLES

Antony G. Brown, Ian Meadows and Simon D. Turner[1]

Abstract

Palynology has the potential to map viticulture at a variety of spatial scales within the Roman world. Viticulture at the northernmost margin of the Roman Empire can be seen as a clear mark of cultural transfer and the importation of a foreign landscape. Despite the textual evidence being ambiguous, palynology has revealed that viticulture was practiced, most likely in a *pastinatio* form in southern and eastern England. Recent work has suggested other areas outside what appears to have been the core area of the Nene Valley in Northamptonshire. Palynology when combined with cultivation evidence is at present the most reliable way of determining the likely presence of a vineyard; however, work is underway on molecular methods such as sedaDNA (sedimentary ancient DNA). Whilst this part of the Empire was never a significant wine producer, this aspect of Roman culture was manifest and had lasting effects on agricultural ecology and landscape.

Introduction

The status of viticulture at the northwestern periphery of the Roman world is archaeologically important as it is a window into the processes of cultural transfer (cf. 'Romanization'), its strength/nature and the relationship between culture and environment it reflects. The approach taken here through palynology has a long and significant place in our understanding of the landscape that the Romans invaded resulting from four centuries of Romano-British agriculture and landscape management.[2] Palynology has also a long history of application to Roman agriculture and horticulture in the Mediterranean, despite its limitations.[3] However, before evaluating the evidence and our methodological approach, it is worth considering the socio-economic and environmental contexts of Roman viticulture in the British Isles. Viticulture, especially to

[1]This work has greatly benefitted from assistance from many palynologists and archaeologists including Jackie Hatton (for pollen data), Rob Scaife, James Grieg, Petra Dark, Pat Wiltshire, Mark Knight and Kevin Trott.
[2]Dark 2000.
[3]See below and Mercuri 2014.

produce wine, is an investment and thus implies some sort of demand.[4] What was the potential demand and how was it distributed in time and space? Various potential sources can be identified: the Roman army stationed permanently to protect the northern frontier of the Empire; Roman administrators and citizens of towns; owners of farmsteads; and the wider rural Romano-British population, at least in southern Britain.[5] We must also consider whether the elite of Iron Age Britain had developed a preference for the consumption and use of wine even before the Roman invasion of 43 CE (e.g. use in burial ceremonies).[6] Given these numerous potential sources of demand, the scale and locations of production might shed further light on the drivers of demand and investment and thus nuance our understanding of a potential viticultural economy across the British Isles.

As they colonized further north of the Mediterranean, the Romans moved into the cool temperate biome that is climatically different from the typical area of grape domestication and early cultivation.[7] However, the potential range of viticulture can also be assessed from the distribution of its progenitor species, *Vitis vinifera* ssp. *sylvestris*, which grows naturally from the coast of the Atlantic (but probably not the British Isles) to North Africa.[8] This geographical range probably reflects an expansion from a climatically restricted distribution (refugia) in the eastern Mediterranean and Balkans during the last Ice Age.[9]

However, the natural climatic envelope for *Vitis vinifera* also allows for growth and grape production in locations with thermally favourable microclimates – whether natural or artificial – such as in southern and eastern England. This is complicated by the considerable genetic variation within *Vitis vinifera*, part of which arises from geographically related genetic variation in its progenitor species *Vitis vinifera* ssp. *sylvestris*.[10] Strong shoot growth and high vigour (high growth rates) are other means for adaptability and survival under sub-optimal environmental conditions. This is well expressed in warm regions where bud opening is limited to apical regions (shoot tips), and which have a higher potential for continuous development. On the other hand, the high vigour and polarity (functionally healthy cells) within buds allows inhibited secondary buds of the grapevine 'eye' to develop after frost damage to the young growth from major buds. Thus, the conglomerate of buds in the grapevine 'eye' is controlled by an interaction between environment and major shoot development. This allows north and westwards establishment, especially in suitable sites across large areas of France,

---

[4]McGovern 2003.
[5]Van der Veen et al. 2008.
[6]cf. Carver 2001.
[7]Now accepted to be both the Eastern Anatolia-Caucasus-Caspian and Levantine regions (McGovern 2003; Dong et al. 2023; McGovern and Dodd and Van Limbergen in this volume).
[8]McGovern 2003.
[9]Butorac et al. 2018.
[10]Bacilieri et al. 2013.

northern Spain and southern England – a distribution that is reflected in modern *Vitis* capability mapping and modelling.[11]

It is not only thermal tolerance which controls the range expansion of *Vitis vinifera* but also its rooting characteristics. On thin calcareous or sandy soils it can develop both deep and wide root systems, whereas in moist clay-rich soils only shallow but wide systems develop.[12] In this respect, south eastern England is similar to northern France being part of the same Mesozoic litho-structural system (the Cretaceous-chalk Paris-London Basin). Using a palynological modelling approach Macias-Fauria and Willis have shown that the temporal-spatial distribution of *Vitis vinifera* is the most difficult woody species to predict climatically due to cultural/anthropogenic aspects, including planting and evolution under Muslim or Christian rule in southern Europe around the tenth century CE.[13]

Justification for a palynological approach and towards a protocol

Several potential sources of indirect evidence can be used to study Roman viticulture in England and elsewhere, including documentary, biological and archaeological remains.[14] Tacitus (*Agr.* 12) in *c.* 98 CE, for example, remarks that 'all plants except the olive and vine could be grown in Britain', but this can be regarded as unreliable because he is comparing the grapevine and olive tree to other more plentiful resources (e.g. metals). The famous edict by Probus (SHA, *Prob.* 18.8) *c.* 280 CE, who 'granted permission to all Gauls and the Spaniards and Britons to cultivate vineyards and make wines', effectively ending Domitian's earlier wine edict (of 92 CE and probably never enforced), also cannot be taken as firm evidence for viticulture in this region. While viticulture was established in southern Gaul by the first century CE, it did not spread to the Seine, Loire, Moselle and Trier until later (second to fourth centuries CE) suggesting a delay in its adoption at the northern fringe of the Roman Empire. On the other hand, strong trading links existed between Roman towns in Britain and France (e.g. Lincoln and Bordeaux shown by the altar inscription blessing the travels of Marcus Aurelius Lunaris) and archaeological remains have been found of barrels and artefacts that suggest wine-consumption in the British Isles.[15] That wine was consumed and in demand is reflected also by direct evidence from at least one letter found at Vindolanda concerning a shipment of wine.[16]

---

[11] Cardell et al. 2019.
[12] Lavee 2000.
[13] Macias-Fauria and Willis 2013.
[14] Williams 1977; Witcher 2013.
[15] Lee 2022.
[16] Birley 1991.

Despite the somewhat contradictory and opaque references in the classical literature, palynological studies have shown that viticulture was indeed practiced in the first to second centuries CE in at least one area of southeastern England – the Nene Valley.[17] The use of palynology is predicated upon the taphonomy of *Vitis* pollen, which is produced in low amounts and does not *generally* travel far from the plant, due to the species being hermaphroditic and also largely insect pollinated.[18] The fact that some *Vitis* pollen can travel long distances is known through pollen analysis from the northern hemisphere ice caps, such as Severnaya Zemla, in the Arctic Ocean.[19] In this record, one pollen grain was found at *c.* 40,000 BP, another single grain at 4000 BP, but two grains at *c.* 1400 CE. This suggests that in the High Medieval period, European V*itis* cultivation was so extensive that it increased the input of *Vitis* pollen into the upper atmosphere. These sites, alongside many others in the European Pollen Database, show that the general background level of *Vitis* in most northern European pollen records is under 0.01 per cent.[20]

A complimentary experimental approach to detecting a vineyard by using pollen is the instrumentation of three vineyards in England with multiple pollen traps, which allowed us to estimate the probability of detecting *Vitis* pollen at specific distance away from a plant or vineyard.[21] Longer-term monitoring (more than one season) has also shown that pollen production and pollen grain size varies with climate and is an order of magnitude higher in more productive areas.[22] This data allows us to postulate that any count of *Vitis* pollen over approximately 0.1 per cent of the total land pollen would probably reflect a vineyard close to the pollen site (Figure 10.1). This indicates that pollen counts need to be unusually large – in excess of 1,000 grains – and the samples should be from a pollen-preserving sediment from as close to the grapevines as possible and well-dated enough to be correlated with the on-site activity.

From these investigations, both ancient and modern, we can generalize a protocol for the collection and investigation of pollen. If the aim is specifically to text the hypothesis that cultivation features or a context was used for viticulture the samples should come from that context, even if the soil or sediment is not highly organic or anaerobic. Failing that an adjacent anaerobic sediment type can be used such as a palaeochannel ideally under 10 metres from the putative vineyard site. The processing of the pollen sample is standard as are the keys used to identify the pollen[23] along with a pollen collection and some dedicated keys to the Vitaceae family.[24]

---

[17]Brown et al. 2001.
[18]Erdtman 1952; Turner and Brown 2004; Kvavadze et al. 2010.
[19]Andreev et al. 1997.
[20]Fyfe et al. 2009. See the European Pollen Database: http://www.europeanpollendatabase.net/
[21]Turner and Brown 2004: Figure 1.
[22]Kvavadze et al. 2010.
[23]E.g. Moore et al. 1991.
[24]Punt et al. 2003; Cartaxo-Pinto et al. 2022.

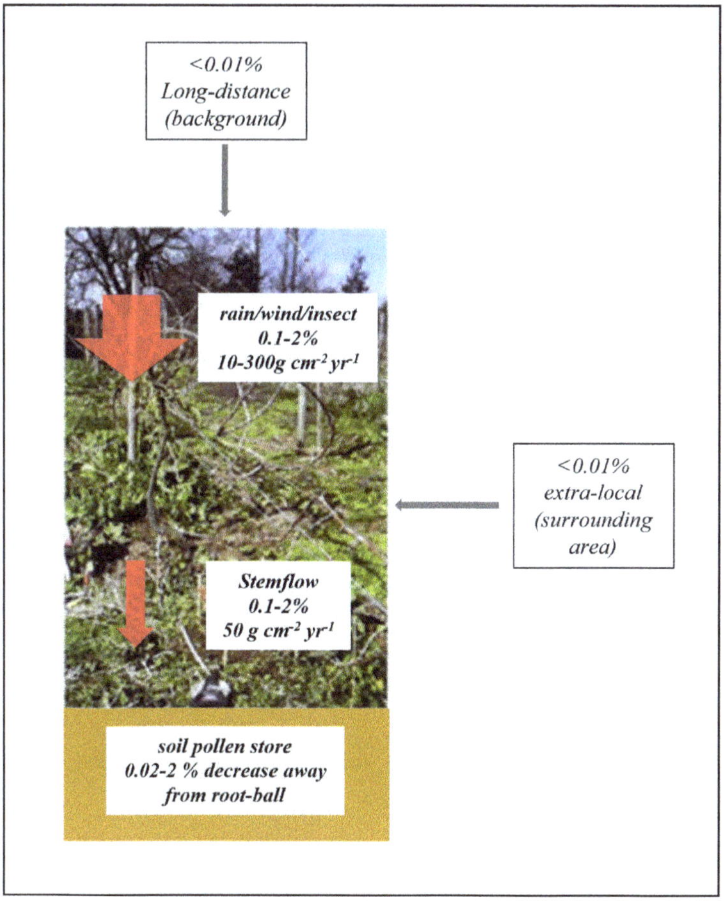

Figure 10.1 An approximate budget for the distribution of pollen around a vineyard. The percentages given are percentages of the Total Land Pollen (TLP) sum. Based on the instrumentation adapted from Turner and Brown 2004.

Site distribution in the British Isles

Excavations from 1993 to 1998 at the Wollaston site complex revealed a low terrace surrounded by three villas with several settlements, and two to three areas of vineyard trenches.[25] The trenches had vertical sides, flat bottoms and clear root boles with stake holes at distances typical of Roman vineyards.[26] *Vitis* pollen was recovered from two of these areas. The first was from a middle Roman period well in the settlement, and the second from buried trenches in the northern part of the site. The *Vitis* pollen, along with the form, dimensions and presence of post holes in the northern vineyard, identified the

[25]Atkins and Meadows, forthcoming.
[26]Meadows 1996; Brown et al. 2001.

Roman trenches as *pastinatio*, as visualized in Figure 10.2. In the Roman world there were numerous methods of training or growing vines, including training them to grow up trees, but *pastinatio* represents the most labour-intensive method of cultivation. It required the excavation of bedding trenches within which the vines were planted, which were in turn supported by a wooden trellis arrangement to aid growth through sunlight exposure of the full plants (especially important on level sites), reduce disease and facilitate harvesting the crop. The presence of this style of cultivation at Wollaston (Figure 10.3) suggests not only a familiarity with Mediterranean cultivation methods but also a preparedness to invest in viticulture, labour and infrastructure.

A palynological approach has provided further evidence to suggest that these were not the only Roman vineyards in the British Isles. There are now four additional sites which have provided Roman *Vitis* pollen as well as five sites that have *Vitis* in medieval levels, including one in Ireland (Figure 10.4). The Roman sites are also described in Table 10.1

Figure 10.2 A visualization of *pastinatio*. Courtesy MOLA and drawn by A. Thorne.

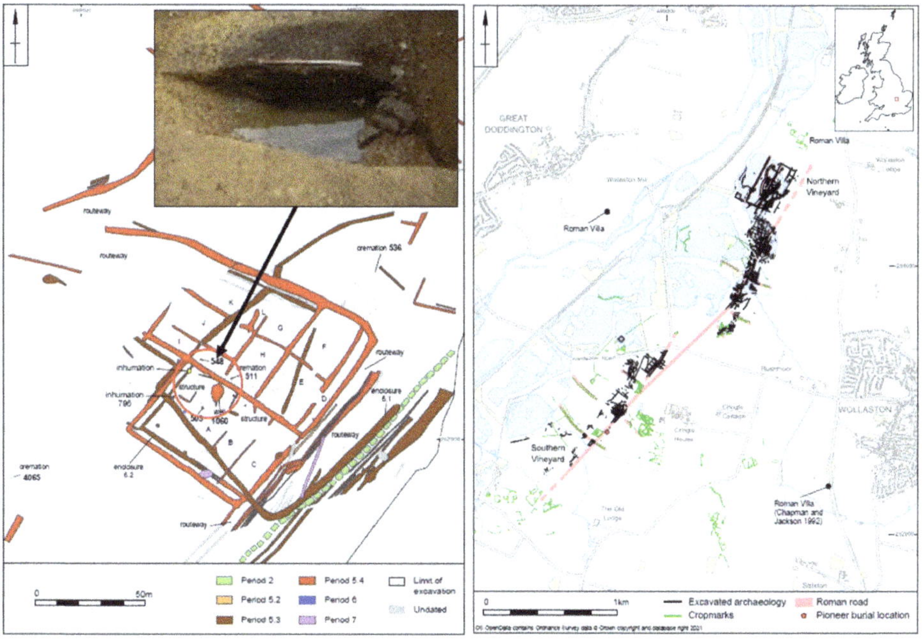

Figure 10.3. The first *Vitis* find from a Middle Roman well at settlement 2 at Wollaston (*left*); Wollaston site complex and location of the Roman vineyards and related archaeology (*right*). Adapted from Atkins and Meadows in Museum of London Archaeology, forthcoming.

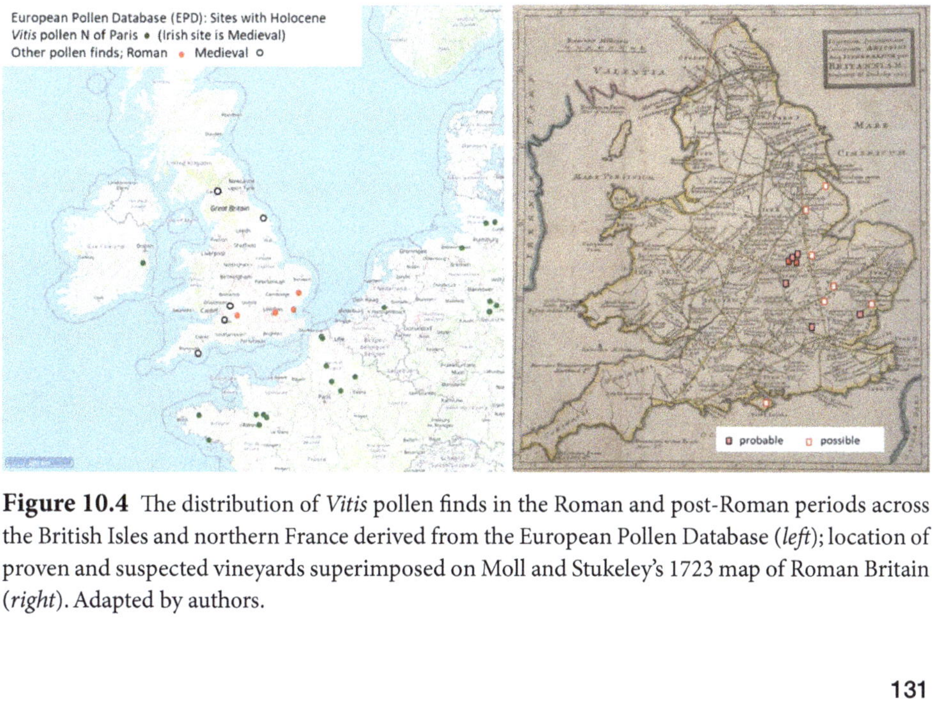

Figure 10.4 The distribution of *Vitis* pollen finds in the Roman and post-Roman periods across the British Isles and northern France derived from the European Pollen Database (*left*); location of proven and suspected vineyards superimposed on Moll and Stukeley's 1723 map of Roman Britain (*right*). Adapted by authors.

Table 10.1 Sites with pollen, macrofossil and archaeological evidence of viticulture and key associated plant records in the British Isles. Other crops/arable weeds in bold are derived from an ongoing sedaDNA study (www.terrace.no)

County	*Site (masl)*	*Evidence and environment*	*Est. age (context)*	*No of grains*	*Pollen %*	*Other crops / arable weeds (pollen/sedaDNA)*	*Nearby archaeology*	*Reference*
Northamptonshire	Appleby Lodge, Wellingborough (110 masl)	Nene Valley Large number of parallel flat-bottomed ditches aligned SW–NE on gravel terrace	Second century	-	-	-	Roman settlements SW of Finedon, Roman town at Irchester	Clarke and Meadows 2014
	Earls Barton (48 masl)	Cultivation trenches and ditches, sand and gravel terrace of the River Nene	LIA–Second century	-	-	**Hordeinae (barley subtribe of Poacaea), *Avena* (oat), *Triticum aestivum* (wheat), Chenopodioideae (goosefoot family), *Vicia fabia* (broad bean), *Rubus idaeus* (raspberry)**	Roman settlement, ditches pottery, tiles	Meadows (unpub.)
Oxfordshire	Symmetry Park, Bicester (74 masl)	4,500 metres of linear trenches covering 3.1 hectares, sand and gravel over mudstone	Early Roman	-	-	-	Close to Akeman Street (Alchester to St Albans)	Nowak 2018

Bedfordshire	Tavistock Av., Ampthill, (72–76 masl)	Small stream valley slope, numerous parallel cultivation rows over 9 ha, Amptill clay and Lower Greensand with calcareous soils	Late First or early Second century	-	-	Pollen very similar to Wollaston but no *Vitis* (despite count >1000 grains)	Nearby domestic settlement	Brown 2010
Essex	Heybridge, Elms Fm, Area R (3 masl)	*Vitis* pollen, floodplain/terrace with a palaeochannel	Fourth century	1	0.4%	Cerealia, *Avena* (oat), *Pisum sativum* (pea)	Large Roman settlement, temple high-status burials	Greig 2015
	Takerley (101–103 masl)	Large number of parallel flat-bottomed ditches aligned E–W with numerous pits	Early Roman	-	-	-	Off Stane Street (Roman road from Braughing to Colchester)	Wilkins and Roberts 2003
	Colchester (8 masl)	Military latrine	44–*c.* 49 CE	seeds	-	*Ficus* (fig), *Papaver somniferum* (opium poppy)	Major Roman town (Camulodunum)	Holloway 2019
Lincolnshire	North Thoresby (40 masl)	Regular network of ditches, and six trenches excavated in 1959	Late Third century	-	-	-	Roman road from Cadeby to North Coates (coast to Lincoln)	Webster et al. 1967

(*Continued*)

Table 10.1 ***Continued***

County	*Site (masl)*	*Evidence and environment*	*Est. age (context)*	*No of grains*	*Pollen %*	*Other crops / arable weeds (pollen/sedaDNA)*	*Nearby archaeology*	*Reference*
	Washingborough near Lincoln (5 masl)	Irregular large parallel ditches with pits	First–Fourth century	-	-	-	High-status Roman settlement with burials and kilns	Rylatt 2004
Buckinghamshire	Stanton Low (58 masl)	Regularly spaced parallel ditches, tripod used for preparation of wine	Second century	-	-	-	Roman port and settlement, villa, ritual activity	Woodfield and Johnson 1989
Wiltshire	Market Lavington (91 masl)	Anglo-Saxon cemetery	820–980 CE and later	-	1%	*Secale* (rye), other cereals, *Ribes rubrum* t. (redcurrent), *Linum* cf. *usitatissimum* (linseed), and *Cannabis* t. (hemp)	Probably agricultural dump near cemetery associated with Saxon settlement	Wiltshire 1997
Cambridgeshire	Godmanchester, Rectory Fm (13 masl)	Well fill, ornamental ponds	Early Roman	seeds	-	*Ficus* (fig), *Foeniculum vulgare* (fennel), *Rosa* p. (rose), *Daucus carota* (carrot)	Roman town (Durovigutum)	Carruthers and Hunter Dowse 2019

with all, except North Thoresby, in southeastern England in lowlands and on river gravels or calcareous geologies.

There is also archaeological evidence of viticulture, predominantly from bedding trenches, at a further seven sites clustered in east midland and southeast England (Figure 10.3). As was first recognized at Wollaston, these trenches are parallel, *c.* 0.85 metres wide and *c.* 0.3 metres deep, with flat bottoms and typically spaced *c.* 5 metres apart. The dimensions, spacing and form of these trenches are characteristic of *pastinatio* and align closely to dimensions of those that have been archaeologically detected and excavated elsewhere, especially in France, Italy and Sicily.[27]

There are other sites not necessarily recognized by the excavators as vineyards but where the form and location of ditches makes a vineyard possible or even likely. The site at Stanton Low, Buckinghamshire, provides a good example, where parallel trenches of comparable dimensions to Wollaston were excavated near buildings of villa quality and status, having a mosaic, rib vault voussoir tiles, stamped tegulae, a luxury bronze folding tripod, a lead coffin, an anvil and a coulter.[28] There are also indications of viticulture at Rock on the Isle of Wight, where a two-story building with an abnormal number of flagons, possible winepress and an undercroft is recorded.[29] Excavations on the south slope of Mersley Down in the centre of the Isle of Wight have also been interpreted as a Roman vineyard.[30]

A survey undertaken of excavation reports that include Roman planting trenches in the east of England with a rectangular cross-sectional form (*c.* 0.6–0.8 metres wide, 0.25 metres deep) and spaced 4.5–5.5 metres apart recorded fifty-two sites, mostly dating between the first and second centuries CE.[31] The authors argue that most were used for vegetables (e.g. peas, beans, carrots) rather than for vines, although very few of the sites had any ecological data making their use still open to question. Interestingly, there was no statistical relationship of these sites to settlements or villas but there was a statistically significant proximity to Roman roads, as is illustrated on a larger scale by the proximity of possible vineyards to major Roman roads depicted on Stukeley's classic eighteenth-century map of Roman Britain (Figure 10.3).

More widespread but also non-definitive evidence comes from the distribution of artefacts that could be associated with grapevine management, such as the pruning knife (*falx vinitoria*), which are not found in the UK. Small reaping hooks that might be used in a similar manner for grapevines, however, are widely distributed at over sixty sites in southern England.[32] Although these are more confidently associated with the cultivation of grapevines in areas like the Moselle[33] they could also be used for a variety of pruning

[27]Columella, *Rust.* 4.25.424–27; Boissinot 2009; Boissinot and Brochier 1998; Volpe 2009; Dodd 2022a.
[28]Woodfield and Johnson 1989.
[29]Lyne 2006.
[30]Trott pers. comm.
[31]Wiseman et al. 2020.
[32]Brown et al. 2001.
[33]Gilles 1995.

tasks on other trees and crops. Other tools, such as hoes, are less likely to be vine specific. The cited lack of winepresses or vinicultural production facilities in Roman Britain, with the possible exception of Rock on the Isle of Wight,[34] while problematic, is inconclusive because certain types of presses and treading vats can be made almost entirely from wood and thus are rarely preserved.[35] This includes modified barrels and box-like constructions for smaller-scale production, which remain a gaping *lacuna* in our understanding.

Palynology, associated crops, weed flora, and management

To verify the occurrences of *Vitis* at the Wollaston site, samples were counted by palynologists over several years. *Vitis* concentrations were found to vary from fifty to 350+ grains per cm^3 of sediment (Table 10.2). There does not appear to be a statistical relationship between the total pollen concentration and the concentration of *Vitis* grains, which suggests that the *Vitis* occurrence is due to sample location rather than the concentration of pollen in a sample (i.e. pollen influx and preservation) supporting its value as an accurate indicator of cultivation.

The pollen assemblages from Wollaston also provide data on crops grown nearby or perhaps intercropped, as well as other plants, including those typical of arable and disturbed land (Table 10.2). Intercropping with vines or the *arbustum* (training plants up trees) technique was common in the Roman Mediterranean and is well attested both in archaeology and texts.[36] As well as trees, crops grown between and around grapevines could include fruit, vegetables and the keeping of animals, such as chickens.[37]

From the large pollen counts, which are all dominated by herbaceous types, especially grasses (*Poaceae*) and weeds such as mugworts (*Artemisia*) and dandelions (*Lactuceae*), a diverse ruderal flora is revealed, including typical arable weeds (e.g. *Polygonum aviculare*, *Chenopodiaceae* and *Galium*) (Table 10.2). Three of these ruderal species – summer pheasant's eye (*Adonis aestivalis*), prickly poppy (*Papaver argemone*) (both identified independently by two palynologists), and opium poppy (*Papaver somniferum*) – are believed to be Roman introductions.[38] Several other species are climbers (e.g. *Convolvulus*, *Lonicera*), and other species are typical of nutrient-enriched grazing land (*Plantago*, *Solidago*, *Rumex acetosa*, *Trifolium pratense*) and ditch flora (*Apium nodiflorum*, *Filipendula*, *Cyperaceae*). Given that the Roman lowland environment was virtually treeless (under 2 per cent of the total land pollen being trees), this presence of climbers could be due to abandonment of training poles or perhaps hedges. Otherwise,

[34] Lyne 2006.
[35] cf. Italy in Dodd 2022a.
[36] Pliny, *HN* 14.14; Van Limbergen et al. 2017; Dodd 2022a.
[37] Brun 2003; Dodd 2022a.
[38] Van der Veen et al. 2008; Preston et al. 2004; Witcher 2013.

Table 10.2 Pollen Summary for *Vitis* and other possible crop and arable weed types from the Wollaston Sites: Hard Water Road (HWR) and Wollaston Trenches (Woll) with Concentrations. Key for pollen analyst ID: J. Hatton (JH), S. Turner (ST), A. Brown (AGB)

ID	*Sample ID*	*Total*	*Vitis*	*Vitis*	*V conc.*	*TP conc.*	*Apiaceae*	*Brassicaceae*	*Chenopodiaceae*	*Cerealia*	*Solanaceae*	*Others/Comments*
	Context		*No.*	*%*	*g cm³*	*g cm³*	*Apium t.*	*Sinapis t.*	*C album*	*A-T t.*	*S nigra*	
							Monolith (399)					
JH ST	HWR 399.00	1039	1	0.1	72	75612	144	432	1514	721	-	*Adonis* (1.4%), *Filipendula, Galium, Cyperaceae, Planatgo, Rumex, Trifolium, Polygonum aviculare*
	HWR 399.04	1021	4	0.39	377	97035	283	283	1792	849	-	*Adonis* (1.4%), *Filipendula, Galium, Cyperaceae, Plantago, Rumex, Trifolium, Urtica, Solidago*
	HWR 399.08	1010	-	-	-	160347	5398	476	4763	2540	-	Poss degr. *Vitis, Filipendula, Galium, Cyperaceae, Plantago, Rumex*
	HWR 399.12	1006	-	-	-	23321	3113	137	732	458	-	*Filipendula.Trifolium, Rumex, Plantago*
	HWR 399.16	1050	1	0.1	53	55801	1701	213	159	159	111	*Filipendula, Cyperaceae, Plantago, Rumex, Trifolium, Urtica*
	HWR 399.20	1036	1	0.1	63	65623	1584	-	1267	127	127	*Polygonum convolvulus, Filipendula, Plantago, Rumex, Trifolium*

(*Continued*)

Table 10.2 ***Continued***

ID	*Sample ID*	*Total*	*Vitis*	*Vitis*	*V conc.*	*TP conc.*	*Apiaceae*	*Brassicaceae*	*Chenopodiaceae*	*Cerealia*	*Solanaceae*	*Others/Comments*
	Context		*No.*	*%*	*g cm³*	*g cm³*	*Apium t.*	*Sinapis t.*	*C album*	*A-T t.*	*S nigra*	
								Well (1064)				
JH ST	HWR 1064.00	1028	4	0.3891	140	35914	594	559	419	70	175	*Cyperaceae, Filipendula, Galium, Plantago, Rumex, Polygonum aviculare, Trifolium, Urtica*
	HWR 1064.04	1015	-	-	-	30750	485	303	212	364	61	*Cyperaceae, Filipendula, Galium, Plantago, Rumex, Trifolium, Urtica*
	HWR 1064.08	1006	-	-	-	33379	332	796	299	66	66	*Cyperaceae, Filipendula, Galium, Plantago, Rumex, Polygonum aviculare, Trifolium, Urtica*
	HWR 1064.12	1007	-	-	-	35677	709	886	213	248	-	*Cyperaceae, Filipendula, Galium, Plantago, Rumex, Trifolium, Urtica*
	HWR 1064.16	1014	-		-	32693	355	613	193	161	-	*Papaver argemone, Cyperaceae, Filipendula, Galium, Plantago, Rumex, Trifolium, Urtica*
	HWR 1064.20	1077	-	-	-	37429	355	869	278	69	69	*Papaver argemone, Cyperaceae, Filipendula, Galium, Plantago, Polygonum aviculare, P. persicaria, P. convolvulus, Rumex, Trifolium*

						Well Monolith						
ST	HWR964032 (2.5)	522	-	-	-	11881		137	410	68		*Filipendula, Plantago, Rumex, Urtica*
	HWR964032 (6.5)	517	-	-	-	10492	41	162	101	41		*Filipendula, Plantago, Polygonum aviculare, Rumex, Trifolium, Urtica*
	HWR964032 (9.5)	502	-	-	-	21270		254	127	169		*Filipendula, Plantago, Rumex, Urtica*
	HWR964032 (11.5)	1009	-	-	-	18886		374	56	337		*Filipendula, Plantago, Polygonum aviculare, Rumex, Trifolium, Urtica*
	HWR964032 (17.5)	529	-	-	-	8393	16	63	126	142		*Linum bienne t., Filipendula, Plantago, Rumex, Urtica*
	HWR964032 (20)	518	-	-	-	13294		205	103	103	319	*Filipendula, Galium, Plantago, Rumex, Urtica*
						Ditches						
AGB	Woll 3/8532a	1041	5	0.47	78	16484		140	248	155		*Adonis, Cyperaceae, Filipendula, Galium, Plantago, Rumex,*
	Woll 3/8532b	1059	8	0.7	113	14991		326	269	170		*Adonis, Cyperaceae, Filipendula, Galium, Plantago, Rumex, Fallopia convolvulus, Trifolium*

(*Continued*)

Table 10.2 ***Continued***

ID	*Sample ID*	*Total*	*Vitis*	*Vitis*	*V conc.*	*TP conc.*	*Apiaceae*	*Brassicaceae*	*Chenopodiaceae*	*Cerealia*	*Solanaceae*	*Others/Comments*
	Context		*No.*	*%*	*g cm*^{3}	*g cm*^{3}	*Apium t.*	*Sinapis t.*	*C album*	*A-T t.*	*S nigra*	
	Woll 3/8530a	856	1			17205		543	181	20		*Adonis, Cyperaceae, Filipendula, Plantago, Rumex*
	Woll 8407.1	965										*Adonis, Cyperaceae, Plantago, Rumex, Filipendula*
	Woll 3/8530c	1108				22633		572	245	143		*Cuscuta europaea, Cyperaceae, Filipendula, Galium, Plantago, Rumex, Polygonum persicaria*

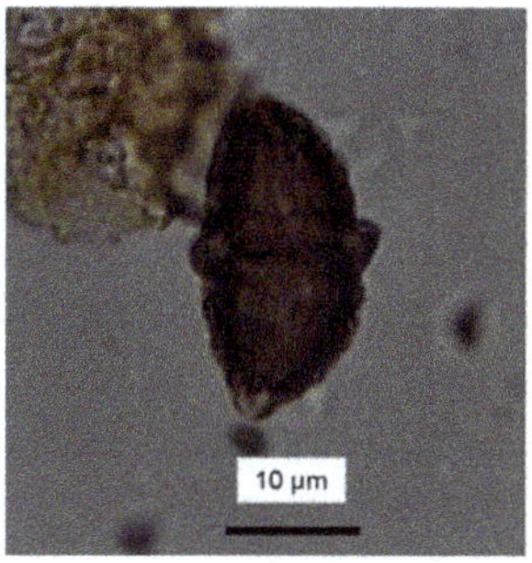

Ascospore of genus ***Valsaria*** (HdV-263) common in Roman deposits and is probably a thermophilous bark fungi (one *V. insitiva* associated with dieback of fruit trees and vines)

Found at:
WOLL8407.1 (3)
HDR523.4 (1)
WOL8383.1 (4)

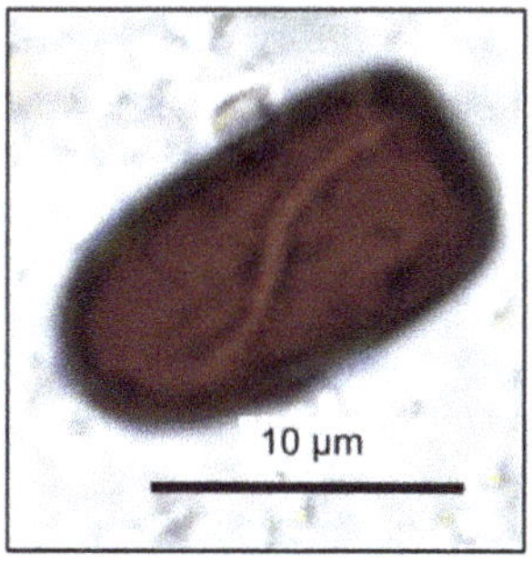

***Sporomiella* t.** (HdV-113) – coprophilous spores (indicates grazing)

Found at:
WOLL8407.1 (16)
WOL8383.1 (2)

Figure 10.5 Some key non-pollen palymorph (NPP) types (*Valsaria* and *Sporomiella*) found in vineyard pollen slides. Courtesy S. Richer.

the assemblage suggests disturbed ground with nutrient enhancement – perhaps both tillage and occasional grazing.

The presence of grazing is also supported by the presence of coprophilous fungal spores, including *Sporomiella* t. (HdV-113)[39] and *Podospora* t. (HdV-368). The ascospore of genus *Valsaria* (DdV-263), which is common in Roman deposits in Italy, is probably a thermophilous bark fungi, and *Valsaria insitiva*, which is associated with warmer climatic conditions[40] and dieback of fruit trees and vines, is a possible 'pest proxy' (Figure 10.5). Concentrations of the fungal rust/smut spores were also found in Anglo-Saxon deposits which had *Vitis* pollen at Market Lavington, Wiltshire.[41] The exploration of fungal spores in archaeological studies of viticulture is just one of many promising potential avenues of further research in the use of pollen and non-pollen palynomorphs (NPPs).

[39]Catalogue numbers given as per Shumilovskikh et al.2022.
[40]Bosi et al. 2019.
[41]Wiltshire 1997.

Conclusions

An important future avenue for the detection and possibly determination of cultivated vines, is the identification of pollen dimorphism.[42] Dimorphism (meaning two forms) is quite common in *Vitis vinifera*, because pollen produced by hermaphroditic and male flowers is trizonocolporate (with three furrows and colpi) but the pollen produced by female flowers is inaperturate (with no pores). Inaperturate pollen has been found in Bronze Age deposits on the Po Plain, northern Italy, and a working hypothesis has been developed stating that the presence of inaperturate grains could be an indicator of cultivation, because domestication produced morphologically hermaphrodite plants which were functionally female and inaperturate.[43] Study of the grains recorded at the vineyards of the British Isles described in this chapter has thus far failed to detect inaperturate *Vitis* pollen grains. They are, however, hard to identify and work is ongoing.

A new research direction combines palynology with sedaDNA. So far this has mostly been used successfully from lakes in relatively cold climates, such as the Boreal and Arctic zones, but also alpine areas in the European Alps.[44] From sediments in Lake Anterne at 2,063 metres in the northern French Alps, Roman grazing and vegetation history were illuminated, and work is underway on sites of lower altitude from which *Vitis* sedaDNA might be recorded.[45] Preliminary analyses of historical terraced vineyards in Soave region, in northeast Italy, as part of the pan-European TerrACE project has produced sedaDNA of *Vitis*, but this work is at an early stage.[46] Where soil or sediment conditions are favourable (e.g. clay-rich) there is no theoretical reason why *Vitis* sedaDNA should not be recoverable, even in the British Isles.[47] Although no *Vitis* sedaDNA was found at Earls Barton Roman vineyard in the Nene Valley, the spectrum of weeds is very similar to the Wollaston sites and more work is needed on these samples.

It is clear, therefore, that palynology still presents one of the most promising avenues to detect the presence of ancient vineyards in more challenging and ephemeral archaeological contexts. This is shown here through the growing body of evidence for Roman viticulture in Britain; however, evidence of widespread viticulture remains limited and Roman Britain still appears to be a consumer rather than producer. Nonetheless, there is evidence of medium-scale viticulture with vineyards concentrated in the Nene Valley and eastern midland England – was it the ancient British wine region? The distribution of sites suggests that production may have been for Roman towns (perhaps again a symbol of cultural transfer) or the Roman army further north. A

---

[42]Mercuri et al. 2021.
[43]Mercuri at al. 2021.
[44]Parducci et al. 2017; Giguet-Covex et al. 2014; Garcés-Pastor et al. 2022.
[45]Giguet-Covex et al. 2019.
[46]Brown et al. 2021a, 2021b; Ficetola pers. comm.
[47]Brown et al. 2022.

comprehensive analysis of all the archaeological data is long overdue. In the interim, palynology can be used to verify whether a potential vineyard site has *Vitis* pollen concentrations above the background value. Pollen and non-pollen microfossils (NPPs), which are predominantly algal and fungal spores, can also inform us about vineyard conditions and management, including the presence of disease. Future research pathways are currently being laid out, including the identification of domesticated pollen forms and the detection of *Vitis* from vineyard soils using sedaDNA.

PART III
VINEYARD AND LANDSCAPE ARCHAEOLOGY

CHAPTER 11
STUDYING ROMAN VITICULTURE IN *BAETICA* WITH GIS MODELLING AND GEOPHYSICAL SURVEY

Pedro Trapero Fernández, Isabel Rondán Sevilla and Lázaro Lagóstena Barrios

Abstract

This chapter discusses a series of techniques applicable to the study of Roman viticulture and wine production, including the modelling of viticultural conditioning factors and the use of non-invasive technologies such as geophysics and remote sensing. We deal with several specific cases in the Roman province of *Ulterior Baetica* that are of wider relevance for future studies of *amphora*-producing workshops and grapevine plantations. In recent years, the History Laboratory of the Institute of Viticultural and Agri-food Studies (IVAGRO) at the University of Cádiz have carried out several surface investigations to locate places of grapevine cultivation and the production of wine *amphorae* in the region of *Gades* at the *Hasta Regia* colony, which today lies in the modern winegrowing region of Jerez-Xerez. Through these discussions, we address the use of non-invasive techniques for the study of Roman vineyard traces.

Introduction

Researchers have always been interested in studying ancient viticulture because of wine's great socio-economic importance and its key place within the so-called Mediterranean triad, thus generating a deeper understanding of ancient societies.[1] The development of wine research has traditionally been divided according to its various stages of cultivation, production, transformation, transport, distribution and consumption – all part of wine's *chaîne opératoire* (see Chapter 2). To guide our reconstructions of the production and transformation phases, we have combined studies of ancient literary sources, especially Roman agronomic texts, with archaeological datasets, such as cultivation traces, productive areas (or *pars fructuaria*), and ceramic workshops.[2] The distribution of wines

[1]Tchernia 1986; Brun 2003.
[2]Martín 1971; Kolendo 1980.

is generally studied through their transport containers, above all, *amphorae*.[3] Finally, regarding consumption, literary sources in particular allow us to enter wine-infused social contexts. Such studies often combine literary and archaeological sources, with several sub-branches developed for the latter in the last century, including epigraphy, the study of ceramic materials and clay fabrics.[4] To these should be added recent contributions from contents analysis and pollen studies among other analyses that increase our understanding regarding types of grapevines, containers and wines.[5]

In recent years, we have focused on the study of viticulture in the south of the Iberian Peninsula, mainly in the region of *Baetica* for which written sources inform us about the export of wines (Strabo 3.2.6; Just., *Epit.* 44.1-2; Pliny, *HN* 3.7.1). We also know a substantial amount about wine production in the area from the agronomists Varro – who was praetor in the *Ulterior* province for a number of years and refers to a dozen specific practices – and Columella, who was local to this region. These authors point to the existence of a mixed production model, perhaps combining Punic or Turdetan influences with Roman techniques.[6]

The distribution of wine from this region is attested archaeologically through finds of Haltern 70 *amphorae*, which appear across two different circuits. On the one hand, a Mediterranean circuit connected to Rome, and, on the other, an Atlantic circuit that supplied the legions of Britannia and Germania.[7] In contrast to this considerable research base, the productive areas in Hispania itself have hardly been studied, which is why our study is focused on a specific wine-producing area of *Baetica*, the Lower Guadalquivir, specifically the territory of the Roman colony of *Hasta Regia*.[8]

In this chapter, we explore the application of new tools to the study of the first phase in wine production – vine-growing. The History Laboratory of the Institute of Viticultural and Agri-food Studies (IVAGRO), from the University of Cádiz, has undertaken several studies regarding the knowledge of viticulture mainly in the province of *Baetica*.[9] The laboratory is composed of historians, archaeologists, engineers and computer scientists who carry out non-invasive studies, generally using a combination of traditional surface survey, remote sensing and geophysical prospection. Since its inception six years ago, the laboratory has built up experience in the application of a series of techniques that advance our knowledge of viticulture and the associated landscape. We focus here on two areas of research: first, a summary of the models combining historical sources and Geographic Information Systems (GIS); and second, the exploration of several case studies using non-invasive and geophysical research techniques to study Roman grapevine plantations and ceramic workshops. We aim to illustrate how these sources

[3]E.g. Peacock and Williams 1986.
[4]For an overview, see Dodd and Van Limbergen in this volume.
[5]Drieu et al. 2020.
[6]Trapero Fernández 2021b.
[7]Brun and Gilles 2001; Ehmig 2003.
[8]Revilla 1995; Lagóstena 2016.
[9]Trapero Fernández 2018; Ruiz Gil et al. 2019; Lagóstena 2021.

and techniques can be used, what results or interpretations are gained, and how our methodologies might be further developed.

The application of GIS modelling

Territorial modelling using GIS offers great potential for understanding production processes in the ancient world. In the case of viticulture, we have developed a series of scenarios combining geographical information with Latin agronomic knowledge.[10] Our main area of study has been the Lower Guadalquivir, for which we have specific references from the local agronomist Columella.

The processes and geographical determinants taken into consideration for this study are summarized in Figure 11.1. We modelled insolation, slope, wind direction and soil types to identify optimal areas for the ancient vineyard.[11] It is possible to relate current geographical factors to ancient sources; for example, we are told that the ideal location for a vineyard was the middle of a hill (Varro, *Rust.* 1.6.5), and that areas with a southerly exposure and high amounts of sunshine were preferred (Columella, *Rust.* 4.22.8). At the same time, these areas needed to be protected from dangerous winds – which, in our specific area, implies an easterly orientation – and have a specific type of soil (Columella, *Rust.* 5.5.15, 2.2.1-21; Pliny, *HN* 17.3-4). A certain value can then be assigned to each of these factors. In our case, for example, the optimal soils described by Columella coincide with those that are best for present-day vineyards.[12]

These indicators can then be translated into a GIS, using basic tools such as generating slopes or orientations and reclassifying. At this stage, we used data from the ancient agronomists, weighting importance according to their preferred indications. Finally, all values are combined to check where all or most of the indicators are met. However, the difficulty lies in weighting the importance of each criterion. Several articles have already been published on the use of these criterion,[13] but generally a correct weighting is done according to the specifics of the study area. By understanding the Roman mentality and seeing which conditions are more important for a certain area, it then becomes possible to establish a ranking. For example, in the Lower Guadalquivir it is very important not to place grapevines to the east or south because the winds can be harmful.

The method used was compared with archaeological sites that possess evidence for the production of wine and in most cases a match was found. Furthermore, the experience allowed us to extend the study to other activities, such as the proportion of riparian land necessary to cultivate the vineyard, as advised by the surveyors (Columella, *Rust.* 4.30.1). Using this method, it is possible to calculate the number of workers required for the

---

[10] Lagóstena and Trapero 2019; Trapero Fernández 2021a.
[11] For the specific methodology used, see Trapero Fernández 2016, 2021a.
[12] Trapero Fernández 2021a.
[13] Trapero Fernández 2016; Stubert et al. 2020.

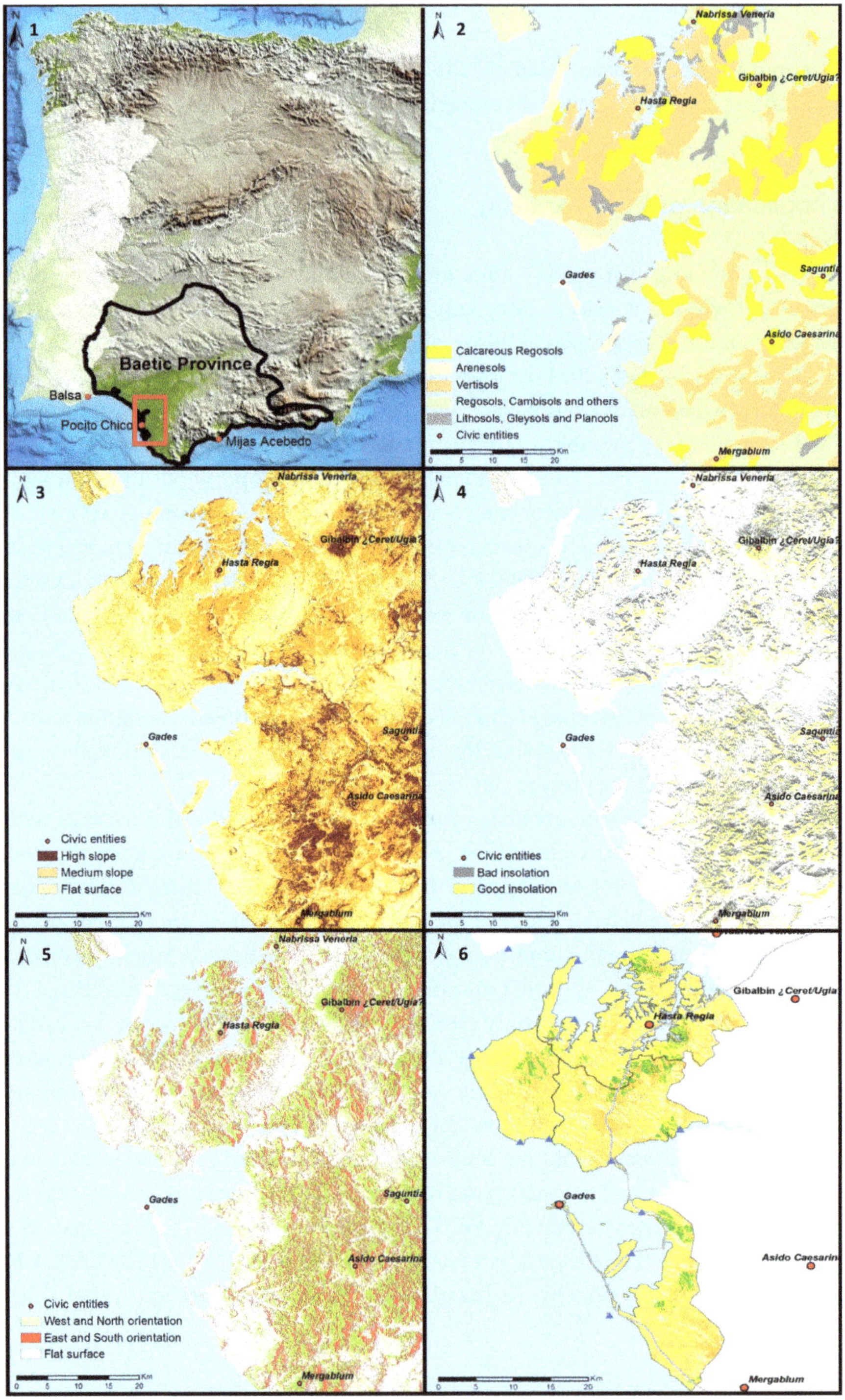

Figure 11.1 (1) location of the Lower Guadalquivir in red and other case studies in the *Baetica* and its surroundings; (2) soil classification; (3) slope classification; (4) wind orientation classification; (5) insolation orientation; (6) final results of suitable areas for vineyards in green, good areas for vineyards in light green, principal ports and harbours indicated by blue triangles. Adapted from Lagóstena and Trapero 2019.

vineyard, as well as to infer land dedicated to cereals necessary to feed these people, therefore generating a theoretical estimation of area and number of workers per productive unit. We aim to improve this method in the future to obtain a predictive model, which will allow us to locate potential deposits in the field. The model should also be expanded to include other agricultural activities to establish the different potential uses of the land.

This method serves in the first place as one useful to study the landscape, but it can also be used to pinpoint possible areas that have not been prospected. The challenge is to locate the places that a Roman farmer would have preferenced on a map, to which we can add other questions such as proximity to communication routes, cities and waterways. This method has been tested in other case studies, and there are examples of similar applications in other European areas.[14] Our contribution has the advantage of considering the precepts of the Roman agronomists as the basis of the study, as other approaches use statistical correlations, with a careful reading of classical sources prioritized.

Geophysics applied to a pottery workshop

The use of GIS is growing as a methodological tool in ancient viticulture studies, which allows for our models to be contrasted with other techniques.[15] Geophysical prospection, however, despite its great potential is not as often utilized in this field. These techniques consist of technological tools that allow us to visualize possible archaeological remains without excavation. Two of the most commonly used techniques are ground-penetrating radar (GPR) and magnetometry.[16] The former measures the amplitude of reflections caused by transitions between layers or structures with different dielectric permittivity, while magnetometry registers variations of the local magnetic field.[17] Both techniques have their advantages and disadvantages. GPR allows the visualization of depths as well as a ground plan, giving in ideal conditions a lot of information, but it is limited by the type of soil, being challenging to interpret if the conductivity of the soil is poor. The magnetometer on the other hand does not allow a visualization of different depths, but it is quick and cheaper to use, and especially useful to detect burnt structures (because of a different magnetic signal). However, it is limited in efficacy if there are metallic elements on the surface. In theory, winemaking structures and storage areas should be detectable by these methods, but in what follows we focus specifically on the traces of pottery workshops and kilns.

---

[14]Stubert et al. 2020.
[15]Bernigaud et al. 2021 and the contributions of Bernigaud et al. and McLean in this volume.
[16]See Chapter 3 in the present volume for extended discussion of these techniques and their use.
[17]Goodman and Piro 2013.

The production of *amphorae* and other containers, such as *dolia*, for the storage and transformation of wine is a major research branch within studies of ancient viticulture.[18] Geophysical techniques can be particularly useful when applied to the study of artisanal spaces dedicated to the production of *amphorae*. Applied geophysics can be used for predictive planning of archaeological excavation as well as to gain knowledge on the organization of production spaces, such as kilns and ancillary structures. Additionally, in a recent publication we have carried out a study on volumetric calculations applied to production remains (Figure 11.2).[19] This kind of analysis allows us to contribute to the knowledge of craft spaces and quantitative economic history.

This is precisely one of the objectives of our work on production areas, especially for wine *amphorae*, as a line of knowledge on the agricultural economy of *Baetica*. In this territory, several examples exist of *figlinae* on the banks of the Guadalquivir,[20] although the areas of the territory *Hastensis* and *Gaditanus* have yet to be discovered. As we currently lack excavation data, the application of geophysical methods offers opportunities to explore these productive spaces. For this reason, we have experimented with different techniques in the study of Roman pottery workshops.

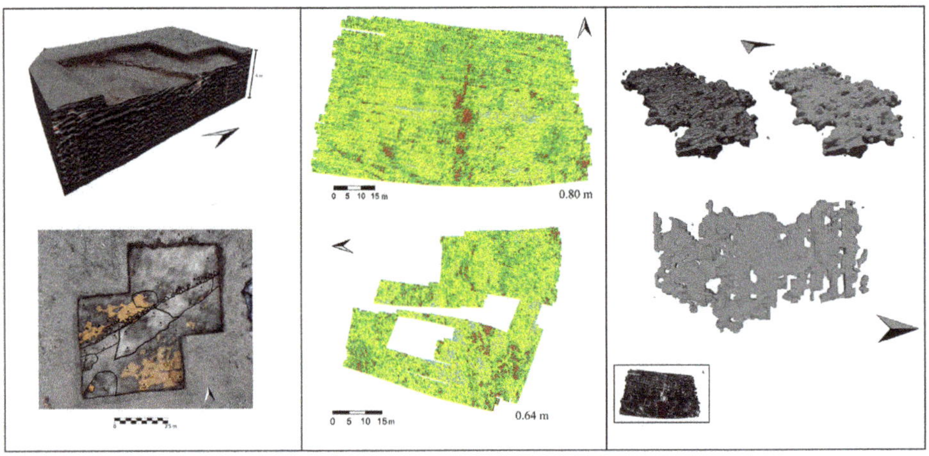

Figure 11.2 Processes and models to study kilns and ceramic dumps using geophysical techniques. Results of GPR exploration and excavation (*left*); Miramundo and La Rana ceramic production sites with processed data (*centre*); volumetric modelling of the *amphora* dump from the Miramundo production site at Puerto Real, Cádiz (*right*). Courtesy L. Lagóstena and the Heritage Geodetection Unit, University of Cádiz, and after Lagóstena and Molina 2019.

---

[18] E.g. Carandini 1989.

[19] Lagóstena and Molina 2019.

[20] Lagóstena and Trapero 2019.

Figlinae were complex working spaces, especially when active for long periods, changing and being reused in different periods, which makes recognition by using geophysics difficult.[21] While these workshops consisted of well-defined functional elements, they are difficult to locate through surface survey. The use of geophysical techniques enables the identification of these elements and allows us to plan other archaeological interventions more precisely. A multichannel 3D GPR from IDS Company and a multichannel fluxgate magnetometer from Sensys company was used to investigate these structures and some of the results are given in Figure 11.2. Above all, these techniques were applied to three *figlinae* in 1) the suburbium of the municipio of Arva (Alcolea del Río, Seville); 2) La Rana (Gatas de Gorgos, Alicante); and 3) Miramundo (Puerto Real, Cádiz).

Processing the GPR results allows us to effectively understand the spatial organization of these ceramic production facilities.[22] It becomes possible to model the volume of material produced or study the size and typology of kilns, even defining certain phases such as abandonment. This ordinarily facilitates a quicker and more cost-effective way to understand these facilities without large-scale, expensive excavations. A serial set of time- or depth-slices is of great interest to know the historical evolution of these production spaces (even if we should remain careful about drawing historical conclusions on the basis of time-slices, as a deeper slice does not always indicate an older structure or layer), and we aim to apply it to the wine *amphora*-producing workshops of the Lower Guadalquivir. This methodology could then be expanded to other case studies in different regions.

On the dumping areas of ceramic production facilities, through the data provided by the 3D GPR it is possible to model the volumes documented in *amphora* production contexts. The data on accumulated waste from kilns producing wine *amphorae* is of great interest for estimating the total production of the site, and for ranking the different installations according to their relative importance. GPR offers the distinct advantage of accurately measuring subterranean volume, and thus, in this case, facilitates volumetric studies relevant to possible discarded materials.

It is important to recognize that *amphorae* serve as indispensable instruments for comprehending one aspect of the winemaking process, especially storage and distribution. Information regarding the production of *amphorae* is generally limited due to the requirement of laborious and costly excavation, thus geophysical survey provides an important and sometimes more accessible alternative to investigate production sites. By investigating sites in this manner and comparing the results to excavated contexts, we can begin querying features by volume and explore connections between waste and production scale.

---

[21] Lagóstena 1996.

[22] Lagóstena and Molina 2019.

Geophysics applied to plantation systems and agricultural traces

Geophysical techniques have been applied mainly to the study of what we could consider positive features and structures, for example walls. However, they can also be used for negative features, such as pits, holes or trenches.[23] The equipment and technical setup developed by the current team allows us to explore a larger ground area in a shorter time, which facilitates the efficient exploration of these negative features across landscapes (Figures 11.3.1–3).

We illustrate here different cases in and outside *Baetica* of specific studies where Roman vineyard cultivation pits can be observed. The method for determining that they are indeed from this period – a recurrent challenge in geophysical prospection – is explained in each case study. Generally it is achieved by association with nearby structures. We also analyse the visualized patterns and agricultural traces through measurements, based on the available information on grapevine spacings by ancient agronomists. The case studies selected are particularly relevant because of the traces analysed and their historical peculiarities. The first is located within the Roman municipality of *Balsa* (Luz de Tavira, Portugal), and the second probably within the territory of the *Gades* and *Hasta Regia* colony, in Pocito Chico (El Puerto de Santa María,

Figure 11.3 (1) Sensys MX V3 magnetometer; (2) IDS Stream X geo radar; (3) 3D-Radar MK IV geo radar; (4) areas surveyed at Luz de Tavira (visualized here at a depth of -0.5 metres) with remains of a periurban villa in red, geological substratum in blue, and salted fish production *cetariae* in green; (5) cultivation traces indicated with arrows at the eastern corner of the *cetariae* indicated in the green rectangle. Photographs and diagrams by authors.

[23]Bitella et al. 2015.

Cádiz, Spain). We have developed other geophysical surveys with similar results at the Villa de Acebedo (Mijas, Málaga, Spain), but this remains under study.

Balsa (Luz de Tavira, Algarve)

This site is located in the south of *Lusitania*, in the Flavian municipality of *Balsa*, (Luz de Tavira, Algarve), and was studied through collaboration with the University of Algarve. The geophysical surveys were carried out on the plots bordering the estuary. The composition of the soils is based on consolidated sands and gravels, usually a good soil on which to use GPR if it is not salinized. The equipment used is the 3DRadar Mk IV GPR, a multichannel, multi-frequency system, with a scanning width of approximately 2 metres. The area presents at least three groups of anomalies (Figure 11.3.4): 1) to the north, the remains of a periurban/suburban villa (currently under excavation); 2) in blue, what is probably the geological substratum, although it cannot be excluded that the stone is worked; and 3) to the south, a salted fish production *cetariae*.

It is at the southeast corner of the field that a set of homogeneous and orderly arranged point reflections were documented (Figure 11.3.5). These points appear at a depth of around 30 centimetres and are set 4 and 2.5 metres apart respectively, likely coinciding with grapevine row distances, and the predominant orientation is northwest–southeast. In this particular case, we do not yet have archaeological evidence that allows the identification of these patterns as Roman cultivation traces. However, the archaeological context – that is, within the production area of the Roman town of *Balsa*, and without traces of other structures with a different chronology – perhaps suggests that they are indeed traces of Roman vineyards.

Pocito Chico (El Puerto de Santa María, Cádiz)

The second case study is the Pocito Chico farmhouse, in El Puerto de Santa María, Cádiz. The area is the subject of an archaeological study that has revealed evidence from the Chalcolithic, Late Bronze Age, Roman and Andalusian medieval occupation periods.[24] It is located *c.* 10 kilometres from *Portus Gaditanus* and the colony of *Hasta Regia*. The geological composition of the area is notable for the presence of calcareous soil, 'albarizas' – a soft rock formed by sediments with traces of diatomaceous algae. In this case GPR was used (IDS Stream X multichannel georadar with a 200 MHz centre frequency) as the soil shows a clear conductivity contrast. The most visible group of signals shows a pattern of homogenous geophysical disturbances oriented northeast–southwest and spaced 4 and 2 metres apart respectively (Figure 11.4). We have chosen two signals and analysed their radargram (profile with signals) view. The reflections have a thickness of about 0.5 metres from –0.40 metres and –0.90 metres in depth (Figure 11.4.2). When compared to the results obtained by magnetometry (using a Sensys MX V3), the anomalies follow the same pattern (Figure 11.4.4).

---

[24] López Amador and Ruiz Gil 2007.

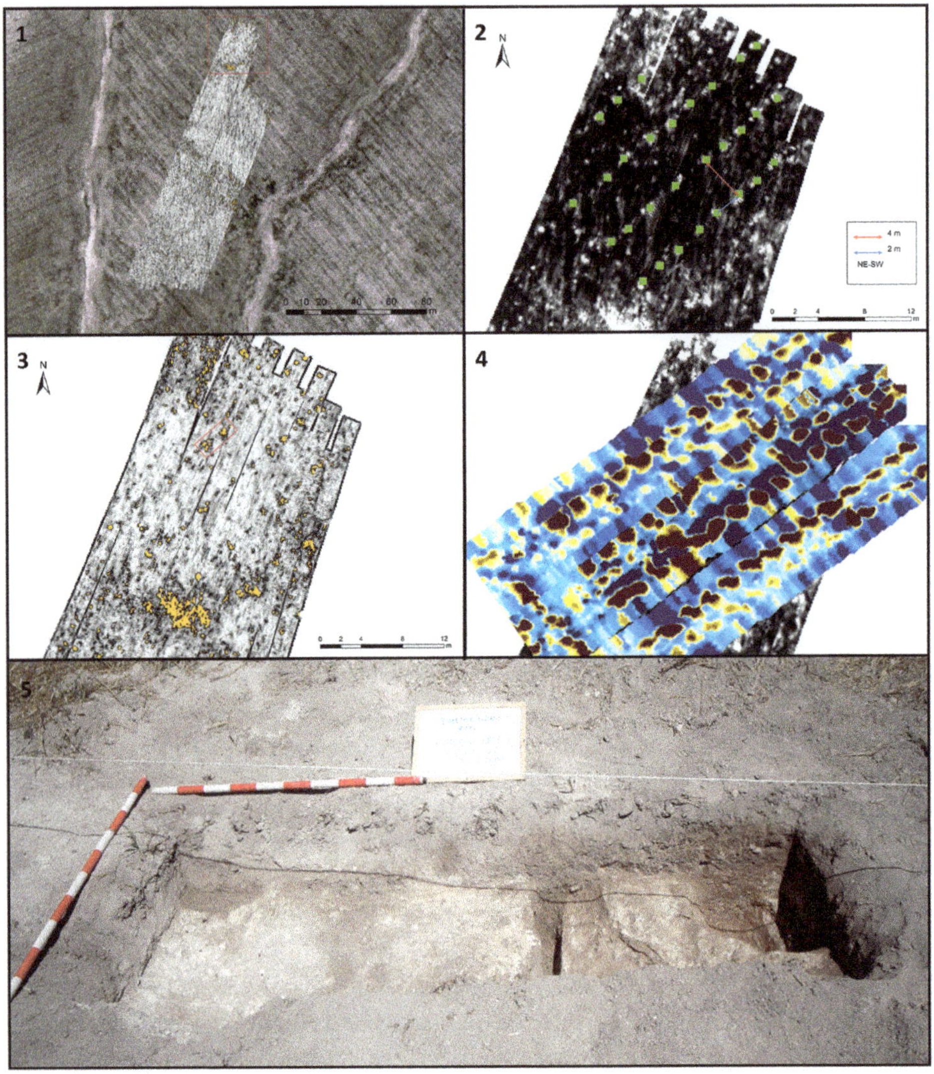

Figure 11.4 (1) area surveyed at Pocito Chico (El Puerto de Santa María, Cádiz) with an indication of areas with cultivation pits highlighted in the red square; (2) regularly spaced anomalies indicated between a depth of -0.40 and -0.90 metres; (3) processed image with cumulative overlapping of all depths where anomalies were detected; (4) results of the magnetometry survey; (5) archaeological excavation designed to test the anomalies detected using geophysical prospection at Pocito Chico. Images and diagrams by authors.

A small excavation was carried out in the area highlighted in red in figure 11.4.3 to compare the geophysical data with archaeological data (Figure 11.4.5). It was found that these were artificially excavated holes in the calcareous bedrock, which included the remains of roots. These could then be analysed to determine whether they belong to grapevines or other plants. So, in this case, we have been able to verify that the signals detected with geophysics do indeed correspond to cultivation pits. Another example of crop marks is also shown at this site, but it is not clear that they date to the Roman period, but rather may be earlier or later given the broad chronology of the site.[25]

Acebedo (Mijas, Málaga)

The last site, currently under study, is at the Acebedo Roman villa (Mijas, Málaga), which has existing excavated areas and is located next to the Fuengirola river. The GPR system used was the same as that in the previous case and the survey covered a large area, documenting several buildings and structures. Treatment and post-processing of the data was undertaken to improve their visualization and interpretation. In this case, they have been documented at a depth of 60 centimetres, with the best visualization at 80 centimetres, and are 4.5 metres and 2.5 metres in spacing, following a northwest–southeast orientation. Part of this area has been excavated and the remains of roots and planting pits associated with Roman pottery have been documented.[26]

Discussion

The application of GIS to the study of viticulture is growing, being one of the main tools now used to understand past landscapes of production. The examples presented here all show results that appears largely consistent with ancient literary sources or have been verified by archaeological excavation and additional study. For the development of this type of technique in the future, it is possible to proceed in two ways. On the one hand, the work should be tested in diverse environments and contexts to untangle the peculiarities of the territory and reliability of the model. On the other, the model should also be applied to locate potentially interesting and as yet unrecognized sites for viticulture. The latter should be a priority in areas that have not yet seen large-scale systematic survey and can help to plan and organize such efforts. Each study area has unique constraints that must be taken into account. A region with poor agricultural soils is not the same as a region with good soils; for example, in the latter, the pedology criterion might not be as important as slope or orientation

Geophysical techniques, while used widely across archaeological fieldwork and research, are usually applied to the historical study of rural sites and remain largely underutilized for examining production facilities and even less so agricultural traces. As illustrated in this chapter, volumetric analyses can be carried out that facilitate estimations of the amount of pottery that a certain production dump site may contain. This allows decisions to be made regarding further fieldwork, including excavation, but also strategies

[25]López Amador and Ruiz 2005.

[26]This work was carried out by the heritage department of Mijas municipality, which kindly lent the unpublished data.

to understand the economic importance of this type of production. Thanks to projects such as the geophysical surveys in the Roman municipality of Arva – which has an artisanal *suburbium* where we have detected several *amphora* kilns, in this case for oil[27] – these methodologies are being developed further. In the future, they will undoubtedly be of great use to study the historical relationship between kilns and subsidiary installations.

What is perhaps most promising, however, is the still underdeveloped study of negative agricultural features using geophysical techniques. One of the problems in the study of ancient viticulture is that we possess relatively few examples of the cultivation process, making it difficult to locate the areas where they are found and challenging to excavate. Geophysical prospection encourages easier and more efficient detection. Of course, there are certain limitations. The first is that not enough studies on this subject exist to be able to explore possible interrelationships between the size of a plantation with chronology. Hopefully, these new studies in *Baetica* will allow other researchers to raise and share their data in this regard. Some caution should also be exercised in the use of the parameters. Because of the GPR data acquisition itself, some uncertainty exists regarding the depth, or Z-axis, since we need to know perfectly the type of soil being analysed. This can be corrected by additional laboratory or comparative study if areas are excavated.

In Table 11.1 the data that can be obtained with the use of GPR are summarized, such as the orientation, width and depth of the hole, and vineyard spacing in several directions. Even if it is unlikely that the recommendations of the ancient agronomists were followed consistently by Roman farmers,[28] these data can be used to compare various historical information and thus explore chronological possibilities. The table provides examples of sources that apply to the area of study, especially the Roman agronomist Columella, together with other authors and chronologies, such as Theophrastus or Palladius, who are not as directly related to the region under study here. However, comparing data from the ancient literature remains complex, especially considering the diversity in purpose, audience and context between each author.[29] Additionally, from a practical perspective, the 'when' and 'how' was central to ancient agronomic literature, making potential comparisons to archaeological data potentially more difficult across regions and periods. *Baetica* provides an opportunity to contextualize these cases, as the texts of authors that spent considerable periods of time in this region, such as Columella and Varro, are comparatively detailed. Similarly, local changes brought about by medieval Muslim and Christian forms of agriculture are visible in later authors including Al-Awam and Alonso de Herrera, Andalusian and Castilian agronomists respectively. These later sources reveal that their fundamental basis is drawn from the Roman agronomists, in particular Columella. Medieval agricultural forms thus appear to be based largely on Roman

[27]Under publication, see Ruiz et al. 2021.
[28]Terrenato and Becker 2012.
[29]Martín, 1971.

practices.[30] Regional continuity of some of these practices in present-day agriculture has even been recognized. This makes it possible to cautiously draw comparisons between sources and across broader chronologies.

Table 11.1 makes it clear that the pattern between the case studies of *Balsa* and Mijas is very similar, and, when considered in light of other contextual archaeological remains, is likely associated with the Roman world, even if an attribution to other periods cannot be excluded. If the measurements given by the agronomists are compared, they most closely resemble those described by Columella (Table 11.1). Pocito Chico may be Roman, but could also be later given that the spacing between lines is greater. The study of these theoretical values, compared with other cultivation frameworks in the future, will be used to determine approximate chronologies.

Conclusion

The case studies presented here exemplify the successful application of new analysis and modelling techniques to understand Roman viticulture. Although these are in many cases incipient research lines, their potential is enormous with substantial benefit, especially with further technical development and equipment that provides better results and is more accessible. In addition, the techniques illustrated here are cheaper and more efficient than other modes of analysis with the potential to ascertain greater quantities of

Table 11.1 Measurements of the pits in the GPR case studies and historical sources

Source	*Orientation*	*Hole width (metres)*	*Hole depth (metres)*	*Hole spacing (metres)*	*Line spacing (metres)*
GPR in *Balsa*	NE–SW	0.4–0.5	0.5	2.4–2.5	2.4–2.5
GPR in Pocito Chico	NE–SW	0.5–0.6	0.9	2.2–2.4	4.0–4.2
GPR in Mijas Acebedo	NE–SW	0.5–0.6	0.6–0.9	2.0–.2.2	2.4–2.5
Theophr., *Hist. pl.* 2.5	-	-	0.3	-	2.7–?
Columella, *Rust.* 3.13.3	-	as wide as the hoe	0.6	2.1–3.0	2.1–3.0
Columella, *Rust.* 3.15.1	-	-	0.6–0.9	1.5–2.1	1.5–2.1
Columella, *Rust.* 5.5.1	-	0.6–0.9	0.9	-	1.2–3.0
Palladius, *Op. Agr.* 2.10	-	0.75–0.9	0.9	-	3.0–3.3
Al-Awam 7.45.186–190	-	0.6	0.6	-	4.5
Alonso de Herrera 2.8	-	-	0.8	-	4.5–6.0

30 García Armendáriz 1995.

data with further development and comparative study. While one does not have the same chronological and typological granularity compared to data derived from an excavation, if enough is accumulated to be able to make historical comparisons, in different periods and places, the reliability of the method will eventually increase. We believe that non-invasive methods should be used as a means of studying viticulture, and are especially useful for their contribution to the generation of economic models.

CHAPTER 12
NEW HIGH-RESOLUTION APPROACHES FOR VINEYARD ARCHAEOLOGY: EVIDENCE FROM THE REGION OF POMPEII

Florian Seiler[1]

Abstract

Pompeii is one of the few places in the Roman world where excellent preservation, due to the eruption of Mount Vesuvius in 79 CE, means that ancient vineyards can be studied in a highly accurate and revealing manner. Furthermore, whereas previous excavations of vineyards were always limited to the hilly region of Vesuvius, the situation on the plain was little known. The recent excavation of a Roman vineyard at Scafati, near the river Sarno, is the first comprehensive study of viticulture on the alluvial plain. The multidisciplinary analysis and micro-excavations presented here have enhanced our knowledge of Pompeian viticulture and Roman agriculture more broadly. Evidence of more vineyards in the area, along with a dense network of farmsteads (*villae rusticae*) specialized in wine production, show that grapevines were grown on the plain on a large scale. This flourishing agricultural sector provides further support for the orientation of the Vesuvian region's economy towards viticulture in the Roman period.

Introduction

In the field of viti- and vinicultural archaeology, the excavation of vineyards still plays a marginal role. This is usually for the simple reason that the physical substance of ancient vineyards (i.e. the surface soil) is not normally preserved due to the erosion and decrease of the palaeosols caused by various destructive effects, such as subsequent agricultural

[1]The author thanks colleagues of the SALVE excavation project at Scafati for the multidisciplinary collaboration that led to the results presented here in condensed form. Special thanks are owed to the editors Emlyn Dodd and Dimitri Van Limbergen for the invitation to the conference *Vine-growing and Winemaking in the Roman World* and to contribute a chapter in this volume. Thanks to their critical comments and suggestions, as well as the careful editing of the manuscript, the final version was greatly improved and gained much clarity. Any remaining errors are my own.

and building activity or manual removal.[2] Organic materials, such as vines, supports and other vegetation of a vineyard, seldom survive. In contrast, Pompeii presents a rarely accessible opportunity to study well-preserved ancient vineyards in a largely authentic state, due to the eruption of Mount Vesuvius in 79 CE. The rapid deposition of huge quantities of volcanic material hermetically buried towns, villas and the surrounding landscape across almost the entire Sarno Plain. The latter, in particular, resulted in a well-preserved surface of cultivated fields, features of agricultural activity, and even some remains of vegetation.

This important, but hitherto dormant potential was first exploited scientifically in the 1970s and 1980s by Wilhelmina Jashemski, who excavated gardens in townhouses and villas as well as vineyards within Pompeii, introducing new techniques of excavation, conservation and documentation.[3] Her pioneering archaeobotanical field research set the foundation in vineyard archaeology from which we build. Villa Regina at Boscoreale on the eastern slope of Vesuvius (Figure 12.1) is so far our only example illustrating the entire vinicultural lifecycle, from grapevine cultivation in fields around the villa, through wine production using treading and pressing, and storage in a central *cella vinaria*.[4] The vineyard planted around the villa, also excavated by Jashemski and now replanted, has since been considered a characteristic example of Vesuvian viticulture.

Based on archaeological data, such as wine production and storage infrastructure, and study of *amphorae*, it is evident that viticulture was widespread in the hilly area of Mount Vesuvius as well as in the Lattari Mountains at the southern edge of the Sarno Plain. The ratio of wine-producing *villae rusticae* related to the total number of villas in these regions is about 40 per cent, and it was likely much higher considering the number of unidentified cases.[5] On the other hand, the situation in the river plain is disputed, with some supporting use for the cultivation of cereals, and others vegetables.[6] Based on the fact that large parts of the plain were densely populated by *villae rusticae*, many of them involved in wine production, the agricultural exploitation of this area must be revisited.

Excavation of a vineyard: data and methods

To provide new data and reassess the agricultural use of the Sarno Plain, our approach combined comprehensive (re)study of all *villae rusticae* in the area with the excavation

[2] Another case, mainly found in rocky environments, are pits or trenches dug into the rocky subsoil to provide more space for planting, which are often interpreted as vineyards. Numerous examples of these trenches dug into the tuff bedrock exist in the *suburbium* of Rome (Volpe 2009; Santangeli Valenzani and Volpe 2012), as well as at Megara Hyblaea, Sicily (Boissinot 2009.), Burgundy, France (Garcia et al. 2010), and *Lugdunum*, modern Lyon (Poux 2011). On this planting method also see Columella, *Rust.* 3.13.

[3] Jashemski 1968, 1973, 1979.

[4] De Caro 1994.

[5] Vogel et al. 2016a; Seiler, Vogel and Esposito 2019.

[6] Cf. De Simone 2017: 32–43.

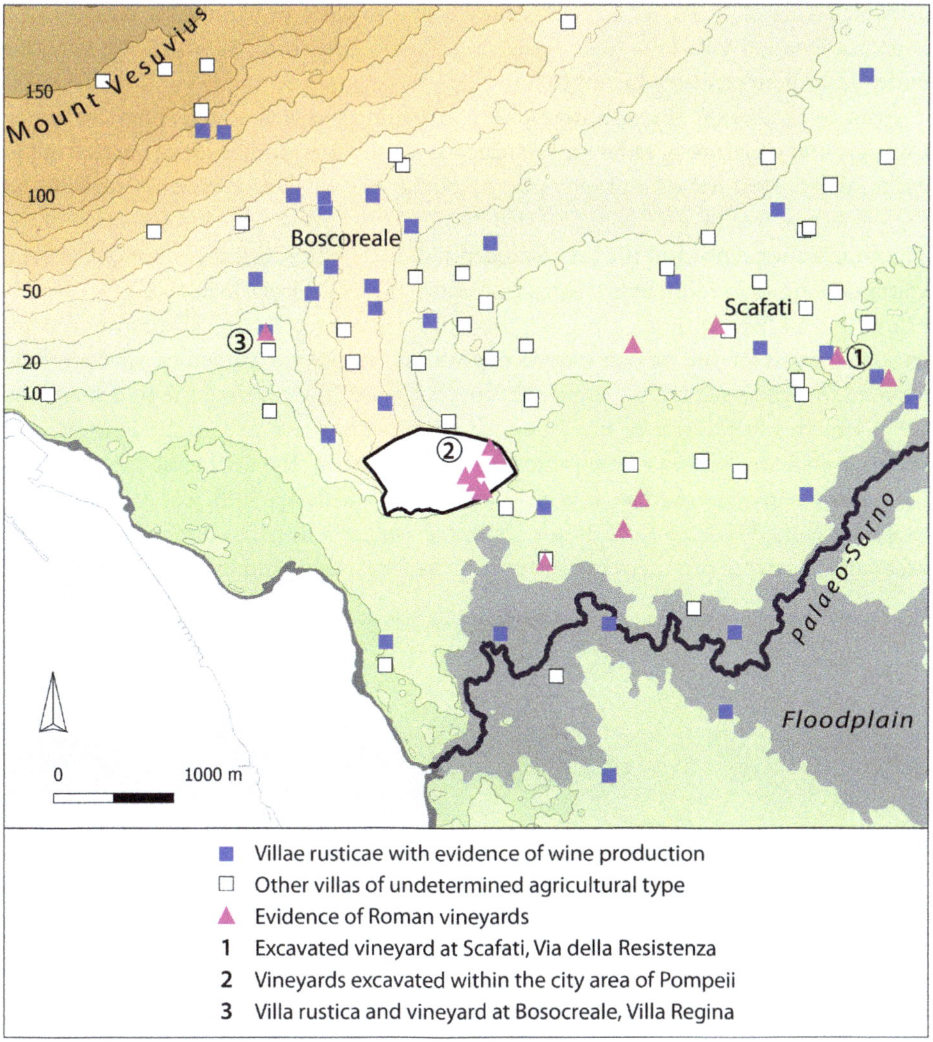

Figure 12.1 Map of the northwestern Sarno River plain with reconstruction of the pre-79 CE topography from a digital elevation model. Courtesy F. Seiler.

of a section of Roman farmland in the river plain, undertaken in 2012.[7] The site chosen for the latter study is located 3 kilometre east of Pompeii on Via della Resistenza in the district of Scafati, and on the edge of the floodplain, 1 kilometre from the course of the river Sarno (Figure 12.1). Close by are five Roman villas – among them the systematically

[7]The final study of the *villae rusticae* in the Sarno River plain by Domenico Esposito will be published shortly as a monograph. The farmland excavation was organized within the research project SALVE of the German Archaeological Institute (DAI), Berlin and funded by the Deutsche Forschungsgemeinschaft (German Research Foundation). Vogel et al. 2016b.

excavated Villa of N. Popidi Narcissi Maioris, and in close proximity the remains of a large unidentified villa (Via della Resistenza, Fondo La Feltra), to which the vineyard under study here probably belonged.[8]

Prior to excavation, three stratigraphic core drillings as well as a geophysical survey using ground penetrating radar and geoelectric resistivity soundings were carried out to ensure that the natural stratigraphy was intact, the 79 CE palaeolayers undisturbed, and the subsurface free of archaeological structures. Coring also enabled observation of the stratigraphic sequence and the expected depth of the 79 CE soil. Excavation of the thick volcanoclastic and sedimentary deposition above the 79 CE layer, located 3.7–4.1 metres below the modern ground level, required the use of a mechanical excavator. After removing most of the volcanoclastic deposition, the 79 CE palaeosol was carefully exposed by hand so as not to damage the intact surface. The depth of the trench and the 1.9–2.4 metres thick layer of loose pumice made it necessary to step the long sides of the excavated area and to create ramps on the short sides. The final exposed surface of the Roman field comprised an area of 12 × 4 metres, which was sufficient to conduct the analyses planned (Figure 12.2). In the southwest corner, a deep 2 × 2 metres trench was excavated to a depth of 6 metres to investigate the Palaeolithic and Iron Age stratigraphy.[9]

Figure 12.2 View of the excavated trench of the vineyard at Via della Resistenza, Scafati, Italy. Courtesy F. Seiler.

[8]Villa of N. Popidi Narcissi Maioris: De' Spagnolis 2002. Villa in Via della Resistenza (Fondo La Feltra, ex Cartucciera), Villa in Località Castagno, Villa in Via Passanti: De' Spagnolis 1991–2; De' Spagnolis 1995. Villa in Via Tricino (Fondo Acanfora): De' Spagnolis 1989: 44.

[9]The excavation resulted in a stratigraphy of *c.* 6 metres deep, which consisted of a multi-layered sequence of repeated volcanic depositions and soil formations spanning at least 10,000 years. It provided clear evidence of repeated anthropogenic agricultural activity (Vogel et al. 2016b).

These field analyses were conducted by specialists from disciplines including geoscience, archaeobotany, palynology and archaeology.[10]

After removing the upper grey pumice stratum, yellowish, regularly distributed stains were visible towards the top of the white pumice deposit, which turned out to be caused by decaying organic material left by the stakes to which vines were fixed. In fact, they matched the position of stake holes eventually located on the 79 CE ground level. Immediately after cleaning the palaeosurface it was sampled for pollen, minimizing the time for contamination. In total, seventy-two pollen samples were taken from the surface (marked by white tags in Figure 12.2). Following this, the entire excavated surface was laser-scanned and georeferenced to create a digital 3D model with 2 millimetre resolution (Figure 12.3). This high-resolution model of the 79 CE ground surface before invasive

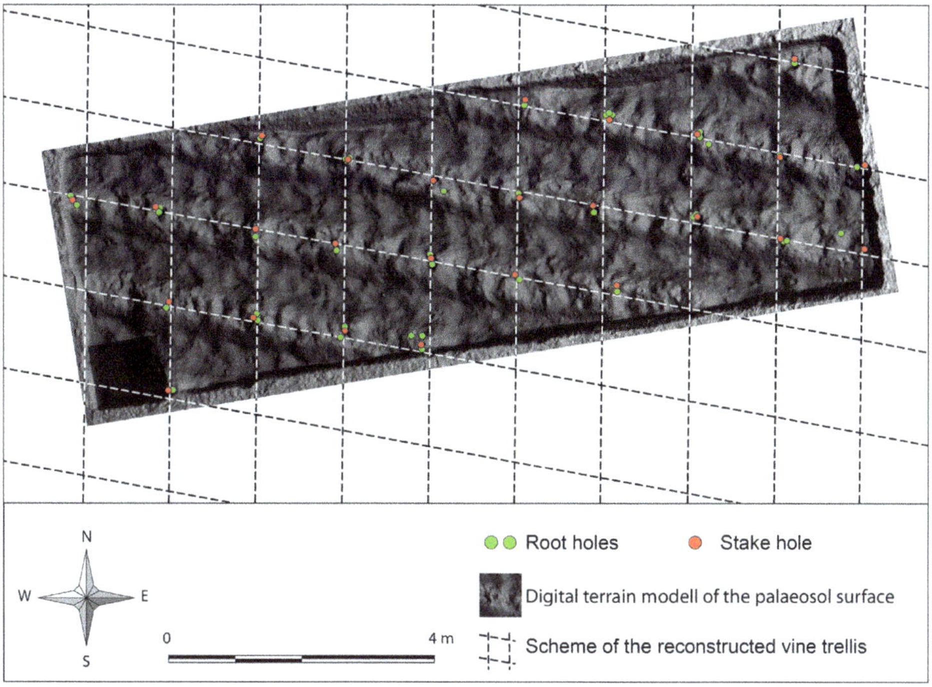

Figure 12.3. Digital model of the laser-scanned Roman palaeosol, organized in ridges and furrows. On top of the ridges the locations of root and stake holes are indicated along with the reconstructed scheme of the vine trellis. Courtesy F. Seiler.

[10]Geology, core drillings, georadar: Giovanni di Maio, Geomed s.r.l., Scafati. Pedology, geoelectrics, modelling: Michael Märker, University of Pavia; Sebastian Vogel, Leibniz-Institut für Agrartechnik und Bioökonomie e.V. Potsdam. Micromorphology: Ivano Rellini, University of Genova. Archaeobotany, Archaeozoology: Mark Robinson, Jenny Robinson, University of Oxford. Palynology: Linda Steinhübel, Universität Göttingen. Archaeology: Pia Kastenmeier, Terzigno; Catello Imperatore, Pompei; Domenico Esposito, DAI Berlin. Topography and laser-scanning: Ascanio D'Andrea et al., Akhet s.r.l., Roma. Excavation: Vitiello s.r.l., Pompei. Scientific direction and archaeology: Florian Seiler, DAI Berlin.

further excavation and sampling provides a source of data that can be used for future research; for example, to reproduce traces of the soil cultivation and to accurately measure the position of plants.

The exposed palaeosol was exceptionally well-preserved. The topsoil was dark brown, soft and very homogeneous, had a sandy loam texture moderately enriched in humus and nutrient elements, and was moderately alkaline and free of carbonates according to the pedological analysis.[11] There was chemically and macroscopically no significant difference between the soil on the ridge and in the furrow (see below), and it can thus be characterized as a nutrient-rich, loose, fine soil, highly suitable for agricultural use, with a thickness of 90 centimetres, and cultivated to a depth of 60 centimetres. The ground was organized in ridges and furrows, orientated roughly east–west, 100°ESE, showing clear signs of agricultural use. Six ridges were located in the excavated area, each with a height of 20–25 centimetres, width of 75–95 centimetres, and cross-sectional distance from one ridge to the next of 120–135 centimetres, in average 130 centimetres. The terrain sloped slightly down from east to west in the direction of the furrows with a gradient of about 2 per cent. This would have aided the drainage of the field as abundant surface water could easily run off; alternatively, rainwater could be held by making small embankments. The furrows show fresh traces of tilling the ground with an 18–20 centimetres broad hoe, which appears to have been used to lift portions of soil from the bottom up onto the ridges. The hoeing marks, at regular intervals of 50–70 centimetres one behind the other, can be seen in all furrows and, according to the marks visible on the palaeosol, entailed the worker stepping backwards. In the same action, the furrow could also be cleared of harmful roots and weeds. Use of the hoe is not surprising, as it is the typical agricultural tool in this region, with examples found in Pompeii as well as in the surrounding Vesuvian villas.[12]

As was already well-known from Pompeii, most of the organic material from the vineyard had decomposed and the surrounding geological material solidified. This created cavities revealing negative impressions of grapevine stems, roots and stakes (Figures 12.4.1–6). Some of the larger voids were filled with pumice lapilli, which further improved preservation (Figures 12.4.3 and 12.4.5). Smaller, more deeply ramified roots left perfectly preserved negatives, despite being free of lapilli (Figure 12.4.2). Study of these cavities along the top of the ridges revealed a total of twenty-six clusters of holes, with each cluster consisting of two or more holes. Stake holes were identified as those that were vertical without lateral elements, whereas the others were interpreted as plant roots. Small depressions in the soil were located around almost every plant cluster (Figure 12.4.1), perhaps indicating a particular agricultural practice; however, it is not clear whether they were made to hold rainwater or to expose and trim harmful head roots that developed within a year.[13]

[11]Vogel et al. 2016b: 165–7.

[12]Cf. Kolendo 1985; Borgongino 1999: 89. Harvey 2010. Almost identical hoes remain in use today in this area.

[13]On the latter, see Columella, *Rust.* 4.24.11.

Figure 12.4. Examples of the micro-excavations of root complexes. (1) cluster of three root holes and one stake hole, located in a small depression; (2) horizontal exposed side root and a small stake hole beside a larger root hole; (3) and (4) vertical sections showing cavities of a stake and a crooked vine stem, filled with lapilli from the eruption; (5) exposure of an old knobby rootstock beside a stake cavity filled with lapilli; (6) horizontally exposed tubular root cavity, an example of propagation by layering, beside an old rootstock. Photographs courtesy F. Seiler.

Overall, fifteen stakes had a single plant hole, seven had two holes, and four had three holes. The diameter of the plant holes varies from two to seven centimetres, although more than half (56 per cent) are of a small to medium size (2–4 centimetres). It appeared, therefore, that we were observing plants of different ages. When compared to modern grapevines, we estimated the vines of this Roman vineyard to be between two and ten years old when Vesuvius erupted, and they were buried. Furthermore, in many cases – though not all – the older plants had one or two younger plants beside. This combined planting of vines of different ages seems to be a typical feature, at least in this exposed area of the vineyard, which suggests that it was, in large part, in the process of being rejuvenated by young vines.

While plaster casts were made of root cavities at Pompeii,[14] in this case the soil was too soft and root voids of such small dimensions that it was not possible to remove the lapilli to create spaces into which plaster could be poured. Instead, we developed techniques in micro-excavation so that the root systems could still be studied. Three different methods were tested and applied (Figure 12.4):

1. A method of vertical sections, which first set a large section perpendicular to the ridge to clear the structure of the main root. By means of several small vertical sections, set in varying angles according to the root structures, we were able to uncover large parts of the root system (Figures 12.4.1, 3 and 4).
2. Gradual removal of horizontal strata around a hole or group of holes filled with lapilli, to record the spreading of lateral roots; however, using this method it proved difficult to examine the vertical extent of roots.
3. Combining the application of vertical and horizontal sections, following the main root structures, to foster a more detailed understanding of a root system in depth and width. In this way, we produced complex excavation structures, with varying levels and depths wherever the roots came to light. (Figures 12.4.2, 5 and 6).

By using these excavation techniques a better understanding of the root systems was gained as well as observations on the training and propagation of the vines. Two propagation methods were visible from the micro-excavations. The first, illuminated by using deep vertical sections, used cuttings and inserted a long shoot vertically into the ground, which remained there after roots took hold. This left long tubular voids in the soil, measuring up to 75 centimetres in length and 1.5–1.8 centimetres in diameter. The second used layering. A low branch or summer shoot from the stem was bent down and inserted into the ground to take root, after which it was cut off from the parent vine and replanted elsewhere. Archaeologically, this appeared as a long, horizontal tubular hollow, with a hole near the stem but no physical join (see Figure 12.4.6). The propagation of vines is an effective practice for rejuvenating the vineyard, which is also apparent in the relatively large number of young vines.

[14]Jashemski 1968: 72; 1973: 30.

Typically, side roots have a tendency to grow in all directions where there are good growing conditions and space; however, in this instance, roots were observed to grow mostly along the ridges, where they could be traced up to a length of 75 centimetres from the stem (Figure 12.4.2). Apparently, only few roots grew transversal to the furrow's direction. It appears that they were prevented from growing in this manner and were continuously pruned or damaged when the furrows were hoed. The main vertical roots (e.g. the subsurface stems) were mostly crooked in growth (Figure 12.4.5). They were found to be often 4–6 centimetres thick, rarely as large as 10 centimetres, and could be traced down to a depth of about 30–40 centimetres. Compared to modern grapevines, the well-developed root structure of the vines suggests an age of the vineyard of at least ten years; furthermore, given the large number of small vines, it is apparent that it is an older cultivation with continuously rejuvenated plants.

The root growth patterns revealed by excavation also shed light on the way the vines were shaped and trained using stakes. Where multiple stems belong to the same plant, the vines were probably trained in a branching manner on a stake with crossbars that had an X-shaped arrangement. This grid layout also derives from the pattern of the intervals between stake holes, which have relatively homogenous spacing of 106–154 centimetres along the ridges and 115–145 centimetres perpendicular to them. The average spacing of 130 centimetres in both longitudinal and transversal directions results in the reconstruction of a rhomboid grid arrangement, visible in Figure 12.3. An additional argument for this solution comes from the fact that, due to the relatively small diameter (3–4 centimetres) of the stakes, it was probably necessary to stabilize them with crossbars to prevent them from collapsing. This model corresponds with a quadrilateral trellis in the form of a pergola - a light but stable construction of crossbars supported by four stakes (Figure 12.5).

The stakes, according to the excavated holes, were pointed at the base, 3–4 centimetres thick and set 20–35 centimetres deep into the ground (Figures 12.4.3–12.4.5). The stains of decayed organic material found on the surface of the white pumice at the beginning of the excavation (see above), and 90–100 centimetres from the top of the ridges, give a reference on the original minimum height of the stakes. They were probably not much higher and give headroom in the furrows of about 120–130 centimetres.[15] Only a few of the stains do not match perfectly with excavated stake holes, most likely because these stakes were leaning sideways, an effect perhaps caused by forces from the eruption or other impacts. It appears that there was a tendency to plant vines on the south side of the stake (twelve times), while the north, east and west appear less common (three times each), and the distribution around the stake occurs mainly with multiple stems (five times) (Figure 12.3). The preference for a southern position is probably related to the intended exposure of the vines to the sun.

A series of scientific analyses were undertaken to investigate the environment of the vineyard as well as to reveal further information about cultivation practices. Root cavities

[15]According to Columella (*Rust.* 4.12.2), the support (for young vines) should have a height of no more than four feet (*c.* 120 centimetres).

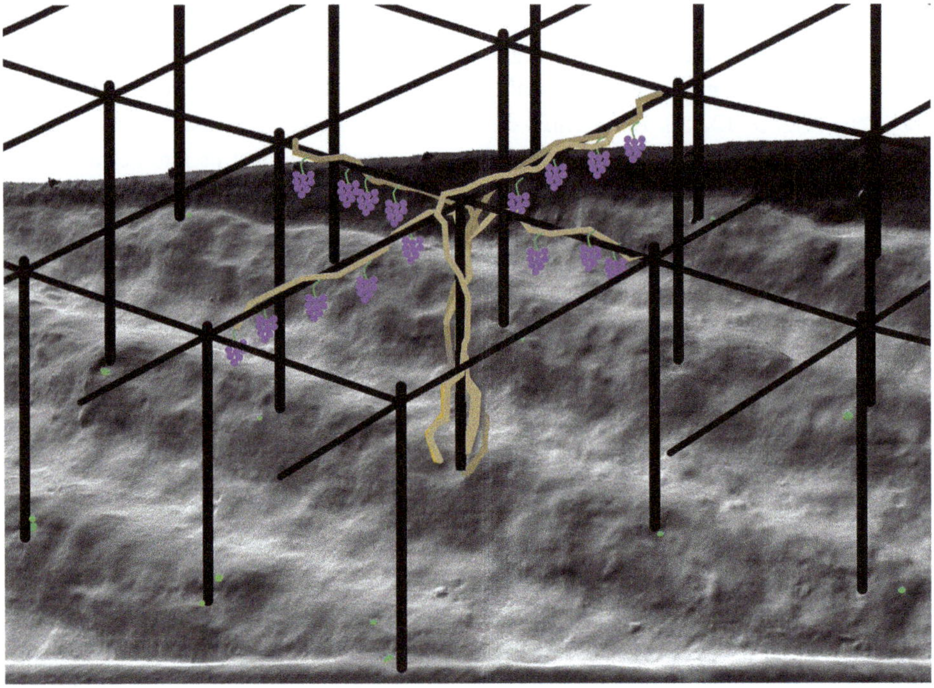

Figure 12.5 Theoretical graphic reconstruction of the *vitis compluviata* trellis system based on the data from Scafati. Courtesy F. Seiler.

were carefully examined for organic plant remains with three providing residues of decayed vine wood, observed in different voids. When studied using macroscopic botanical analysis they were found to be preserved in the form of white fibres filled with calcium carbonate, showing the characteristic large vessels of *Vitis* (grapevine), which definitively confirmed the attribution of a vineyard. Land snail shells, including *Zonitidae* (eleven specimens), *Trochoidea trochides* (three specimens) and *Cochlicella acuta* (twenty-nine specimens), were also found in some of the root cavities. *Zonitidae* are a species that prefers a shaded habitat with high atmospheric humidity, like that under fallen leaves, in tussocks of vegetation and next the trunk of the vine. The other two species favour sunny habitats on well-drained soil where ground vegetation is sparse.

Several bulk samples of the Roman palaeosol, taken from different areas and depths, were analysed using flotation and subsequent macroscopic examination. Remarkably, the samples resulted in little material – only a charcoal fragment showing the typical large vessels, broad rays, and radial files of small multiple vessels in the latewood of *Vitis* (grapevine), small fragment of a carbonized olive stone, and a fragment of adult molar tooth enamel from a sheep or goat. An assemblage too small and heterogeneous to draw any useful conclusions. The notably few macrofossils and sparse organic material present in the palaeosol raise questions about fertilization regimes and whether it took place at all. Micromorphological investigations sought to interrogate this question further.

Analysis of samples taken in the Roman palaeosol from different soil depths did not reveal any typical signatures for the presence of manure, such as organic material from decomposed animal remains and faeces. Thin section analysis confirmed the presence of artefacts, for example particles of pottery, plant residues and vine charcoal, but the composition is in line with the macro remains (above). Furthermore, no difference in soil consistency was found between samples from the furrow and the ridge, perhaps since it was repeatedly dug up by hoeing thus becoming mixed and relatively homogeneous.

Palynological analysis was also integrated with the aim of beginning to reconstruct the general vegetation composition of the environment before the eruption of 79 CE, specifically of the vineyard. A total of seventy-two surface samples from fifteen units were taken from the excavated area immediately after removing the last layer of pumice, supplemented by sixteen control samples from the pumice layer covering the Roman palaeosol. Various control methods showed that the pollen samples have a very low degree of contamination.[16] However, pollen grains were mostly in a poor state of preservation, which made the identification of species challenging and limited subsequent interpretation. A sufficient number of pollen grains in each sampling unit illuminated a low degree (*c.* 8 per cent) of arboreal pollen compared to non-arboreal pollen, among them *Pinus* (pine tree), *Picea* (spruce), *Quercus* (oak), *Alnus* (alder) and *Corylus* (hazel). This was interpreted, together with other evidence, such as the high diversity of non-arboreal pollen, to indicate an environment of mostly open vegetation in the wider area. The tree species identified that prefer a humid environment, such as alder, may have been located nearby the river Sarno, whereas the other species were probably found in diverse habitats across the wider surroundings including the slopes of the mountains. A variety of non-arboreal pollen was identified, dominated by *Pocaceae*, *Cyperaceae*, *Asteracea*, all of which are grasses and herbs and characterize environments of open vegetation. Agricultural species identified include varieties of vegetable plants, for example *Brassicaceae*, *Apiaceae* and *Allium*, which were less dominant in the total pollen count, but may indicate growth in the more immediate vicinity. The occurrence of *Cerealia* pollen was low and inconsistent, interpreted to be due to a few individual cereal plants that accidentally entered the vineyard.[17] The most surprising result was an extremely low count of *Vitis* pollen. In fact, only one pollen grain was actually identified with certainty as *Vitis*, though this may be related to the poor state of preservation of the pollen recovered and difficulties of identification.[18]

Overall, the pollen analysis provides a diverse image of vegetation both in the wider surroundings and more immediately in the vineyard. The broader landscape is

[16]The control samples were taken directly from the pumice covering the 79 CE palaeosol. They consist of sixteen sampling units. Six supplementary single samples were taken, one from modern soil, four from earlier palaeosols, and an additional one from the pumice (Steinhübel 2013: 7 and 15).

[17]One might also consider whether the *Cerealia* pollen represents the residue of an earlier cereal field in this location before it was transformed into the vineyard.

[18]Compare the discussion in Chapter 10 of the present volume regarding *Vitis* pollen levels and preservation.

characterized as areas of mostly open vegetation, where sporadic tree habitats alternated with agricultural fields and fallow land, and the areas closer to the river probably attracted tree species that preferred humid environs. The apparent contradiction between low *Vitis* pollen recovered and archaeologically evidenced grapevine cultivation remains difficult to explain. One plausible interpretation could again be recurring soil cultivation, which may have had a negative impact on pollen preservation. Isolated cereal plants may represent the residue of previous growing or may come from a nearby grain field.

Discussion

The data and results obtained from the excavation and scientific analysis allow us to contextualize the vineyard within the viticultural ecosystem of the Sarno River plain. This can be nuanced further by including data from earlier landscape research,[19] allowing a more precise morphological and environmental characterization. Although an area of only forty-eight square metres (4 × 12 metres) was investigated in what was originally a much larger field, significant results were obtained regarding Roman grapevine and soil cultivation. Aligned with descriptions by ancient sources regarding the importance of soil quality for agricultural production, the loose soil at this site, with its homogenous consistency and nutrient-rich composition, offered conditions suited to the growth of grapevines.[20] Since the soil in the entire area of the plain developed from the volcanic substrate of previous eruptions, its properties do not differ significantly from palaeosol found on the slopes of Mount Vesuvius.[21]

Topography also plays a notable role in viticulture. The site's relatively flat location on the fluvial plain at the edge of the flood zone does not immediately seem to be a favourable place for vine growing compared to a location on a slope (Figure 12.1). A vine's exposure to the sun on flat terrain is less in comparison to that on a slope. Moreover, given the influence of the nearby river on the microclimate, higher groundwater table and risk of flooding, the potential for poorer growing conditions is present.[22] Some of these less favourable conditions were, however, compensated by specific measures implemented by those that laid out and worked the vineyard. The level gradient of the terrain was deliberately used such that the field, organized into furrows and ridges running east–west, was laid out in the direction of the incline so that surface water could run off. Furthermore, the east–west orientation of the planting rows was optimal to get maximum sunlight from morning to evening. Recurrent hoeing loosened the soil, which reduced moisture in the wet season, and allowed more moisture infiltration in the dry season. This evidence suggests that the orientation and layout of the field took advantage of

---

[19]Seiler et al. 2011; Vogel, Märker and Seiler 2011.

[20]Columella, *Rust.* 3.1, 11–12; Vogel et al. 2016b: 165–7.

[21]Vogel and Märker 2012, 2014.

[22]The groundwater table was determined to be at a depth of 7.7 metres below the modern surface at this site, equal to a depth of *c.* 3.7 metres below the 79 CE level.

natural factors, moulded and activated through human intervention, rather than by planned measures such as land division and centuriation. Related to the latter, there is no correspondence in the alignment of the main axes of the field to neighbouring *villae rusticae*.[23]

The use of ridges and furrows to shape a field is a well-known agricultural practice in the Vesuvian region, but is less attested (or, more likely, preserved) elsewhere in Italy. Numerous examples of exposed arable land show that this method of tilling was practised both on the slopes of Vesuvius (e.g. Boscoreale, Terzigno, and Pompeii) and on the slopes of the Lattari Mountains (e.g. Casola di Napoli), as well as on the plain (e.g. Scafati and San Marzano).[24] Only in a few cases is more precise data available on the measurement and shape of the soil as well as botanical features. Examples of agricultural areas that are likely vineyards provide a range of spacing between ridges, from 120 centimetres (Insula II.5, Pompeii), 130 centimetres (Via della Resistenza, Scafati), 135 centimetres (Villa Regina, Boscoreale) to 175 centimetres (Cimitero, Scafati). Instead of a standard distance between rows, different measurements were clearly used, which may correspond to different descriptions provided by the ancient agricultural writers.[25] The furrows were not only used for drainage but also provided a method of channelling rainwater to moisten the vines, the control of which is important in the ability to manage grape growth.[26] The trunk and head of the vine were kept dry as they were raised up being dug into the soil of the ridge, while the deeper vertical and lateral roots received sufficient moisture by running lower than the level of the furrows. By regularly hoeing the furrows, as advised by the Roman agrarian sources, the soil was cleared of weeds and excess lateral roots could be cut.[27] The soft and loose soil was easy to work with the hoe, which was used in antiquity as it is today, and this may also have aided the formation of ridges and furrows.[28]

As we can see, the planting and care of grapevines was subject to rational planning and careful cultivation. This observation can be extended to almost all the known Roman vineyards in this area. At the site discussed in this chapter, at Scafati, each vine tended to be grown on the south side of a stake to take advantage of maximum sun exposure. One or two younger vines adjacent to an older vine demonstrate efforts to continuously rejuvenate the vineyard through cycles. Comparable evidence of young vines, based on root cavity diameter, have also been found in the vineyards in Pompeii and the

[23]The villas within a radius of 500 metres are all oriented approximately the same way, about 130° SE, while the orientation of the field is 100°SE. For villa names and references, see n. 8.

[24]Villa Regina at Boscoreale (De Caro 1994: 95–114); Boccia al Mauro, Cava Ranieri at Terzigno (unpublished); Insula II.5 at Pompeii (Jashemski 1973); Monticelli at Casola di Napoli (Miniero 1991–2: 225–8); Cimitero at Scafati (unpublished); San Marzano (unpublished). In general, see Borgongino 1999. A large number of other examples remain unpublished.

[25]Pliny, *HN* 17.171 recommands four feet, whereas Columella, *Rust.* 3.13.3 gives five feet.

[26]Cf. Borgongino 1999: 89.

[27]Cf. Jashemski 1973: 37–8.

[28]White 1967: 36–43; Borgongino 1999: 89 and 134 with Figure 112.

rejuvenation of an older vineyard with young vines is also documented in the vineyard of Villa Regina at Boscoreale.[29] The process of rejuvenation is also evident at Scafati in the two methods of propagation using either cuttings or layering. The method of layering is also attested twice in a vineyard at Pompeii, with tubular root cavities similar to those at Scafati (cf. Figure 12.4.6).[30] These methods illustrate how older or dead plants could be constantly replaced with young plants and provided further evidence of the highly developed and intensively practised nature of Roman viticulture, especially in this region during the last decades before the eruption of Vesuvius in 79 CE. Propagation in one's own vineyard might be seen as a valuable investment, since the cost of buying young vines, as visible in the ancient sources, was relatively high.[31]

The suggested reconstruction of the quadrilateral supporting framework on which the vine was trained in four lateral directions (Figure 12.5) – the so-called *vitis compluviata*[32] – is said to be a very successful training method to grow grapevines in Italy.[33] The thin stakes that are also widely used in the vineyards of Pompeii gain stability when they are connected with crossbars.[34] The vertical supports for the vines, called *pedamenta* according to ancient terminology, were preferably made from hardwood.[35] The crossbars on which the vine runs transversally are called *iuga*, literally yokes, and were made of lighter material.[36] This type of trellis might have been better able to withstand forces such as strong winds.[37] Another benefit of this type is that shading is provided by the leaf canopy, which may have been particularly useful in the dry season.[38]

---

[29]Ninety per cent of the root cavities uncovered in the vineyard in Insula II.5 at Pompeii had a diameter of 2–5.5 centimetres, which indicates young plants (Jashemski 1973: 30). The vineyard Insula III.7 provided very small root holes of 1.5–5 centimetres diameter, indicating vines that were not yet old enough to be staked (Jashemski 1979: 228–30). At Villa Regina, judging by the size of the root cavities, some vines appeared to be twenty years old or more, while others were about three years old (De Caro 1994: 98).

[30]Jashemski 1979: 206 with Figure 299.

[31]Columella, *Rust.* 3.3.11–15.

[32]Varro, *Rust.* 1.8.2; Columella, *Rust.* 4.24.14, 4.26.3; Pliny, *HN* 17.164–66. The correct reconstruction of the *compluviata* type is given in Jashemski 1973: 34 with Figure 3. A different reconstruction, in which the vine is planted in the centre of the *compluvium*, is shown by Ruffing, Renger and Gutsfeld 2002: 429–31. However, this is contradicted by Columella's instruction (*Rust.* 4.13) that the vines should be tied close to the stakes.

[33]Varro, *Rust.* 1.8.2. Cf. Jashemski 1973: 34–6.

[34]Cf. Columella, *Rust.* 4.16.4, 4.17.6–8.

[35]Varro, *Rust.* 1.8.1. Columella, *Rust.* 4.26.1 and Pliny, *HN* 17.174 describe various types of durable wood used for stakes, such as olive tree, junipers, cypress, chestnut. We were not able to detect the type of wood used for these supports. A preserved organic fragment of a pointed stake, found together with residues of grapevine in an excavation near the River Sarno south of Pompeii, is of cypress wood (cf. De Carolis, Senatore and Ciarallo 2012). A large number of stakes were found in a *villa rustica* known as S. Maria la Carità in Messigno. They were temporarily stored in the courtyard, probably for use in a vineyard. Their wood was almost perfectly preserved, but neither the type of wood nor the measurements and the number were noted (cf. Della Corte 1923). Chestnut stakes were found stacked in the *villa rustica* called Cività Giuliana, Fondo Brancaccio near Pompeii (cf. Della Corte 1921a), and a graffito from the villa Fondo Prisco refers to the preparation of 1,300 stakes for the vineyard, of which 840 were pointed and 460 not pointed, the latter perhaps for use as crossbars (Della Corte 1921b; *CIL* IV 6886).

[36]Reeds tied in bundles or willow rods are recommended for the so-called *iuga* (Varro, *Rust.* 1.8.1–2; Columella, *Rust.* 4.16.4; Pliny, *HN* 17.166.174).

[37]Cf. Columella, *Rust.* 4.17.7–8.

[38]Cf. Columella, *Rust.* 4.17.8.

The main advantage, however, may have been the higher yield of grapes obtained by the four-sided branched growth.[39] Jashemski's hypothesis of recognizing in this type the traditional Vesuvian form of viticulture receives support by these inter-site similarities. A striking iconographic confirmation is also provided by the well-known wall painting from the Casa del Centenario, which clearly represents quadrangular vine trellises of this type occupying the slopes presumably of Vesuvius.[40]

Conclusion

The first overarching conclusion to draw from the comparison of the vineyard at Scafati and other examples in the urban area of Pompeii and at Boscoreale is the high degree of similarities in quantitative and qualitative features. It is perhaps more convincing to say now that the ridge and furrow system was the most common form of viticulture in the Vesuvian region, with advantages to vineyard management now clearer, from monitoring growth, care of the soil and vegetation, to the harvest and pruning. As far as we can tell from the cavities of decayed plants and stakes, the method of vine cultivation across the region was also the same, using a quadrilateral, *vitis compluviata* framework for training, as recommended by the ancient sources. Whether on the slopes of Mount Vesuvius or on the plain, one can generally speak of a common traditional form of viticulture throughout the northern side of the Sarno Valley. Pompeii and the *villae rusticae* formed a large network deeply involved in the profitable business of wine. Pompeii might be considered a hub of the regional wine trade, as evidenced by relevant infrastructure and archaeological finds.[41] Well-known families of wine producers settled in Pompeii,[42] such as the *Lassii*, *Clodii* and *Holconii*, whose famous wines, such as *Vesuvinum* and *Pompeianum*, were distributed to many parts of the Roman Empire.[43] Geographically, the wine region is defined by the course of the Sarno River on the south side, and on the north it includes the slopes of Mount Vesuvius, where ancient viticulture is archaeologically attested in the area of Boscotrecase, Boscoreale and Terzigno.[44]

Nevertheless, certain differences between the two microregions of the Pompeiian viticultural territory are also evident – one on the slope, the other on the plain. Of the three main factors that can influence the quality of a wine – soil, climate and location – the palaeosol is of a high quality everywhere, while the latter two see variable conditions from the plain up towards Vesuvius. Due to the proximity of the river, different

[39] Emphasized by Pliny, *HN* 17, 164.
[40] Jashemski 1979: 220 with Figure 318; Ciarallo and Carolis 1999: 56 with Figure 2.
[41] Cf. Seiler 2023. The harbour of Pompeii has not yet been discovered. Warehouses in the southern *suburbium* at Bottaro indicate commercial activities in the wider area of the harbour. Stocks of wine *amphorae* were found in the so-called Villa B at Oplontis, which was a warehouse (Clarke, De Caro and Lagi 2015: 142–55).
[42] Cf. Day 1932: 203–8 with tables D and E; Purcell 1985: 8.
[43] Cf. Tchernia 1986: 176–7.
[44] Cf. Seiler, Vogel and Esposito 2019: 190–7.

microclimatic conditions, including increased humidity, can be assumed even though no data has yet been gathered. In addition, sun exposure on flat terrain is lower in the morning and evening, especially in winter. The soil moisture might also have been higher. The effects on viticulture of these even minimal deviations should not be underestimated and might dictate the need to grow different grape varieties in these different microclimates. Therefore, grapes grown in the Scafati vineyard were perhaps not the same as those cultivated at Villa Regina and on the slope of Vesuvius. They may have been of varieties such as *Vennu(n)cula, Murgentina* (also called *Pompeiana*) or *Holconia*, which are graded by Columella as third class.[45] Certain vine varieties, not specified by name, were adapted to humid climatic conditions, therefore suitable for cultivation in the lowlands.[46] Grape varieties grown on flat terrain are said to produce a higher yield, but were not of such high quality as the varieties of, for example, *Amineae*.[47] While speculative at present, if this assessment is correct, viticulture in the plain might have specialized in producing larger quantities of wine of average quality, probably intended for mass use or consumption.[48]

Historically, this investigation was limited to a relatively short period of a few decades before the eruption of Vesuvius in 79 CE. While we do not know when the vineyard excavated in Scafati was established, the root cavities indicate ages that do not exceed a decade (shortly after the earthquake of 62 CE).[49] However, it cannot be ruled out that the vineyard was established earlier and replanted recently with young plants. Here, and considering the palynological data, it is interesting to speculate whether the field was not cultivated with cereals before the vineyard was planted.[50] This raises the question of whether the preponderance of viticulture in the plain was a phenomenon of the early Imperial period, while in the hilly locations it was already established by the late Republican period.

These intensive investigations of a small part of the vineyard in Scafati demonstrate the potential of such a comprehensive and multidisciplinary research methodology along with the results that can be obtained. It is clear that this approach should be applied more regularly and in other regions, building this field into the future. In particular, arable palaeosols unearthed by modern building projects in the Vesuvian region should

---

[45]Columella, *Rust.* 3.2.25–27.

[46]Columella *Rust.* 3.1.5; De Simone (2017: 33) suggests that the variety *Holconia* was probably cultivated in the *Ager Pompeianus*, which includes the plain.

[47]Columella, *Rust.* 3.2.6–7.

[48]While quantitative calculations of wine production are fraught with difficulty, for examples see De Simone 2017: 35–40. For the socio-economic perspective, see Purcell 1985.

[49]The few ceramic sherds from the Roman palaeosol (late second century BCE to first century CE) are stratigraphically not applicable to the question of vineyard establishment.

[50]Jongman's (1988) hypothesis of predominantly cereal agriculture on the plain has been rightly criticized in favour of viticulture (cf. Purcell 1990; Scheidel 1992; Senatore 1998). However, as the example of Scafati might suggest, there could also be an economic transition in the early Imperial period in which viticulture became increasingly prevalent on the river plain and the cultivation of cereals gradually declined.

be studied just as carefully to collect high-resolution ancient agrarian and archaeobotanical data to develop hypotheses on an even broader basis. Such an evidence-based approach would generate necessary data to reconsider in more depth the current issue of the wine economy of Pompeii.[51]

[51]The debate on Pompeii's agricultural economy, initiated by Rostovtzeff (1957) and Finley (1975), was reconsidered by Jongman (1988). His thesis that the main production of the plain was grain achieved little support and was rigorously criticized by, e.g. Purcell (1990), Scheidel (1992), and Senatore (1998). In a quantitative study, De Simone (2017) re-founded the thesis of wine production in the Vesuvian area including the Sarno plain. The contemporaneous re-study of rural settlement by the SALVE project by Domenico Esposito provides a new, broader basis to reappraise this question from the ground up.

CHAPTER 13
EMPLOYING REMOTE SENSING AND MULTISPECTRAL SATELLITE DATA TO MEASURE THE EXTENT OF GRAPEVINE AND OLIVE VEGETATION: A CASE STUDY IN THE LANDSCAPE OF WESTERN ROUGH CILICIA, TURKEY

Christopher D. Dore and Nicholas K. Rauh

Abstract

This chapter discusses an innovative method for identifying and modelling the presence of grapevines in the contemporary landscape and using this to inform our understanding of the situation in antiquity. A range of multispectral satellite images from IKONOS and SPOT-5 were sourced and processed to assess land coverage of grapevines and olive trees. This involved the algorithmic 'training' of software to identify the spectra of wild grapevines and olive trees. The result was a 20:1 ratio of grapevines to olive trees and, hence, based on the remarkable density of vines in an unusual maquis environment, a compelling, albeit speculative, case for the vestigial remnants and wild survival of a formerly extensively cultivated resource in Rough Cilicia.

Introduction

Significant debate persists regarding the proper identification of the remains of ancient wine and oil presses in the Mediterranean world.[1] Following the testimony provided by literary sources that wine produced in Roman Rough Cilicia enjoyed a high reputation across a wide area of the Mediterranean, the Rough Cilicia Survey Project (RCSP) investigated the remains of local *amphora* production centres (kiln sites) and the associated remains of dozens of press complexes in the survey region of the Antalya Province, Gazipasha district, south coastal Turkey, as a means to verify the existence and

[1] cf. Van Limbergen et al., forthcoming.

the scale of this activity (Figure 13.1). Given the presence of surviving stone elements at nearly every identified press complex in the region, namely olive crushing wheels and hollowed out crushing basins,[2] a lively debate has persisted regarding the true focus of ancient Rough Cilician surplus production.

To address this issue, the RCSP explored alternative procedures that bore potential to shed further light on the agricultural economy and its regional focus. We increasingly turned to the use of multispectral remote sensing data and techniques to identify and measure the percentage of land cover represented by grapevines and olive trees in the survey zone. Aydınoğlu and Rauh display a satellite image that used blue and red pixels to demonstrate the approximate ratio of grapevine versus olive tree vegetation (respectively) in the contemporary ground cover of western Rough Cilicia.[3] When combined with the literary and archaeological evidence, the 20:1 ratio of grapevines to olive trees paints a compelling picture of the predominance of grapevines, and hence the likelihood that wine production stimulated economic development in Roman Rough Cilicia. Here, we describe the procedures employed to generate these outcomes and conclusions.

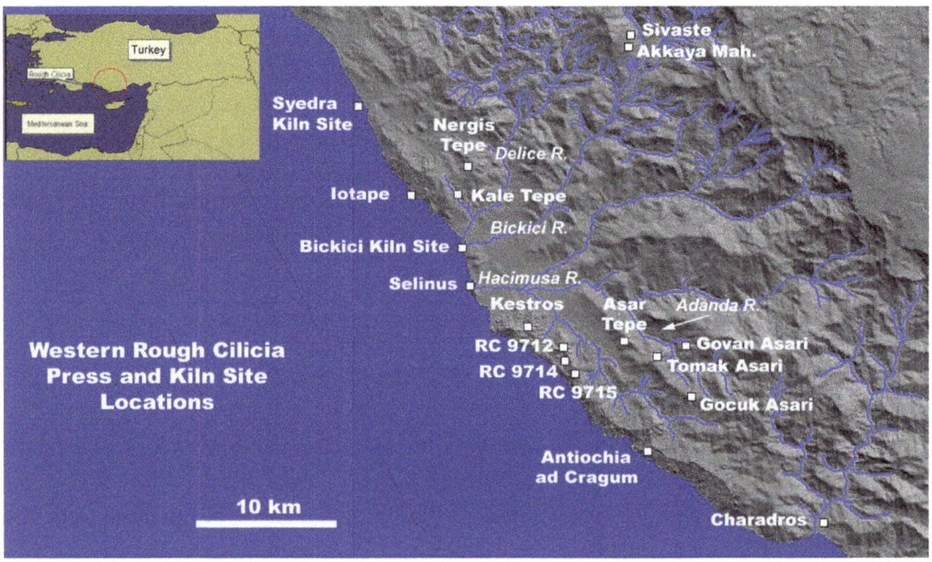

Figure 13.1 Map of RCSP survey area showing locations of ancient press complexes and *amphora* kiln sites identified by the survey. Image by C. Dore and N. Rauh.

[2]For discussion, see Aydınoğlu and Rauh in Van Limbergen et al., forthcoming.

[3]Aydınoğlu and Rauh in Van Limbergen et al., forthcoming.

The multispectral identification of grapevines in Rough Cilicia

Across western Rough Cilicia we had the advantage of surveying a largely rural region where the lowlands (below 500 metres elevation) were covered by dense, nearly impenetrable maquis scrub brush. It has been demonstrated by geobotanists that maquis scrub is secondary vegetation directly consequent to human activity such as land clearance, agriculture and herding.[4] When agricultural landscapes became abandoned and terracing collapsed, maquis scrub colonized the empty terrain and, over decades, achieved a climax of growth. Maquis consists of an alliance of resilient flora that become increasingly resilient by meshing together.[5] These flora regenerate vegetatively from subterranean dormant buds and are essentially resistant to herding, fire and drought. They can not only endure continuous human interference, but also rapidly invade sites of mesic, ruined vegetation at neighbouring elevations. Prior to urban development and land clearance over the past fifty years, maquis formed the ground cover for most of the low-lying landscape in the survey region. What seems unique about the maquis scrub here is the additional inclusion of vestigial growth of olive trees and grapevines in the floral alliance.

While it is true that both olives and grapes grow naturally in Mediterranean landscapes below 900 metres elevation, the fact remains that grapevines are not usually regarded as constituent elements of the maquis alliance. Given the fact that human population declined significantly as ancient sites were abandoned at the end of the Roman era, and that local population remained low until recent times, it is likely that the maquis ground cover colonized the landscape at the end of antiquity and remained in a steady state until relatively recently. Under these circumstances, we can argue that the maquis ground cover in western Rough Cilicia represents a palimpsest of vegetation that incorporates the vestiges of flora that were cultivated in Roman antiquity meshed with plants more typically representative of the maquis alliance. This is significant because it enables us to conduct remote sensing in a manner that might not lend itself naturally to investigations elsewhere, particularly in regions such as Italy where the agricultural landscape has remained in continuous use for millennia.

The principles governing the spectral analysis of satellite imagery are as follows. When sunlight strikes the earth, plant life and other materials absorb and reflect light differentially. The reflected light that is not absorbed by the atmosphere and land cover is captured by the sensor of the orbiting satellite. Different wavelengths of reflectance are captured, and their averages displayed in bands each having a certain spectral range. Most materials have unique reflectance, and the number and thickness of bands allow these differences to be identified and visualized in analysis. This process of classification can be supervised or unsupervised. Unsupervised classification determines *a priori* the

---

[4]Naveh and Dan 1973; Zohary 1973: 356, 384, 403; Di Castri et al. 1981; Bottema et al. 1990.

[5]These flora include *Quercus calliprinos*, *Phyllyrea media*, *Rhamnus alaternus*, *Laurus nobilis*, *Pistachia palaestina*, *Styrax officinalis*, *Cercis siiliquastrum* and *Arbutus* (Zohary 1973: 356, 404, 500).

number of classes desired, and the algorithm places individual pixels into classes during analysis. In this case, classes may not align with classes (objects or items) of interest. Supervised classification is used where known objects of interest are first identified in the imagery, and then analysis calculates the spectral signature of these items. A second step of analysis searches through the imagery to find other locations with the same spectral signatures.

To conduct the remote sensing analysis, we first utilized the project's IKONOS 4-band multispectral satellite imagery to analyse the vegetation pattern of grapevines in the survey region, based on the spectral signature of a large vineyard visible in the satellite data on the northern slope of a mountain in the Adanda canyon, just below the fortified acropolis of the ancient site of Lamos (Figure 13.2). The survey team encountered the vineyard while conducting systematic pedestrian survey at Lamos during the 2003 season. Although team members had previously encountered grapevines in the field, this was the first time a visible and particularly dense patch of vines could be located in the satellite data. Using IKONOS, we were able to capture the spectral signature of the vineyard in the 4-band multispectral data and used this to estimate the extent of wild grapevine vegetation throughout the survey region. This investigation formed the basis of our initial publication.[6]

Analysis of the spectral signature of the vineyard in the satellite data used both unsupervised and supervised classification. Unsupervised classification was initially used for exploratory data analysis and to assess the spectral variability and the archaeologically meaningful classification potential of the imagery. When these were better understood, supervised classification was incorporated. Field surveyors identified active vineyards, the vineyards were found in the image, vineyard pixels were used to create a spectral signature for grapes, and this signature was identified in other areas of the imagery where the vegetation was unknown.

As Figure 13.2 illustrates, our operation revealed the existence of widespread landscape features emitting the same pattern of spectral reflectance as the known vineyard. This vegetation seems particularly pronounced along south and southeast facing slopes of the mountains that surround Gazipasha. It is especially visible in areas now dominated by maquis scrub. Although one must allow for the possibility that plant growth, such as vines spectrally resembling grapevines, have a similar reflectance and are represented in the results, the most likely conclusion is that grapevines grow wild and are abundant throughout the survey region. Moreover, based on the results, it was possible to suggest that grapevines thrive amid areas of dense maquis scrub far more than in cultivated land parcels, such as the vineyard identified near Lamos.

Following this discovery, team members made several attempts to ground truth areas of the coastal ridge for the presence of wild grapevines, including those in the immediate vicinity of Nephelion. In one instance, we found the ancient theatre of Nephelion entirely engulfed by grapevines (Figure 13.2). Based on preliminary attempts to verify the results

[6]Rauh et al. 2006.

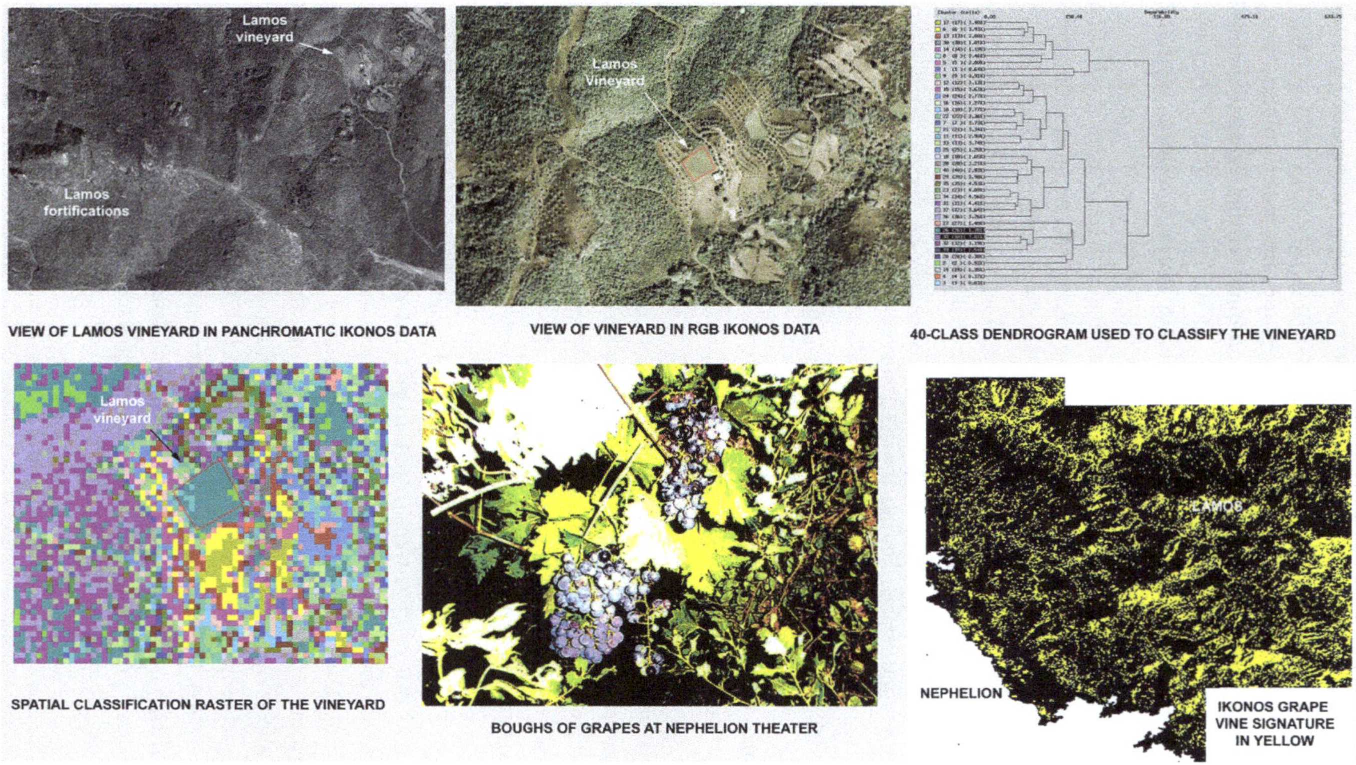

Figure 13.2 *Top*, left to right: view of the modern vineyard near the Late Roman fortification at Lamos in the IKONOS panchromatic data; view of the vineyard in the RGB IKONOS data; forty-class dendrogram used to spectrally identify grapevines based on three combined classes (26, 30 and 39). *Bottom*, left to right: spatial classification raster employing the combined classes to identify the Lamos vineyard; boughs of grapes in the maquis scrub shrouding the ancient theatre at Nephelion in August 2003; extent of the captured Lamos vineyard spectral signature when applied to the southern portion of the IKONOS multispectral data. Images by C. Dore and N. Rauh.

of analysis, grapevine growth appears to be as extensive as indicated by the multispectral data. Remote sensing analysis thus appeared to offer an accurate indicator of the presence of this natural resource in the survey region and did so in a manner unmatched in scale and efficiency by pedestrian archaeological investigation. In fact, the impenetrable nature of the region's maquis scrub rendered the existence of grapevines indiscernible to team members in the field, and it was only because of the remote sensing exercise that the pedestrian team began to investigate additional areas of the maquis scrub. What was revealed was that grapevines exploited the surrounding brush as a form of trellis, climbing to the top of the scrub canopy and spreading out above.

By this time, the presence of olive-crushing wheels and basins in the region's production installations, not to mention the debate this was generating among team members, stimulated an increasing need to identify the extent of olive trees versus grapevines in regional land cover. It took some time to identify an olive grove suitable to serve as the reference target for multispectral remote sensing analysis – a task made more challenging by the fact that in this region mature olive trees tend to stand in mixed forests making an accurate sampling of the chromatic signature questionable. During the 2005 survey season, a newly planted olive grove in open terrain in the village of Beyrebucak was located and georeferenced (Figure 13.3).[7] Classification training sets were thus able to be generated for both olive trees and grapevines.

To identify these two vegetation types and compare their distribution, four bands of the SPOT-5 data were used with unsupervised classification using a self-organization neural network algorithm. The classification resulted in 131 classes, which was a deliberate overclassification to divide the vegetation into more than their ideal number of classes. This is an easy, but simplistic, way of dealing with intra-class spectral variability. Following the classification, the classes that spatially occurred within the training polygons could be visually identified then combined. This resulted in two sets of pixels: one for grapevines and the other for olive trees.

The results proved substantially more dramatic than the original analysis using the IKONOS data. In Figure 13.4, the extent of the land cover represented by gravevines appears in blue and that of olive trees in red. In fact, the presence of the red pixels is barely perceptible in the larger pan (Figure 13.4, top) and requires a close-up view (Figure 13.4, below). Based on a count of pixels, the ratio of grapevines to olive trees (in area) stands in excess of 20:1. It is necessary to repeat that this result is based on current land cover and reflects the existence of these types of flora as seen today. To what degree they reflect vestigial remains of ancient viticultural and oleicultural cultivation remains a matter of speculation. However, it is important to remember that while olive trees are sometimes classified as components to the maquis alliance, grapevines are not.[8] The presence of such dense growth of grapevines in the maquis scrub (and forests) throughout

[7]The grove is visible in Google Earth at the base of a mountain (Azı Tepe) beside the D400 Mersin-Antalya National Highway 36.211659, 32.392352.

[8]Zohary 1973: 511.

Figure 13.3 View of the olive grove at Beyrebucak in the IKONOS panchromatic data (*above*); preparing training set of individual olive trees in the IKONOS panchromatic data (*below*). Images by C. Dore and N. Rauh.

the region of western Rough Cilicia likely points, therefore, to the survival of a formerly extensively cultivated resource in the wild.

The development of multispectral satellite data to identify vegetation types

Our reliance on remote sensing using multispectral satellite data to investigate contemporary ground cover in Rough Cilicia was to some degree an accidental outcome to our primary agenda to construct a project geographic information system (GIS). From its inception, the RCSP based its research procedures on georeferencing all forms

Figure 13.4 Wide view of grapevine signature (blue) and olive tree signature (red) in the SPOT-5 Multispectral Data (*above*); close-up view of grapevine signature (blue) and olive tree signature (red) in the SPOT-5 Multispectral Data for Beyrebucak (*below*). Images by C. Dore and N. Rauh.

of survey data for the purpose of organizing and presenting data in spatial terms. At the outset in 1996, the project lacked a suitable base map for the region. To build one we acquired spatial data in a wide variety of forms. These included 1:5000 cadastral maps of the immediate coastline, 1:10000 Turkish military topographical maps of interior regions, 1:50000 topographical maps of an extended survey area that were available in Russian, and ultimately reliance on 1:100000 Shuttle Radar Topography Mission data (made available in 2002) to round out the periphery of the survey area. These sources proved instrumental to our construction of the RCSP Digital Elevation Model (DEM, see Figure 13.5). It must be stressed that these sources were topographical maps that had to be digitized and were purely for visualization purposes.

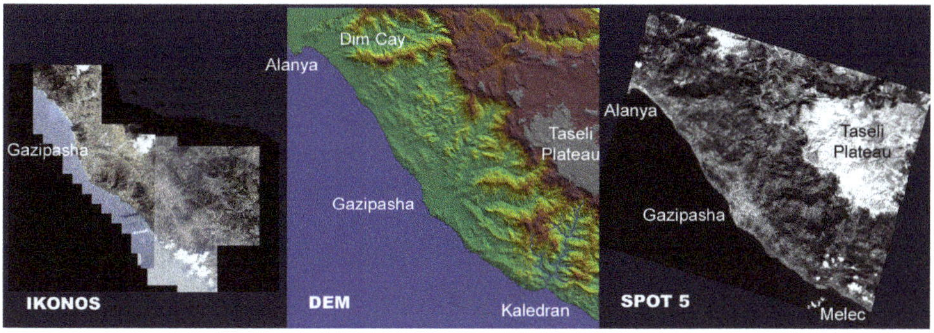

Figure 13.5 Left to right: IKONOS RGB bands for the RCSP survey area; digital elevation model of the survey area; SPOT-5 panchromatic image for the survey area. Images by C. Dore and N. Rauh.

In 2001, we secured generous support from Space Imaging Inc. and its subsidiary, SI Eurasia, to acquire some 200 square kilometres of 1 m^2 resolution georeferenced IKONOS satellite data for the survey area (Figure 13.5).[9] As opposed to topographical maps, these are pixelated matrices of sun-lit reflectance captured by the sensors of the orbiting IKONOS satellite. The IKONOS data consisted of two sets of five bands: one panchromatic and four multispectral.[10] Multispectral data is commonly used in agricultural and vegetation analyses across landscapes, so we soon realized that the data at our disposal could also be used to identify, map and measure various forms of vegetation in our survey area.

Simultaneously, our research focus was shifting increasingly to ancient resource utilization in all its forms, including the exploitation of forestry resources in the Taurus mountains.[11] By 2003, the team was investigating landscapes in the cedar zone (*c.* 1500–1800 metres elevation), where highland Roman-era settlements encountered by the pedestrian team were tentatively identified as lumbering camps. We also modelled river basins to determine rates of erosion as a consequence of deforestation and the possibility that alluvial deposition in the river basins could be dated using pollen and carbon data.[12]

[9]The project expresses its gratitude to Space Imaging and Space Imaging Eurasia, Inc.

[10]Data were acquired from the Space Imaging IKONOS earth imaging satellite consisting of one panchromatic band (0.45–0.90 μm) with one-metre spatial resolution and four multispectral bands (blue 0.45–0.52 μm; green 0.51–0.60 μm; red 0.6–0.70 μm; and near infrared 0.76–0.85 μm) of four-metre resolution. Eleven-bit data were obtained providing 2,048 levels of measurement in a pixel. Two scenes were necessary to cover the project area, and both were acquired on 17 September 2001, the first at 8:53 and the second at 8:55. The data were prepared using a cubic convolution interpolator rather than a nearest neighbour algorithm. The nearest neighbour algorithm generally results in better-looking imagery but can result in pixel value shifting between bands that can create problems in mathematical classification.

[11]In addition to the mounting evidence for oil and wine production, literary testimony indicated that the region was important for its timber resources, particularly including highland cedar forests that were useful for shipbuilding (Akkemik et al. 2012).

[12]Rauh 2006, 2008, 2012; Akkemik et al. 2012; Karlioğlu et al. 2014, 2016.

Identification of specific kinds of vegetation in the satellite data helped to approximate the locations, elevations and density of specific forest types in the current landscape to furnish this discussion with quantifiable metrics. In addition, by 2003 the expanding boundaries of the survey regions had grown to incorporate highland terrain in the Taşeli Plateau to the north, as well as river basins such as the Dim Cay to the northwest and the Kaledran basin to the southeast (Figure 13.5).

As noted above, in 2003, the survey team located and georeferenced an active vineyard on the north slope of Lamos. Using IKONOS, we were able to capture the spectral signature of the vineyard in the 4-band multispectral data and used this to estimate the extent of wild grapevine vegetation throughout the survey region. Following this analysis, the project extended field survey into areas that were beyond the edge of the IKONOS imagery. In order to continue testing the use of multispectral imaging to detect certain types of vegetation on a wider scale, imagery was acquired from Satellite Pour l'Observation de la Terre 5, or SPOT-5 (Figure 13.5).[13] The spectral bands were lower resolution than the IKONOS bands and had 256 levels of measurement compared to 2,048 for IKONOS. However, the 'higher' energy orientation of the bands in the red/infrared parts of the spectrum promised to be better for vegetation classification and mapping (to identify grapevines, olive, cedar, juniper or fir trees). The SPOT-5 data also had the practical advantage of including the entire project area in a single image. The image scene extends from the ancient acropolis of Korakesion (modern Alanya) in the west, most of the Taşeli Plateau to the north and the Kaledran river basin, and the coastal villages of Meleç and Tenzile in the east (Figure 13.5). The SPOT-5 image was also virtually cloud free, which, combined with the single scene, eliminated significant pre-processing steps to ready data for classification.

By the time the project had acquired the SPOT-5 data in 2006, we had already located the recently planted olive grove mentioned above. Like the vineyard near Lamos, the olive grove at Beyrebucak was situated on exposed terrain that was visible in the satellite data (Figure 13.3). It was then possible to use the SPOT-5 data to spectrally identify both the Lamos vineyard and the olive grove in Beyrebucak and to generate the image displayed in Figure 13.4. The point to this digression is twofold: first, it is intended to show how serendipitous the reasoning behind acquiring satellite data for the survey project was; and second, it explains the degree to which the trajectory of the project evolved over time. In essence, a combination of growing familiarity with the survey landscape and the evolving needs of the survey tended to lead to more suitable, bespoke and innovative methodologies.

---

[13]Although the imagery was purchased in 2006, it was obtained by satellite in October 2002. Data consisted of a panchromatic (0.48–0.71 μm), visible green (0.50–0.59 μm), visible red (0.61–0.68 μm), near-infrared (0.78–0.89), and short-wave infrared (SWIR) (1.58–1.75 μm) band. The ground resolution of the bands was five metres for the panchromatic band, 10 metres for the multispectral bands, and 20 metres for the SWIR band, which was resampled to 10 metres for vegetation classification. The collection was on 31 October 2002 at 9:04, approximately one year after the IKONOS imagery.

Conclusion

The use of remote sensing satellite imagery and multispectral analysis to delineate areas of sustained grapevine and olive tree vegetation offers tremendous insight to our reconstruction of the accumulating archaeological evidence for Roman-era wine and oil production in Rough Cilicia. As recognized elsewhere,[14] the results of the remote sensing analysis are a substantial step forward toward redressing the balance between otherwise conflicting information about ancient wine and oil production and indicates that wine was perhaps the region's primary surplus commodity. Remote sensing of grapevines and olive trees in western Rough Cilicia demonstrates not only that these florae grow naturally throughout the region, but more importantly that they have probably done so stretching back into antiquity at considerable scale. Depending on the energy applied by local inhabitants, this resource was available for utilization either as a ready source of food or as raw material for the production of wine and oil. The presence of press installations and *amphora* kiln sites merely indicates the manner and degree to which the inhabitants of this region harnessed these resources for purposes of wine and oil production in the Roman era. The bulk of the literary and archaeological evidence points to wine as the principal commodity exported from Rough Cilicia in the Roman era and how the high ratio of grapevines versus olive trees in the contemporary land cover of western Rough Cilicia appears to confirm this. Population decline during the seventh century CE alongside further evolutions with the advent of Islam reduced wine production considerably during the Middle Ages. This raises the distinct possibility that the extensive growth of uncultivated grapevines visible today represents the survival of formerly cultivated vines that merged with the maquis alliance and expanded their habitat as the Roman-era agricultural landscape was abandoned.

[14]cf. Aydınoğlu and Rauh in Van Limbergen et al., forthcoming.

PART IV

MODELLING AND EXPERIMENTAL ARCHAEOLOGY

CHAPTER 14
QUANTIFYING ROMAN WINE PRODUCTION: NEW APPROACHES TO VINEYARD SIZES AND WINE YIELDS IN THE *AGER BARCINONENSIS*

Antoni Martín i Oliveras[1]

'Agri bis centum colo jugera; vinea centum.
Jugeribus colitur, prataque dimidium.
Silva supra duplum, quam prata et vinea, et arvum.'

Decimus Magnus Ausonius, *De herediolo* 6.21-22.[2]

Abstract

This chapter illuminates the different types, shapes and extensions of vineyards documented by the Roman agrarian writers including diversity across the provinces. It also extends our knowledge regarding the processes, factors and variables involved in the supply chain and production function of Roman viticulture to quantify its main economic values. These processes can be assessed, quantified and yields obtained by applying different methods of calculation. From these yield ratios, and knowing the quantity of winemaking facilities, we can obtain an estimated scale of the vineyards managed, crop achieved and wine made by a property. Applying this methodology as a case study to a hypothetical *fundus* located in the ancient Laetanian region of the *Ager Barcinonensis* (Catalonia, Spain) we can reconstruct and model the socio-economic dynamics of this territory for a set chronological period during the Roman era.

[1]Special thanks to Emlyn Dodd and Dimitri Van Limbergen for accepting our contribution in the conference *Vine-growing and Winemaking in the Roman World*, held in Rome from 27–9 October 2021, and for their confidence in commissioning one of the chapters for this book. All possible errors of course remain my own. This study was funded through the research projects: *Cella Vinaria* Research and Heritage Project (2003–22) and CLT009/18/00045 '*Dinàmiques socioeconòmiques del món rural romà: formes d'hàbitat i cultura material al litoral central català (I)' – Projectes Quadriennals de Recerca en matèria d'Arqueologia i Paleontologia (2018–2021) – Departament de Cultura, Generalitat de Catalunya.*

[2]'I cultivate two hundred *iugera* of land; A hundred are cultivated as vineyards, half are cultivated as meadows. The woods are more than twice (of the property surface) as much as the meadows, the vineyards, and the farmland' (author's translation).

Introduction

Establishing the scale and dimensions of viticultural properties – and rural properties more broadly – during the Roman period is not an easy task. Generally, viniculture does not typically develop large estates (*latifundii*).[3] In the first phases of settlement, it occurred in the so-called economic 'villa system' and afterwards through developing intensive or specialized vinicultural production where tenants dominated the rural environment rather than further specialization processes that in some situations suppose a progressive concentration of properties in one or few landlords.[4]

The supply chain of Roman viticulture extends from the cultivation and production processes through to storage, transport and commercial activities and delivery to the end customer or consumer. The production function is the global system that characterizes a productive activity.[5] Production as a whole is controlled by the inputs of a system, which is worked by, for example, raw materials, labour, machinery, tools and facilities, to foster optimal productive performance and result in outputs, which can be finished products, goods or services. In the Roman world, the processes for making wine and other wine or grape derived products were impacted by such factors, conditions and variables at different scales, including macro-, meso- and microeconomic.[6]

These production processes can be quantified, and estimated crop and yield ratios obtained by applying different methods of calculation. From these values, and by knowing the number and various features of winemaking facilities (wineries), such as the number of wine presses (*torculares*), collection vats (*lacus*) and storage containers (e.g. *dolia defossa*), we can obtain an estimated vineyard scale managed in a hypothetical agricultural plot of land and the potential grape crop.

Viticulture was central in the economy of *Hispania Citerior Tarraconensis* between the first century BCE and third century CE. A large number of vineyards, wineries and pottery workshops are clustered in specific territorial areas, including the Laetanian region located on the northeast coast of the Iberian Peninsula. Traditionally, this has been interpreted as evidence for the existence of an intensive and specialized wine economy associated with large-scale production and trade of wine, packaged in *amphorae* or other containers (e.g. *dolia*, *cupae*, *cullei*), and orientated mainly towards

[3]Tchernia 2013: 153.

[4]Martín i Oliveras and Revilla 2019: 146–7.

[5]In economics, the production function is the equation that expresses the relationship between the quantities of productive factors or inputs used (such as land, labour and capital) and the amount of product or outputs obtained (in this case wine and derivative products). It states the amount of product that can be obtained from every combination of factors, assuming that the most efficient available methods of production are used (Jones 2014: 9; Martín i Oliveras 2015a: 19–38; 2016: 34–54).

[6]Derivates from must and wine include *defrutum* reduced from must by one-third in volume (Pliny, *HN* 12.19.3–5), *sapa* reduced by two-thirds (Columella, *Rust.* 12.21.12), and *caroenum* reduced by 60 per cent (Palladius, *Op. Agr.* 11.18). Others include sweet wine with the addition of honey and macerated fruits called *mulsum* and vinegar or *acetum* (Brun 2003: 84–93).

an export market. The Roman colony of *Ivlia Augusta Faventia Paterna Barcino* (modern Barcelona, Catalonia, Spain), founded by the emperor Augustus between 15–13 BCE, was one of the main urban centres around which the Laetanian territory was structured. It is the only Roman city in the region that seems to possess a centuriated territory dedicated to intensive and specialized vinicultural production and trade.

An approach to the standard scale of Roman vineyards

The Roman texts provide diverse examples of vineyard sizes. Cato (*Agr.* 11) lists the labour and equipment required to manage a vineyard of 100 *iugera* (*iug.*).[7] However, it remains unclear whether this represents a real or idealized 'standard' Roman vineyard.[8] Other Roman authors describe different scales. Pliny (*HN* 14.48) mentions a sixty *iug. vineis* in the surrounds of Rome ('... *LX jugerum non amplius vineis excultis in Nomentano Agro* ...') where the expression '... *non amplius vineis* ...' seems to refer to a smaller scale. The poet Ausonius (*De herediolo* 6.21.22), at the end of the fourth century CE, speaks about a vineyard of 100 *iug.* ('*vinea centum/jugeribus colitur*') inside an estate of 500 *iug.* and located next the Garonne River in *Gallia Aquitania* where, however, half of the total property is occupied by woods. Pliny (*HN* 14.48–52) and Columella (*Rust.* 3.3–10) both describe the sale price of vineyards and their yields. One, bought for 2,400,000 sesterces (HS) by the philosopher Seneca, was a vineyard replanted ten years ago by the grammarian *Remmius Palaemon* who bought it then for one-quarter of its actual price. Alternatively, Pliny provides a price of 400,000 HS for a harvest from an eight-year-old vineyard with a good harvest yield to wine production ratio of seven *cullei* per *iug.* (148 hl/ha ≅ 1317.19 gal/acre). Columella states the production yield ratio of Seneca's vineyard as eight *cullei*/*iug.* (163 hl/ha ≅ 1450.68 gal/acre), including a sale price of 300 HS/*culleus*, but describes it as unusual.

The resulting scale of these 'standard' vineyards could be between sixty and 360 *iug.* depending on the calculation parameters used. While keeping in mind the fact that the price of land and wine varied across time and region, we can also consider the 1,000 HS provided for the purchase of a one *iug.* vineyard land by Columella (*Rust.* 3.3), or the 400,000 HS sale price of the sixty *iug.* vineyard of Acilius Stenhelus (equalling 6,666 HS/*iug.*; Pliny, *HN* 14.48). Using these examples, Tchernia established a potential vineyard scale between 250–300 *iug.* (62–75 ha ≅ 153.20–185.32 acres), with the price

[7]Vineyard surface correlations: 1 ha = 2.4711 acres = 3.97 *iug.* ≅ *4 iug.*; 100 *iug.* = 25.18 ha = 62.22 acres; 200 *iug.* = 50.36 ha = 124.44 acres (parameters taken from Smith and Anthon 1854: 1024–30).
[8]Tchernia 2013: 154.

for a wine's *culleus* between 166–228 HS and a 'per *iug*.' sale price between 8,000–9,600 HS. It is apparent, therefore, that various parameters create different numerical results, but that it is less different than might initially be thought.[9]

Past attempts have also tried to calculate vineyard scale by using the storage capacity of *dolia* in production facilities, considering both the data given by the Roman agronomists and *dolium* capacities from some well-preserved *cellae vinariae* in Italy and *Gallia*.[10] This remains challenging because correlations between the annual vintage and maximum *cella* storage capacity are not necessarily equivalent. Cato (*Agr.* 11.1) recommends having *dolia* storage capacity for five vintages. Previously, scholars have calculated a storage capacity of only one vintage based on the *cella vinaria* of Villa della Pisanella at Boscoreale, Italy.[11] *Tituli picti* recorded on *amphorae* indicate that some high-quality Roman wines were aged inside *dolia* for four years, but several texts suggest that, generally, Roman winegrowers ought to prepare their *dolia* every year to store the next vintage.[12] Using quantified economic data, Tchernia established a storage capacity catering to one to three vintages.[13]

Different vineyard scales also become evident when calculating backwards from storage capacity, using excavated *cellae vinariae* from Roman villas located in Italy, *Gallia* and *Hispania* (see Table 14.1). This variety of sizes, documented both in Italy and in the provinces, led some to classify the properties as small, medium or large.[14] Great diversity in vineyard scale is thus documented in different regions by the ancient literature as well as by calculating backwards from the capacity of storage containers (*dolia*) and winemaking yields to illuminate potential vineyard extensions.

More accurate and reliable comparative production yields can be obtained by increasing the quantity and diversity of data. For example, by considering the total number of wine presses, and number and capacity of collection vats, cellars and storage containers (*dolia* or *cupae*) so that we might reconstruct the extension of ancient Roman vineyards through different calculation methods. By integrating theoretical or documented Roman grapevine cultivation systems and yield ratios from different grape varietals we can also begin to consider theoretical ratios related to crop weight and winemaking yields in liquid units.[15]

---

[9]Duncan-Jones 1974: 47; Tchernia 2013: 153.
[10]Tchernia 2013: 157–61.
[11]Day 1932: 165–208.
[12]Cato, *Digestum* 18.6.1.3; Pliny, *HN* 180.320.
[13]Tchernia 2013: 157.
[14]See Table 14.2; Galtier 1960; Tirone 1970.
[15]Martín i Oliveras and Revilla 2020: 89–112.

Table 14.1 Estimations of vineyard extension from Roman villas located in Italy, France and Spain

Country	*Site*	*Dolia no.*	*Dolium capacity*	*Storage capacity*	*Vineyard scale (est.)*[16]
Italy	Villa Regina (Boscoreale)[17]	18	± 1 *culleus*	± 18 *cullei*	± 2–6 *iug.*
Italy	La Pisanella (Boscoreale)[18]	80–100	± 2 *cullei*	±160–200 *cullei*	± 60–100 *iug.*
Italy	Settefinestre (Toscana)[19]	80–100	± 2 *cullei*	±160–200 *cullei*	± 60–100 *iug.*
France	Villa du Grand Loou (Var)[20]	68	± 3 *cullei*	± 204 *cullei*	± 60–100 *iug.*
France	Villa de Pardigon 3 (Var)[21]	115	± 3 *cullei*	± 345 *cullei*	± 100–200 *iug.*
France	Villa du Molard (Rhône-Alpes)[22]	200	± 2.5 *cullei*	± 500 *cullei*	± 150–350 *iug.*
France	Villa des Toulons (Var)[23]	200	± 3 *cullei*	± 600 *cullei*	± 250–500 *iug.*
France	Villa de Vareilles (Hérault)[24]	477	± 3 *cullei*	± 1431 *cullei*	± 450–550 *iug.*
Spain	Veral Vallmora (Teià-Maresme)[25]	200	± 2 *cullei*	± 400 *cullei*	± 300–400 *iug.*

Table 14.2 Scale of Roman properties

Scale	*Galtier (1960)*	*Tirone (1970)*
Micro property	< 1 ha ≅ < 4 *iug.*	–
Small property	1–3 ha ≅ 4–12 *iug.*	< 10 ha ≅ < 40 *iug.*
Peasant property	3–7 ha ≅ 12–28 *iug.*	–
Middle property	7–20 ha ≅ 28–80 *iug.*	10–20 ha ≅ 40–80 *iug.*
Big property	20–50 ha ≅ 80–200 *iug.*	20–50 ha ≅ 80–200 *iug.*
Very big property	> 50 ha ≅ > 200 *iug.*	50–100 ha ≅ 200–400 *iug.*
Exceptional property	–	> 100 ha ≅ > 400 *iug.*

Source: Tchernia 2013: 163.

[16]Columella's average wine yield is of ±2–3 *cullei/iug.*
[17]Barnabei 1901; Tchernia 2013: 158.
[18]Pasqui 1897: 397–554; Tchernia 2013: 158.
[19]Carandini and Ricci 1985; Tchernia 2013: 158.
[20]Brun 1986: 193; Tchernia 2013: 158.
[21]Tchernia 2013: 158.
[22]Dechandol et al. 1983; Odiot 1983–4; Lasfargues 1985: 542; Brun 2005: 62–5; Tchernia 2013: 164.
[23]Brun 2005: 62–5; Tchernia 2013: 164.
[24]Maune 2003: 326–7, Tables I-II; Brun 2005: 62–5; Tchernia 2013: 164.
[25]Martín i Oliveras and Revilla 2020: 106.

The Roman viticultural supply chain and production function

A supply chain is a system of resources and processes involved in the production and trade of goods or services from supplier to customer.[26] The supply chain of wine involves a diverse range of actors and processes, from cultivation to eventual use or consumption (see above and Figure 14.1):

1. The production function is the global system that characterizes a productive activity.[27]
2. Production factors are the inputs of the economic system.
3. A specific technology combines these inputs (e.g. raw materials, labour, machinery, tools, facilities) to obtain optimal performance.
4. The outputs are the finished products, goods or services that result from the production activity.[28]
5. The production of goods and services may be in the hands of the state, private producers or set on a diverse spectrum in between.

The Roman vinicultural production process is not alien to these considerations, conditions and microeconomic variables (Figure 14.1). Methodologically, there are two

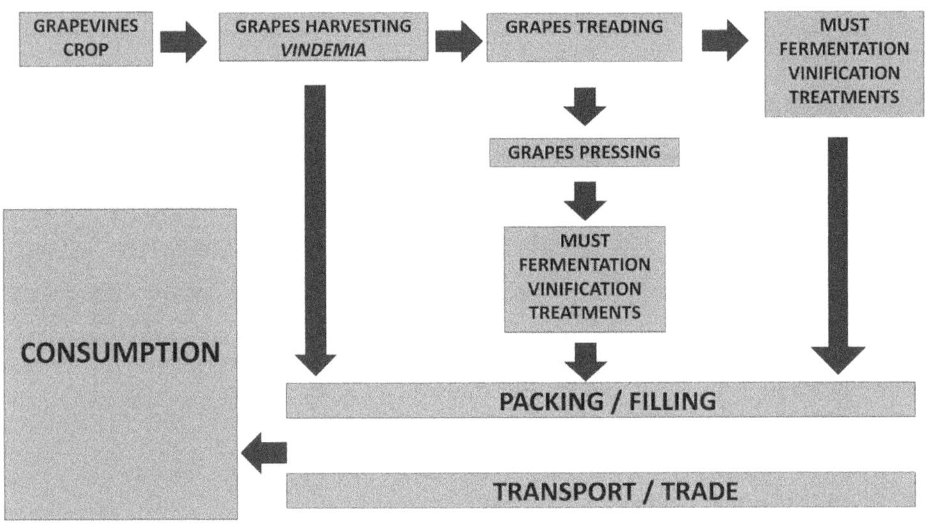

Figure 14.1 Diagram of the Roman viticulture supply chain and winemaking production function. After Martín i Oliveras 2015b: 185, Figure 86.

[26]Jones 2014: 35.
[27]See note 5.
[28]Martín i Oliveras 2015b: 182.

interrelated parameters to consider in order to quantify the Roman vinicultural production function:

1. Determine the estimated vineyard extension or scale; and
2. Determine the estimated viticultural yield.

Vineyard yield calculations

Vineyard yield refers both to the crop yield itself (the productive capacity of the plant), and the total harvest yield (the mass of the grapes harvested prior to pressing from the whole vineyard, property, area or territory). The major parameters to consider in this calculation are (see Figures 14.2 and 14.3):

1. **Planting density**: the ratio of vines per area (e.g. vines/hectare, vines/acre, or *vitis/iug.*) (see Figure 14.3 top and centre).
2. **Yield per cultivar**: the capacity of a grapevine cultivar to produce a specific weight of grapes. This depends on agroecological variables such as type of soil, climate, sunlight, including, for example, the selected grapevine variety, cultivation system, driving and pruning methods adopted, and canopy management (cf. modern examples considering different cultivars in Figure 14.3 below).
3. **Harvest yield**: the total mass of grapes produced in the vineyard, property or territory.

$$PY(t/a) = \frac{ANV \times NC \times CW}{2000}$$

PY(t/a): Potential yield, in tons per acre
ANV: Number of producing vines per acre
NC: Number of clusters per vine
CW: Cluster weight in pounds

$$PY(kg/ha) = \frac{ANV \times NC \times CW}{1000}$$

PY(kg/ha): Potential yield, in kilograms per hectare
ANV: Number of producing vines per hectare
NC: Number of clusters per vine
CW: Cluster weight in grams

$$PY(lib/iug) = \frac{ANV \times NC \times CW}{12}$$

PY(*lib/iug*): Potential yield, in *librae* per *iugerum*
ANV: Number of producing vines per *iugerum*
NC: Number of clusters per vine
CW: Cluster weight in *unciae*

Figure 14.2 Formulae to calculate potential yields in different magnitudes. After Martín i Oliveras and Revilla 2020: 100, Figure 3.

Spacing Feet (m)	Between Rows									
Between Vines	3 (0.9)	4 (1.2)	5 (1.5)	6 (1.8)	7 (2.1)	8 (2.4)	9 (2.7)	10 (3.0)	11 (3.4)	12 (3.7)
3 (0.9)	4840	3630	2904	2420	2074	1815	1613	1452	1320	1210
4 (1.2)	3630	2723	2178	1815	1556	1361	1210	1089	990	908
5 (1.5)	2904	2178	1742	1452	1245	1089	968	871	792	726
6 (1.8)	2420	1815	1452	1210	1037	908	807	726	660	605
7 (2.1)	2074	1556	1245	1037	889	778	691	622	566	519
8 (2.4)	1815	1361	1089	908	778	681	605	545	495	454
9 (2.7)	1613	1210	968	807	691	605	538	484	440	403

	1 m	1,10 m	1,15 m	1,20 m	1,25 m	1,30 m	1,40 m	1,50 m	1,75 m	2 m	2,25 m	2,50 m	2,75 m	3 m
1 m	10.000	9.090	8.695	8.333	8.000	7.692	7.142	6.666	5.714	5.000	4.444	4.000	3.636	3.333
1,10 m		8.264	7.905	7.575	7,272	6.993	6.493	6.060	5.194	4.545	4.040	3.636	3.305	3.030
1,15 m			7.561	7.246	6.956	6.688	6.211	5.797	4.968	4.347	3.864	3.478	3.162	2.898
1,20 m				6.944	6.666	6.410	5.952	5.555	4.761	4.166	3.703	3.333	3.030	2.777
1,25 m					6.400	6.153	5.714	5.333	4.571	4.000	3.555	3.200	2.909	2.666
1,30 m						5.917	5.494	5.128	4.335	3.846	3.418	3.076	2.797	2.564
1,40 m							5.102	4.761	4.081	3.571	3.174	2.857	2.597	2.380
1,50 m								4.444	3.809	3.333	2.962	2.666	2.424	2.222
1,75 m									3.265	2.857	2.539	2.285	2.077	1.904
2 m										2.500	2.222	2.000	1.818	1.666
2,25 m											1.975	1.777	1.616	1.481
2,50 m												1.600	1.454	1.333
2,75 m													1.322	1.212
3 m														1.111

Row width (m)	Number of rows per ha	Number of bearers/ha[1]	Mass of grapes[2] for cultivars with							
			120 g/bunch[3]		150 g/bunch[4]		200 g/bunch[5]		250 g/bunch[6]	
			Per bearer (g)	Per ha (t)	Per bearer (g)	Per ha (t)	Per bearer (g)	Per ha (t)	Per bearer (g)	Per ha (t)
3,0	33	33 000	360	11,8	450	14,9	600	19,8	750	24,8
2,7	37	37 000	360	13,3	450	16,7	600	22,2	750	27,8
2,5	40	40 000	360	14,4	450	18,0	600	24,0	750	30,0
2,2	45	45 000	360	16,2	450	20,3	600	27,0	750	33,8
2,0	50	50 000	360	18,0	450	22,5	600	30,0	750	37,5

[1] With a bearer spacing of 10 cm between bearers
[2] For an average of 3 bunches per spur
[3] E.g. Cabernet Sauvignon, Cabernet franc, Pinot noir
[4] E.g. Merlot, Pinotage
[5] E.g. Crouchen blanc, Morio Muscat, Sauvignon blanc
[6] E.g. Carignan, Colombar, Mourvèdre

Figure 14.3 Vines per acre and vines per hectare tables based on different planting spacing schemes (*top and centre*); mass of grapes for cultivars on different plantation frames and grapevine varieties (*below*). After Martín i Oliveras and Revilla 2020: 97–9.

These three parameters are most commonly used to calculate maximum and average vineyard yields.[29] Using this information, we can calculate the potential yield rates in metric tons/hectare or in tons/acre or in *librae/iug.*

Winery yield calculations

It is also possible to calculate potential yields from the processes of treading and pressing grapes, the transformation through fermentation from must into wine and other grape derivates, as well as the maximum and average production capacity of processing equipment. Thus, we can determine requirements for the quantity and capacity of collection mechanisms and number of containers for ageing and storing the wine (e.g. *dolia, cupae*). Here, we focus on two main operations: treading and pressing yields and collecting and storage yields, with various structures and machinery often found across Greek and Roman winemaking facilities.

Treading and pressing yields

This method attempts to calculate the maximum capacity of must processed during the vintage both by treading in vats or on specially designed floors (*calcatoria*) and by pressing the pomace using presses (*torculares*). It has several advantages compared to the following method:

1. Practically all the harvested grapes ought to be trodden and pressed to obtain the must for wine.
2. Important and relevant values are provided by the ancient texts, experimental archaeology, ethnographic data and modern oenology.
3. Estimations can be tested using mathematical and statistical analysis.[30]

Below, we explore two comparative methods to determine the treading and pressing yields of a winery. The first uses data from Roman agricultural texts, and the second data obtained from experimental archaeology and ethnography.[31]

Treading and pressing yields derived from Roman texts

Various parameters can be inferred from the text of Cato (*Agr.* 11), who states that three presses are necessary to process the harvest from a 100 *iug.* vineyard (≅ 25 ha ≅ 61.77 acres), and from Pliny's (*HN* 18.317) description of a press yield of *c.* twenty *cullei* (≅ 105

---

[29]Martín i Oliveras and Revilla 2020: 97.

[30]From experimental archaeology we only have the data from the Mas des Tourelles experiment, which operated from 1995–8 (Tchernia and Brun 1999: 102–5). Estimates are still debated among scholars but have been accepted by most economic historians due to the scarcity of data and the absence of more reliable information (De Sena 2005: 2 with n. 7).

[31]Martín i Oliveras and Revilla 2020: 106.

hl ≅ 2310 gallons = 73 wine barrels). However, these authors remain vague on several crucial factors: for example, they do not specify how many times the presses operated nor if the quantities given also include the volume of must extracted from treading the grapes beforehand.

To quantify the former variable, we can compare the information given by the Roman agronomists with data from ethnographic and experimental sources to calculate the maximum and average pressing capacities of one so-called Catonian press in a working day (*c.* twelve hours).[32] This calculation integrates the following parameters and variables:

***Vindemia* period (max.):**	≅ 30–44 days (Varro, *Rust.* 1.34.2; Pliny, *HN* 18.319).
***Vindemia* period (av.):**	≅ 12–15 days.
Working hours per day:	12 hours of sun.[33]
Vineyard unit:	1 *iug.* ≅ 2,518 m^2 = 0.2518 ha ≅ 0.623 acres.
Vineyard plots:	100 *iug.* ≅ 25.18 ha ≅ 62.30 acres (Cato, *Agr.* 10.1). 200 *iug.* ≅ 50.36 ha ≅ 124.60 acres (Columella, *Rust.* 2.12).
Modern liquid unit:	1 hl = 100 L ≅ 22 gallons.
Roman liquid unit:	1 *culleus* ≅ 5.24 hl ≅ 524 litres ≅ 115.3 gallons.
Grapes processed per day (av.):	10.6 mt ≅ 10.38 t.
Grapes processed per *vindemia* (15 days):	159 mt ≅ 155.82 t.[34]

From the yields given by Pliny (*HN* 18.317), Brun calculated that 750 hl (≅ 16,497.69 gallons) of must could be processed in a facility with three presses.[35] Supposing that each press could process 250 hl (≅ 5,499.23 gallons), and by applying Pliny's yield per press ratio (20 *cullei* ≅ 105 hl ≅ 2310 gallons = 73 barrels), this results in just three days of work.[36]

---

[32]The Romans divided the day into twelve *horae* (hours) starting at sunrise and ending at sunset. The night was divided into four watches. Sunlight parameters for the months of August, September and October in the Barcelona area are taken from: https://meteogram.es/sol/espana/barcelona/

[33]Assuming a vineyard of Muscat of Alexandria variety in the Laetanian region's agro-ecological conditions (i.e. soil, slope, weather, temperatures, planting system, head-spur pruning) with an extension of 100 *iug.* (≅ 25.18 ha ≅ 62.3 acres) and also considering the average temporal values for processing (treading and pressing) as deduced from Pliny.

[34]Average grape values taken from the Mas des Tourelles experiment (Tchernia and Brun 1999: 102–5).

[35]Brun 2004a: 20. We have not been able to deduce how the total processing yield target of 750 hl was reached. Using the ratio of 33 hl/ha ≅ 263.69 gallons previously established and calculated for 25 ha (≅ 62.30 acres ≅ 100 *iug.*), the result obtained is 825 hl (≅ 18,147.46 gallons).

[36]It is assumed that the period of almost three days is reached by considering that Pliny's yield value of twenty *cullei* for one press in one working day also adds the must produced by treading. In any case, the result seems overly large.

Attempts to calculate production yields by relying only on information given by the Roman agronomists is therefore difficult because the data provided is so incomplete and the results obtained inconsistent. However, we might still conclude that the wineries were designed and built with large harvests in mind and with the aim of processing grapes efficiently.[37]

Treading and pressing yields from experimental and ethnographic sources

Our calculations here utilize the data provided by an experimental archaeology project carried out from 1995–1998 at Mas des Tourelles (Beaucaire, France), which used a reconstructed Roman winery with a replica Catonian press. Helpfully, the study distinguished the must derived from treading (av. 53 per cent of the total yield) as well as the yields from pressing (av. 47 per cent).[38] The experiment provides the following useful quantifications:

- Average must processed per working day = ±63 hl (≅ 1,385.78 gallons ≅ 48 barrels ≅ 12 *cullei* ≅ 240 *amphorae*).
- Average must processed using treading and pressing from a ten-day *vindemia* = 629 hl (≅ 13,836.06 gallons ≅ 439 barrels ≅ 120 *cullei* ≅ 2,700 *amphorae*).
- Average must processed using treading and pressing from a fifteen-day *vindemia* = 943.5 hl (≅ 20,754.09 gallons ≅ 659 barrels ≅ 180 *cullei* ≅ 4,193 *amphorae*).

By applying the Mas des Tourelles yield of 63 hl per day to a 25 ha vineyard (≅ 100 *iug.*), producing an average of three *cullei/iug.*, we arrive at a total of 300 *cullei* (≅ 1,572 hl ≅ 34,645 gallons ≅ 1100 barrels). Taking into consideration the production rate of 63 hl per day using treading and pressing, it would take a single Catonian press *c.* twenty-five working days to process the yield of a 100 *iug.* vineyard (an equivalent grape weight of 265 mt ≅ 260.81 t).[39]

Collection and storage yields

Previous studies based on archaeological remains have tried to calculate production capacity from the data provided by the Roman texts regarding the recommended number of *dolia* required to store the wine produced.[40] In this sense, archaeological material that is well enough preserved to provide the size of collection vats or the number and size of *dolia* in excavated production centres can be used. Even though information from ancient literary sources is often scarce or contradictory and archaeological data incomplete, it is still possible to develop quantification methods when sufficient data is available.

---

[37]Note that we speak of optimum results, not maximum. Achieving a positive balance between crops and processed yields was the main objective pursued by producers.

[38]Tchernia and Brun 1999: 104.

[39]Martín i Oliveras and Revilla 2020: 106–7, table 7. Average yield ± 3 *cullei/iug* provided by Columella, *Rust.* 5.3.5 (Trapero Fernández 2021b: 132, table 7.5). All other data provided by the Mas des Tourelles experiment. Average grapes per working day (12 hours) processed ≅ 10.6 mt ≅ 10.38 t.

[40]cf. Cato, *Agr.* 11.1; Brun 1993a: 307–42; Tchernia 2013: 153–66.

Mathematical methods

These are based on the application of geometric formulae to calculate the volume of different containers (Figure 14.4). The results are expressed in ft^3 or m^3 that can then be transformed into gallons or litres by multiplying or dividing the values obtained with the numeric conversion factor 0.21997. To calculate the volume of a collection vat (*lacus*), the most common formula applied is for the volume of a quadrangular prism: *V=lwh*.[41] Some scholars believe that initial fermentation of the must took place in the *lacus*.[42] Others find this unlikely, given that it would stop the production process for at least one week.[43]

The calculation of *dolia* capacity is more difficult given their irregular shape, which might be considered a regular sphere, irregular spheroid or ellipsoid. There are different formulae to calculate their volume, which include using a regular sphere ($V = \frac{4}{3}\pi$) or irregular spheroid or ellipsoid ($V = \frac{4}{3}\pi abc$).[44]

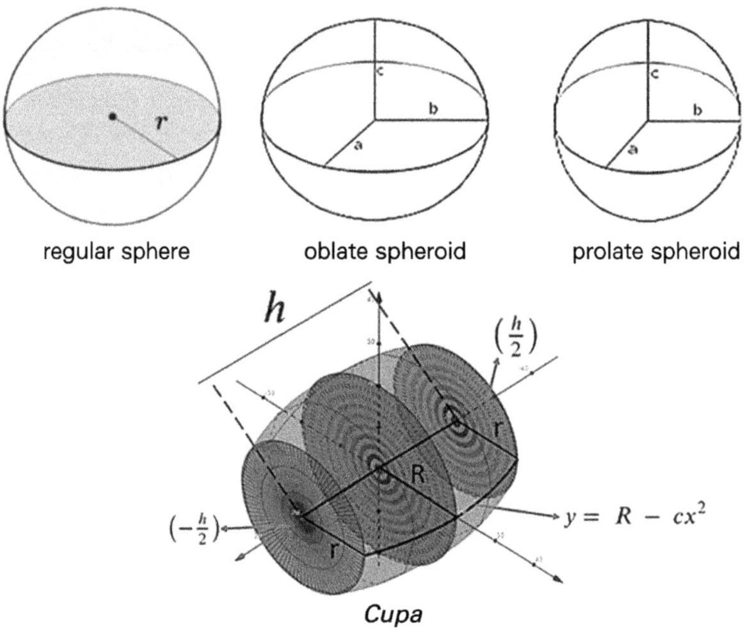

Figure 14.4 Diagrams illustrating mathematical variables used to calculate various shapes that closely reflect storage vessels used in a Roman winery, including a *dolium* and *cupa*.

---

[41]Where *l* is length, *w* is width, and *h* is height.

[42]Carandini and Ricci 1985; Mezquíriz 2004: 135.

[43]Tchernia 1986.

[44]Other methods, based on Archimedes' theory (that the volume of a spheroid or ellipsoid is 2/3 the volume of a circumscribed elliptic cylinder, and π/6 the volume of the circumscribed box) can calculate the volume of a sphere or a spheroid with only an approximate error margin of 0.03 per cent without using the value π, through this formula: $V = \frac{67}{16} r^3$.

Another volume and capacity calculation is proposed by Carrato based on the disc method for calculating the volume of solids of revolution.[45] It consists of dividing the internal *dolium* space in rectangular segments or discs (Figure 14.5). Each rectangle is delimited on the right side by the vertical central axis of the *dolium*.[46] By applying the formula $V = \pi r^2 x h$ we can obtain the volume of each disc with the sum of all discs providing the total volume of the vessel. Different *dolium* shapes are documented across the Roman world and these formulae should be adopted on a case-by-case basis. However, considering the regular sphere's formula and its correlation to the Roman capacity units given by ancient texts we can determine a progressive scale of capacity (in *amphorae* ≅ 26.26 litres ≅ 5.76 gallons; see Figure 14.6).[47]

Wooden *cupae* (barrels) were used as a wine container from at least the late Roman Republic, with a noticeable presence in iconography.[48] *Cellae vinariae* equipped with *cupae* are cited by written sources,[49] and their presence has been interpreted in Roman cellars of Atlantic temperate regions, including northern Portugal, Spain, Aquitaine, Normandy, Belgium, the Netherlands, Rhenania and northern France.[50] Unfortunately,

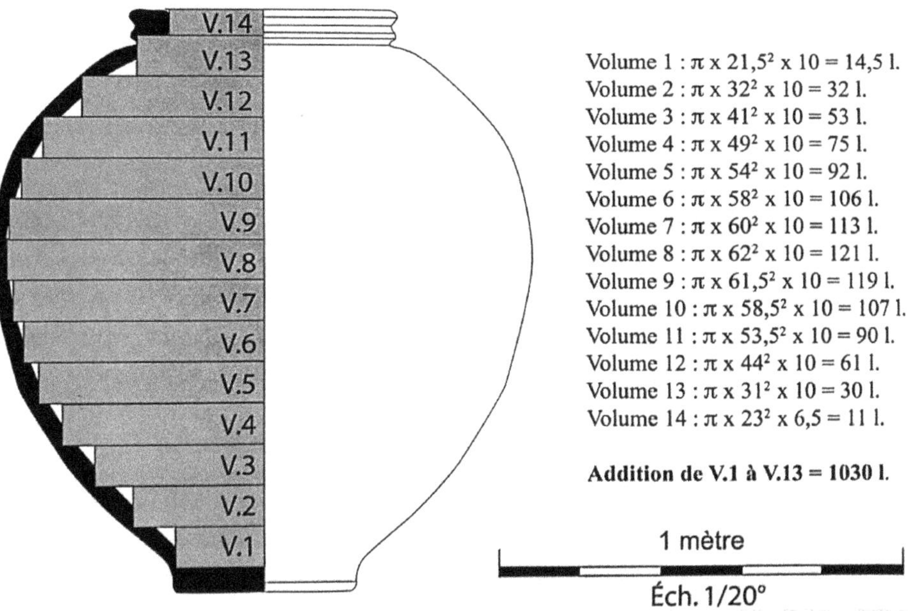

Figure 14.5 Calculation method for the storage capacity of the *dolium*. After Carrato 2017a: 103, Figure 50.

---

[45]See Rice 1987; Senior and Birnie 1995: 319–34.

[46]Carrato 2017a: 103.

[47]'*Amphora*', also called '*quadrantal*', is a Roman standard measure for liquids of 5.76 gallons ≅ 26.26 litres (Lewis and Short 1879).

[48]Strabo, *Geog.* 5.1.12; Pliny *HN* 14.27.122; Baratta 1994a, 1994b, 1997, 2001, 2005.

[49]Palladius *Op. Agr.* 1.18.

[50]Brun 2003: 83; 2005: 66–8 and 131–52.

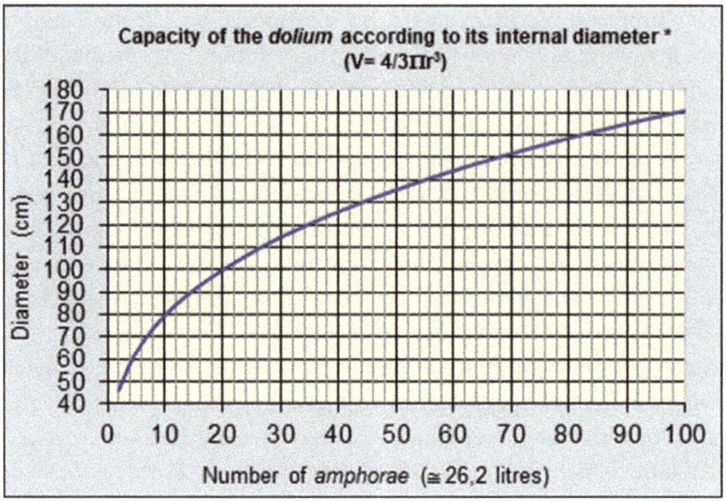

Figure 14.6 Graph illustrating the calculation of the proportional capacity of a *dolium* according to its internal diameter. After Martín i Oliveras 2015b: 105, Figure 29.

due to their construction in largely organic components, archaeological evidence is rarely preserved.[51] It is possible to calculate the capacity of *cupae* by applying the formula: $V = \frac{\pi h}{3}\left(2R^2 + r^2 - \frac{2}{5}(R-r)^2\right)$.[52]

It is also possible to calculate these volumes using 3D design software, such as AutoCAD, Autodesk 3ds Max, Blender 3.0, which incorporate tools (e.g. measure, volume) and utilities (e.g. Blender add-on 3D Print toolbox), to calculate different internal or external measurements as well as the volumes of regular and irregular geometric forms (Figure 14.7).

Empirical methods

One of the most accurate ways to ascertain the capacity of a vessel is to replicate the container, fill it with water, and measure the capacity using a liquid flow counter. It is, however, important to consider the difference in density of water and grape must or wine. One litre of water weighs 1,000 grams whereas one litre of must weighs ± 994 grams.[53] That supposes a difference of 0.6 grams per litre ≅ 0.6 per cent less of the total water filled.[54]

[51]Marlière 2001: 181–201; Marlière and Costa 2005: 228–9, Figure 241; Mille and Rollet 2020: 123–55.

[52]There is also an online geometric form calculator that, among others, includes this shape: https://rechneronline.de/pi/forms.php

[53]Ruíz Hernández 2021. The density of the must can vary depending on the grape cultivar with different concentrations of sucrose, acids, salts and polyphenols.

[54]Formulae: $V\ (must) = (No)Litres\ (H_2O) - \left(\frac{(No)Litres\,(H_2O)x0,6g}{100}\right)$ / $V\ (must) = (No)Gal\ (H_2O) - \left(\frac{(No)Gal\,(H_2O)x0.0013198\,Lb}{100}\right)$.

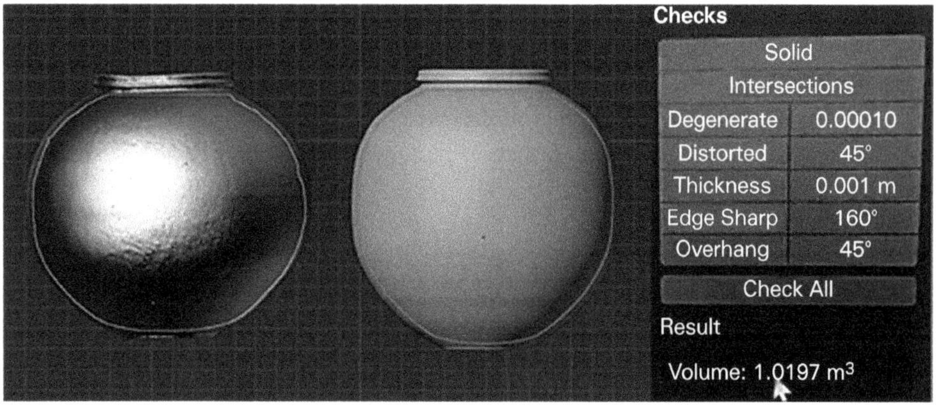

Figure 14.7 Digital methods for the calculation of volume and storage capacity of different *dolium* shapes documented in the Veral de Vallmora–*Cella Vinaria* Archaeological Park (Teià, Maresme, Spain). After Martín i Oliveras and Parés 2020: 947–74.

Modelling intensive vineyard cultivation in the *Ager Barcinonensis*

Before attempting to estimate viticultural scale, it is first necessary to define the scale of an archetypal Roman *fundus* on the centuriated Laetanian *Ager Barcinonensis* plain. The parcel division of an *ager*, which was the extension of arable land attached to a Roman *civitas*, was termed *centuriato*.[55] Written sources mention the existence of undivided areas within the *centuriae* – for example, the integration of woods and grasslands.[56] *Centuriato* were shaped like a systematic grid of parallel planning lines anchored to right-angled axes called *limitatio* and were made by the *gromatici* or *agrimensores*. During the archaic Roman monarchy period (753–509 BCE), only two *iug.* of land were distributed for cultivation to each *colonus* (Varro, *Rust.* 1.10), which were called *haeredium* (Pliny, *HN* 18.2), *festus* or *caespes fortuitus* (Horatius, *Epistulae* 2.15–17). During the Republican period, seven *iug.* were assigned to each inhabitant of Rome and for a long time this size of land was also that distributed among the colonies of conquered territories.[57] Later, during the Imperial period, these concessions were increased and 100 *iug.* were called *sors* and 200 *iug.* were called *sortes* or *haeredia centuria*.[58]

---

[55]According to Florentinus (*Dig.* 50.16.111), the term *fundus* comprised all land and constructions on a Roman country state or property. *Centuriation* is a grid-based parcelling of landscapes for land distribution and field allotment (Palet et al. 2021: 173). *Limitationes* were the paths and the roads between the plots or parcels on a centuriated land (Clavel-Lévêque et al. 1996). The *Gromatici* or *agrimensores* were the Roman land surveyors. One of their tasks was to measure the land for distribution or to allot it among the inhabitants of a new colony (Smith et al. 1890). A *colonus* was originally a peasant or tenant farmer that worked on a country estate but in this case refers to veteran legionaries that were given a plot of land by the Roman state (two *iug.*) in payment for services (Broadhead 2007:148–63).

[56]Varro, *Rust.* 1.6.; Columella, *Rust.* 5.2–3.

[57]Pliny, *HN* 18.3; Livy, *Ab urbe condita* 5.30.

[58]Columella, *Rust.* 1.5.

Fields were usually quadrangular or rectangular in shape and were subdivided into one *iug.* units of *c.* 240 Roman feet long and 120 Roman feet wide. They could also be subdivided into *actus vortus quadratum* – a portion of land corresponding to half a *iug.* – or have other triangular, semi-circular, round or polygonal shapes, depending on the topography of the territory, pre-existing roads, paths and specific agricultural requirements.

Plots were typically organized around two main cardinal axes: one called the *cardo maximus* (north–south) and another perpendicular to this called the *decumanus maximus* (east–west). Auxiliary axes called *decumani* and secondary *cardines* were also used.[59] The most common division of the territory or *cadastratio* in Italy since the third century BCE was made in quadrangular lots of 200 *iug.*, called *centuria*, of about 710 m (≅ 2329.39 ft) on each side (≅ 50.36 ha ≅ 124.19 acres).[60] According to the *agrimensores*, the surfaces could be larger (e.g. 400 *iug.* ≅ 2 *centuriae stricto sensu*), or smaller (e.g. 50 *iug.*, also called *laterculus*).[61] In fact, some authors define the division of *centuriae* differently, including the *cellae*, also called *strigae* or *scamna*, corresponding to internal parallel networks delimited by internal secondary paths, which may or may not be equidistant.[62]

Various approaches have been applied to study centuriated areas with a particularly influential method, developed during the 1980s and 1990s by the French Besançon group, based on two criteria:

1. Constant orientation and modulation based on the *actus* (120 Roman feet, about 35.5 metres ≅ 116.46 feet).
2. Identification of the canonical model of grid squares with sides of twenty *actus*; using methods and techniques of metrology and integration of multidisciplinary approaches, especially fieldwork and surveys that analyse the archaeomorphological sequence from a diachronic perspective.[63]

In Catalonia (Spain), work has been carried out on ancient landscape dynamics in the Plain of Barcelona with a focus on the territorial impact of the foundation of *Colonia Ivlia Augvsta Faventia Paterna Barcino*.[64] Because it is today a busy urban metropolis, much of the Plain of Barcelona has not been able to be surveyed.

---

[59]Roth-Congés 1996: 299–422.

[60]Castagnoli 1984: 7–17; 1993: 805–21; Clavel-Lévêque et al. 1996.

[61]The reliability of data from the *gromatici* or *agrimensores* (Roman land surveyors) and its usability has been debated, see: Smith et al. 1890; Dilke 1971; Campbell 1995: 540–6; 1996: 74–99; 2000; 2006: 173–81.

[62]Ferdière 1988: 133–55; Chouquer and Favory 1991, 2001; Clavel-Lévêque et al. 1996; Conso et al. 2005.

[63]See Chouquer and Favory 1979, 1991; Leveau and Golvin 1979: 817–43; Chouquer 1983: 113–39; Clavel-Lévêque 1983; Chouquer et al. 1987; Berger and Jung 1996: 95–112; Palet 1997; Clavel-Lévêque and Vignot 1998; Menotti 1999; Clavel-Lévêque and Orejas 2002; Labate 2010: 21–30; Leveau 2011: 591–603.

[64]See Palet 1997, with later updates to incorporate information from archaeomorphological, archaeological and palaeoenvironmental data into a GIS (Palet, Fiz and Orengo 2009: 103–26; Palet, Orengo and Riera 2010: 113–29; 2011).

The settlement of *Barcino* was located on a small promontory next to the shoreline. The spatial organization of its adjacent territory through regular parcelled fields (albeit not all necessarily exploited) integrated complex land use dynamics and a characteristic appearance of the ideal territory of a Roman town (Figure 14.8).[65] In this case, centuriation was based on a model grid of 15 x 20 *acti* extending over the littoral plain between the Besós and Llobregat rivers, and from the seashore to the Collserola mountain range, adapted where necessary to natural slopes, promontories and streams. It seems likely that this is why some *decumani* further inland were displaced from the theoretical centuriation grid, in certain cases creating smaller *centuriae* of fifteen *acti*. By basing the southwest/northeast orientation of the grid on the direction of the coastline and littoral range of hills, it seems that the *agrimensores* attempted to ensure efficient land use and good drainage also incorporating the streams to minimize the construction of new *limites*. The parallel axes to the coast form *salti* of three *centuriae*.[66] The effective implementation

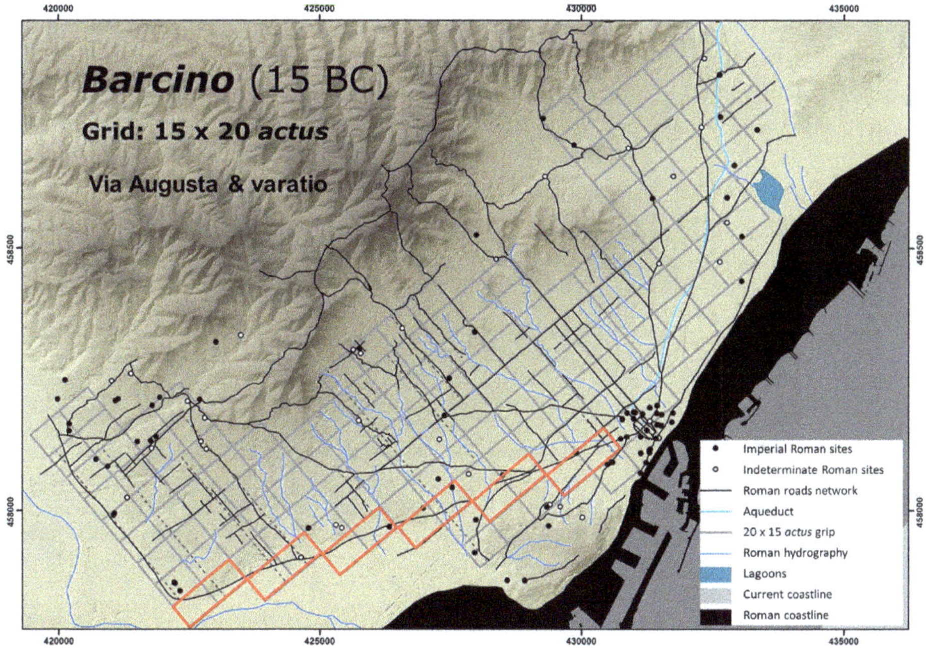

Figure 14.8 Hypothetical reconstruction of the *Ager Barcinonensis* centuriation plain with the axis system of the *varatio*. After Palet, Ortega and Miró 2021: 178, Figure 3.

[65]While centuriated systems organize the land methodologically, they do not necessarily demand agrarian exploitation of the whole system. Recent palaeobotanical studies in the *Ager Barcinonensis* plain show the absence of largescale deforestation and the presence of large grassland areas illuminating a complex, heterogeneous Roman landscape (Palet, Orengo and Riera 2011; Martín i Oliveras 2015b: 45–50). Indeed, new theoretical frameworks indicate socio-cultural and symbolic factors regarding diversity within settlement patterns and land distribution, tenancy and exploitation (Palet, Ortega and Miró 2021: 177–88).

[66]Palet, Ortega and Miró 2021: 177–88.

of these strategies can be seen particularly clearly in the south-west and south-central sectors, where the plain is wider allowing greater implementation and preservation of the grid axes. The grid is delimited by the Roman *Via Augusta* to the southwest and northeast. The construction of this section of the road was dated by milestones to 9–7 BCE.[67] In some places it forms the diagonal axis of several *centuriae* in the theoretical grid, following the system of the *varatio*, and acts as the hypotenuse for various triangles of 40 x 45, 20 x 30, 60 x 45 and 40 x 30 *acti* plots.[68] The correlation between the urban layout of *Barcino* and the centuriated network undoubtedly indicates coordinated planning, with its main axes aligned to the position of the temple of Augustus in the *forum* of the colony.[69] The centuriation appears to have avoided mountainous areas, with traces disappearing in certain sectors, such as the promontory of Montjuïc and the first spurs of the littoral range.

What appears to be the main *Ager Barcinonensis* plot form of 15 x 20 *acti* is roughly equivalent to the more common Roman agrarian field measurement of 150 *iug.*[70] If we subtract 33 per cent of a plot's surface area to cater for household buildings, agrarian facilities, *limitationes*, pathways, grain fields and *horti* for feeding the *villicus*' family, permanent staff and temporary labourers during the *vindemia* season, two-thirds of the agrarian surface remains to be dedicated to intensive viticulture, supposing a standard theoretical vineyard of 100 *iug.* in every parcel.[71]

Quantifying the yields of a theoretical 100 *iugera fundus* in the *Ager Barcinonensis*

The preceding discussion allows us to consider different viticultural yields obtained through the exploitation of a theoretical 'standard' *fundus* of 150 *iug.*, where 100 *iug.* are

[67]Fabre, Mayer and Rodà 1984; *IRC* I.183–184.

[68]Orfila et al. 2020: 127–46; Chouquer and Favory 2001: 89–94; Roth-Congés 1996: 299–422. The *varatio* is a method of land surveying based on the application of right-angled triangles with sides in proportion to integer numbers. The meridian line was the generating axis for each new land allotment, obtained from the indications given by Vitruvius (*De arch.* 6.1; Orfila et al. 2020: 128).

[69]Palet, Fiz and Orengo 2009: 108.

[70]1 *actus* = 120 *pedes* = 35.5 m.
15 x 20 *acti* = 532.5 m x 710 m = 378.075 m^2.
10 m^2 = 37.8075 ha.

[71]White 1970a. An important factor to consider in a *fundus* dedicated to the intensive cultivation of a certain product, such as vine-growing, is that the property remained polycultural, to an extent, to sustain its labour force. Thus, it is estimated that a family of six people would have to cultivate twelve *iug.* (≅ 3 ha), to cover their minimum food requirements for a year. If they also kept animals, twenty *iug.* were required (≅ 5 ha). Farmers, or those in charge of farms, would also be required to allocate the same amount of farmland to ensure the subsistence levels of staff and temporary labour, perhaps 10 to 14 people considering a ratio of one vintner for 10 *iug.* (cf. Pliny, *HN* 18.320), or one vintner for 7 *iug.* (Columella, *Rust.* 3.3.8; Martín-Arroyo and Remesal 2018: 217; Trapero Fernández 2021b: 128–38). All these variables, including palaeoenvironmental data and possible surplus yields of a certain product in a given territory under study, must be taken into account when making inferences to develop spatial, economic and econometric models.

dedicated to vineyards, including the facilities and infrastructure required. Laetanian vineyards included free-trained and head-driven systems, with grapevine trunks between 1.5 *pedes* (1.45 ft = 44.4 cm) and 3 *pedes* (2.90 ft = 88.8 cm) tall, in an *alveus* planting system using spur pruning or *gobelet*.[72] These features offered considerable advantages, including high productivity and lower labour and production costs during harvesting compared to competitors like those in Italy.[73]

To ensure applicability in terms of grapevine variety and its associated productivity relevant to Laetanian wine, we use here the Muscat of Alexandria or Roman Muscat. This is the modern variety closest to the *coccolobis hispana*, described by the Roman agronomists as cultivated in this territory.[74] It is predominantly a white, sweet grape of the fourth epoch with a large vegetative cycle of over 185 days and maturation period of 55+ days that should be harvested later than other similar varieties (between mid-September and mid-October).[75] It is used to make raisins and as table grapes but also produces elegant sweet and dry wines with a powerful floral flavour.

Regarding the vineyard itself, if we suppose a plantation frame of 1.5 x 1.2 m (≅ 4 × 5 *pedes* ≅ 4–5 vines/frame), the estimated number of plants would be 5,555 vines/ha (≅ 2,178 vines/acre ≅ 13,888 vines/*iug.*) (cf. Table 3). These grapevine density values are associated with common spacing in the Laetanian region in Roman times, but a 10 per cent reduction ought to be considered due to the use of land for margins and roads.[76] This leaves 5,000 vines/ha (or 1,960 vines/acre), equivalent in Roman terms to 1,259 *vitis/iug.* using the former and 1,215.2 *vitis/iug.* the latter. Thus, a standard 100 *iug.* Roman vineyard would contain *c.* 120,000 *vitis* per ½ *centuria*.[77]

To estimate the number of grape clusters per vine, so that we can calculate the yield per cultivar, we use the data for the Roman Muscat or Muscat of Alexandria: three branches per vine, with two spurs per branch, resulting in six spurs with two bunches on

---

[72]Columella, *De arboribus* 4.1.2. The Roman agronomists distinguish two ways of planting a vineyard according to the ploughing system used: either in *scrobes* (trenches) or *alveus* (small ditches). The most common plantation system in the Laetanian region consisted of two vines planted on both sides in a rectangular or ovoidal *alveus* of 4 x 1 *pedes* x 1 *pes* (deep). Free-trained and head-driven systems describe vines without any kind of support except, early in their lives, a vertical wooden stake positioned at each vine. When the vine has matured, the support is removed. A common pruning method adopted by Laetanian growers was spur pruning, also called the *gobelet* or open *vase* (see Martín i Oliveras and Revilla 2020: 94–95, Figure 2).

[73]It is possible that this cultivation method was imported to the Iberian Peninsula via Punic colonizers and adopted by the indigenous inhabitants of *Hispania Citerior* (Tchernia 1986: 127; Brun 2004b: 264).

[74]Columella, *Rust.* 3.2.19; Pliny, *HN* 14.29–30; Brun 2004b: 264. The Muscat of Alexandria is one of the most ancient and least hybridized grapevine 'mother' varieties in the world (Myles et al. 2011: Figure 3).

[75]Depending on the soil typology and microclimatic conditions of the vineyard, see: http://www.vitivinicultura.net/moscatel-de-alejandria.html. Some also believe that the ancient variety of *balisca* or *bilisca,* the *vitis Apiana,* cited by Pliny as 'the grapes that attract bees' (*HN* 14.24.81) and the *coccolobis hispana* were the same cultivar (García 1991: 219–21).

[76]Percentage of vine losses taken from: http://www.viverosmacaya.com/plantacion/

[77]Conversion values used here and below: one acre ≅ 0,405 ha; one hectare ≅ 2.47 acres; one *iug.* ≅ 0.620 acres; one *iug.* ≅ 0.2518 ha; one Roman *libra* ≅ 0.7109 lb; one Roman *libra* ≅ 0.32245 kg; one ton = 2,240 lb; one metric ton = 1,000 kg; one Roman wine *amphora* ≅ 81 Roman *librae* (Smith and Anthon 1854: 1024–30).

average.[78] Taking into account a loss of 10 per cent due to fruits used for raisins or table grapes, this results in 250 mt per ½ *centuria* ≅ 275.57 t ≅ 775,314 *librae*.[79] Then it is possible to transform these values into yields represented by *amphorae* and *cullei* per *iug*. The result is ± 9,251 *amphorae* and ± 4–5 *cullei*/*iug*.[80]

It is then possible to work backwards and calculate the minimum infrastructure and facilities that must be in place to process viticultural yields of this scale. Considering the data provided above by experimental archaeology, the treading and pressing infrastructure required to process a 100 *iug. fundus* includes a minimum of one large *calcatoria* and one Catonian wine press working from dawn to dusk for ±40 days.[81] If the infrastructure is doubled, with two *calcatoria* and two presses, the processing time is halved, taking 15–20 working days. Using these maximum production yields, it is possible to calculate the storage capacity required to ferment and age the wine. If we assume that the most common *dolium* capacity was *c.* 2–3 *cullei* (≅ 1048–1572 litres ≅ 230.52–345.79 gallons), then ± 150–250 *dolia* are needed to cater for the yield from a 100 *iug.* vineyard property.

Finally, by comparing these estimates to archaeological evidence from Gaul and coastal *Tarraconensis*, including the Laetanian region (see Table 14.1), it appears that facilities were usually designed to store only one harvest each year. However, the calculations above are designed to reach maximum production yields in a hypothetical 100 *iug.* Laetanian vineyard, but the quantity of grapes cultivated successfully and the number of *dolia* for storage could vary due to a range of intervening factors:

1. Regional diversity in property scales, dimensions, agroecological conditions and land tenancy regimes.
2. Grapevines could be managed by the winegrower to produce more or less fruit to obtain different qualities of wine.
3. Portions of must might be set aside to create grape or wine derivatives (e.g. *defrutum*, *caroenum*, *sapa*).
4. The force and quantity of pressings as well as storage and decanting in other vessels made from organic materials (e.g. wineskins, wooden vats, barrels), all of which leave little archaeological evidence.

---

[78]This is typically estimated based either on the buds remaining at pruning (therefore, potential shoots/vine) or the typical cluster count/shoot for the grape variety. It can also be estimated in season using cluster counts on selected vines within a block. Often, it is done at both times, particularly in years where winter damage has resulted in significant bud damage, or when early frosts have damaged emerging shoots.

[79]One vine = twelve bunches x 200 g ≅ 0.440 lb (average) = 2.4 kg/*vitis* ≅ 5.29 lb (average) x 1,200 *vitis*/*iug.* (average) = 2,880 kg/*iug.* ≅ 6349.31 lb = 288 mt per ½ *centuria* ≅ 317.46 t.

[80]250,000 kg ≅ 551155.65 lb.

0.32245 kg/*libra* ≅ 0.71088 lb/*libra* = 775,314 *librae* per ½ *centuria*.

81 Roman *librae* per wine *amphora* = 9,571 wine *amphorae* x 26.2 L ≅ 5.76 gallons = 250,781 L ≅ 55164.10 gallons.

100 L ≅ 21.99 gallons = 2,508 hl per ½ *centuria* ≅ 22321.08 gal/acre ≅ 25 hl/*iug.* ≅ 549.92 gal/*iug.* ≅ 4–5 *cullei*/*iug.*

[81]2,500 hl ≅ 54992.31 gallons. ± 63 hl ≅ 1385.80 gallons (daily must yield processed per *calcatorium* and Catonian winepress) = ±30–40 working days.

Discussion

Using these hypothetical calculations as a foundation, it is possible to calculate estimations for the total *Ager Barcinonensis* from a macrospatial point of view in maximum values. Considering that a grapevine yield of 2.5 mt/ (≅ 3.08 t/*iug.*) is equivalent to a wine yield of 15 hl/*iug.* (≅ 329.95 gal/*iug.* ≅ 2–3 *cullei*/*iug.*), we can estimate that the maximum potential viticultural yield of the *Ager Barcinonensis* was 37,500 mt of grapevines that could produce between 30,000–45,000 *cullei* of must and 600,000–1,000,000 *amphorae* of wine.[82]

Rates of local wine consumption forms an important, albeit difficult to measure, variable, so that the balance between intra- and extra-regional consumption can be gauged. It is often accepted that approximately 250 litres of wine were consumed per person per year during the Roman period,[83] which, using the above production estimates for the area, caters for ±60,000 to ±100,000 local individuals each year. It has been estimated that the total urban population of the Laetanian region was *c.* 10,052 individuals, which supposes an estimated rural population of *c.* 30,756 inhabitants if we accept the hypothesis that the urban population would be approximately 25 per cent of the total.[84] Alternatively, if a density of 7 or 11–12 people/kilometre^2 is applied, the overall population of the Laetanian region is between 16,408 and 34,567 inhabitants.[85]

This latter population densities provide a total estimated local wine consumption of ±7,828 to ±16,969 *cullei* or ±156,560 to ±339,380 *amphorae* per year for the permanent population and temporary workforce, which leads to excess production of between ±20,000 to ±25,000 *cullei* or 400,000 to 500,000 *amphorae* per year for potential export. It seems, therefore, that these estimations indicate an agency-based intensive and specialized viticultural production model in the *Ager Barcinonensis* and also in the Laetanian region aimed at commercialized overseas trade.[86]

---

[82]±150 parcels of 20 x 15 *acti* (≅ 150 *iug.*) = 22,500 *iug.* 22,500 *iug.*–7,500 *iug.* (considering subsistence crops, housing and facilities) = 15,000 *iug.* of total potential vineyards x 2.5 mt/*iug.* (≅ 2.46 t/*iug.*) = 37,500 mt (max. grape productivity).
15,000 *iug.* x 15 hl/*iug.* (≅ 329.95 gal/*iug.*) = 225,000 hl ≅ 4,949,308 gallons ≅ 157,120 wine barrels. Using 5.24 hl/*culleus* (≅ 115.26 gallons) = 42,939 *cullei* (≅ 858,778 *amphorae*).

[83]Purcell 1985: 1–19; Trapero Fernández 2021b: 133.

[84]Carreras 2014: 53–82.

[85]Martín i Oliveras et al. 2020: 61–2. After the Idescat dataset (https://www.idescat.cat), the territory occupied by the ancient Laetanian region has an extension of 2304.5 km^2 (including the actual shires of Baix Llobregat, Barcelonès, Maresme, Vallès Occidental and Vallès Oriental). In the modern census called 'fogatges', the population densities for these territories, including cities, documented between 11–14 people/km^2 in 1497 and between 16–20 persons/km^2 in 1553 for the Maresme shire (Iglesias 1979). The estimation includes the needs of temporary workforce during the *vindemia* (*c.* 2,000–3,000 adults over two months), so the consumption rate could be increased by ± 1,000 hl/year (Carreras 2014: 53–82; Martín i Oliveras et al. 2020: 61–2).

[86]Martín i Oliveras and Revilla 2019: 152–4; Martín i Oliveras et al. 2020: 61–2.

Conclusion

If trying to establish modern viticultural crop and wine yields is arduous and difficult due to intervening agents, factors and variables that can influence productivity, any attempt to do it for Roman viticulture is even more challenging. This is made more so by gaps in our knowledge and missing data that influences parameters, including the palaeoclimatic conditions, real size of cultivated fields, grapevine training methods and the varieties, vegetative cycles and production yields of grape cultivars. Hence, we must interpret this missing data as best as possible from the descriptions provided by Roman agronomists, which are often imprecise and at times contradictory, the archaeological record, experimental archaeology (sometimes partial and imprecise due to added layers of interpretive bias), or from ethnographic studies that are almost always made in slightly different situations and conditions to that from antiquity. Nonetheless, analysing each aspect of this applied methodology and moulding them to specific regional contexts might allow us to begin to understand, reconstruct and model the socio-economic dynamics of the Roman vinicultural ecosystem in different territories during the Roman period.

Regardless of the difficulties and biases present, it is a worthwhile exercise to attempt to come to grips with these quantifications, as has been previously recognized.[87] Estimations fluctuate in accuracy depending on the quantity and quality of the available datasets, but they almost always provide useful patterns with the potential to be applied to modelling different ancient production and economic processes. As was illuminated in this chapter, one of the most significant of these, with promising future research potential, is modelling the Roman viticultural supply chain and production function in different situations and chronological contexts.

[87]cf. 'les rendements' in Amouretti and Brun 1993: 552.

CHAPTER 15

RECONSTRUCTING VINEYARD GEOGRAPHY, YIELDS AND THE PROFITABILITY OF WINE PRODUCTION IN THE ROMAN EMPIRE: NEW INSIGHTS FROM SPATIAL ANALYSIS, AND AGENT-BASED AND CLIMATE-VEGETATION MODELLING

Nicolas Bernigaud, Alberte Bondeau, Joël Guiot, Laurent Bouby, Frédérique Bertoncello and Marie-Jeanne Ouriachi[1]

Abstract

This chapter presents different types of modelling developed to reconstruct the geography of areas potentially favourable to ancient viticulture, as well as to assess the impact of climate change on yield variations, profitability and the geography of wine production. This work is based on methods of spatial analysis, as well as agro-ecosystem and climate-vegetation models, as developed in the field of geosciences, and agent-based modelling. We focus on Roman Gaul here, but the experimental methods developed can, of course, be used for other study areas and at other geographical scales. The results support the hypothesis of a potentially positive effect of Roman-era global warming on wine yields and profitability, and impact in Late Antiquity of subsequent cooling in Gaul in paradox to the contemporary booming eastern Mediterranean wine industry. Various ways forward considering this and other contradictions are proposed, along with the potential for these models to aid in the strategic targeting of future fieldwork across areas of particularly promising viticultural activity.

Introduction

Historical and archaeological research carried out across the Mediterranean and western Europe gradually allows us to expand our knowledge of the genesis, development and geography of viticulture in the Roman Empire. While it is worth highlighting progress made in recent decades, our knowledge in this area is still lacking and geographically disparate. Viticulture is often difficult to identify based on sometimes controversial or tenuous archaeological evidence (e.g. production facilities, grape seeds or remains of vineyards), and the intensity of research varies greatly from one country to another (see

Chapter 1). The modelling methods developed in the fields of spatial archaeology and geosciences offer the possibility of freeing ourselves from these various archaeological biases of research and interpretation that blur our understanding of ancient viticulture. These methods now make it possible to reconstruct the geographical extension and spatio-temporal dynamics of wine regions, but also to analyse the historical, socio-economic and environmental factors related to their evolutions.

In this chapter, we will present distinct types of modelling that have been developed to reconstruct the geography of areas potentially favourable to ancient viticulture, as well as to assess the impact of climate change on yield variations, profitability and the geography of wine production.[2] This work is based on methods of spatial analysis, as well as other particularly innovative types of modelling such as agro-ecosystem and climate-vegetation models developed in the field of geosciences, but also Agent-Based Modelling (ABM). Most of the work presented here has Roman Gaul as its geographical setting, not only because the production of wine was especially important here in the Roman period but because we have an easy access to enough necessary data. The experimental methods developed for this area can, of course, be used for other study areas and at other geographical scales, including the wider Roman Empire.

Reconstructing the spatial extension of viticulture: spatial analysis and suitability map processing

For over thirty years, tools and methods of spatial analysis derived from geography have been applied in archaeological research to develop models to predict the location of human occupations of different periods in different landscape units.[3] Mainly used from a heritage management perspective, these methods bear potential to illuminate more specific scientific issues such as the reconstruction of the geography of wine production in different provinces of the Roman Empire, such as *Hispania Tarraconensis*, another important western Mediterranean vinicultural region, where recent work has also been undertaken.[4]

---

[1]The research presented in this article was conducted as part of the multidisciplinary project RDMed (Resilience and Adaptation to Droughts and Extreme Climate Events in the Mediterranean: Lessons Learnt from Past on a 1.5°C or Warmer World) and Labex OT-Med both funded by the A*midex Foundation of the University of Aix-Marseille (France). The authors would like to thank the other participants of this project with whom regular exchanges made it possible to advance multidisciplinary discussions and reflections on the issue of the adaptation of ancient societies to climate change over the past three millennia: Loup Bernard (University of Strasbourg), Delphine Isoardi (CCCJ-CNRS, Aix-en-Provence), Philippe Leveau (Centre Camille Jullian, Aix-en-Provence), Alan Kirman (EHESS, Paris), Sylvain Olivier (University of Nîmes), Cees Passchier (University of Mainz), Gül Sürmelihindi (University of Mainz) and Daniel Contreras (University of Florida).

[2]On the importance of climate and variations in temperature and precipitation to grapevines, see Fraga 2019.

[3]E.g. Vita-Finzi et al. 1970; Tobler 1979; Verhagen and Whitley 2020.

[4]cf. Stubert et al. 2020.

We have designed and developed for Roman Gaul a suitability map for vineyards with the R software (and raster package). This cumulative model is based on the addition of six geographical, environmental and climatic criteria considered influential in the establishment of a commercial vineyard in the Gallo-Roman period:

1. Temperatures over 18°C in July;
2. Land oriented to the south and east;[5]
3. Proximity (< 5 kilometres) to the coastline for distribution;
4. Proximity (< 5 kilometres) to a major road for distribution;
5. Proximity (< 5 kilometres) to a river for distribution; and
6. Proximity (< 30 kilometres) to an urban centre, constituting a major place of consumption and redistribution.[6]

The values of these parameters were chosen according to ancient Latin written sources, geographical features of Roman period (e.g. roads, towns) and palaeoclimatic reconstruction.[7] The value defined for each parameter is binary: it can be either good for viticulture (= 1) or not (= 0). There are no mid values (neither good, neither bad). Moreover, we have considered all parameters equally important and decided not to make a weighted model. Obviously, these choices lead to an oversimplification of the reality and a more sophisticated model could be developed on this basis, but the simplicity of this one seems powerful, as indicated by the validation of results.

The suitability map obtained for the High Empire makes it possible to distinguish areas where the variable value of potential (from 0 to 6) is proportional to the number of criteria stacked in each cell. While the high-altitude zones (Alps, Pyrenees, Massif Central) as expected show zero potential (= 0 criteria), as well as some northern plains (Orne, Seine-Maritime, Somme), the most favourable areas to viticulture (> 3 criteria) appear in the southern Mediterranean, the Rhone-Saone axis, the Rhine Valley, the south-west as well as further north in the Paris Basin and the Pays de Loire (Figure 15.1).

The robust nature and validity of this model and method were evaluated by comparison with archaeological data that indicate viticulture and wine production in this region,[8] using the statistical test of relative gain, the simplest and most frequently employed in predictive modelling.[9] The values obtained with this index (from −2.04 to

[5]According to Columella, *Rust.* 3.12.
[6]See the model and input files: https://data.mendeley.com/datasets/j4gtjy9w87/2
[7]See Bernigaud, Bondeau and Guiot 2021.
[8]The archaeological and archaeobotanical evidence that attests the presence of Roman viti- and viniculture in Gaul (e.g. remains of wineries, viticultural tools, traces of vineyards, *amphora* workshops and grape seeds) were mainly extracted from several regional syntheses (e.g. Brun and Laubenheimer 2001; Brun 2005; Brun, Poux and Hervé-Monteil 2011), as well as relevant databases, including the Outagr database dedicated to agricultural tools: http://outagr.huma-num.fr/Outagr/
[9]Verhagen 2008.

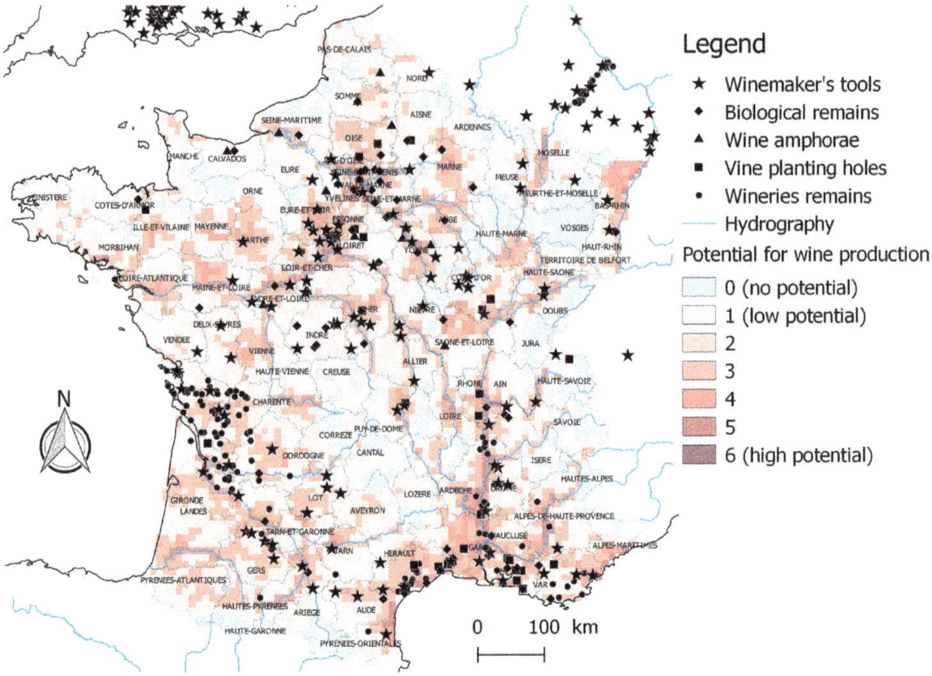

Figure 15.1 Suitability map for the establishment of vineyards in Roman Gaul in the High Empire, overlaid with locations of (bio)archaeological data attesting to viticulture. After Bernigaud et al. 2021.

0.73) for each of the seven levels of potential (from 0 to 6) indicate a very good geographical correlation between the results of the model and the archaeological data.[10]

The main aim of this model is to highlight potentially favourable areas for viticulture where archaeological data attesting such activities are scarce or absent. This is particularly the case in several areas of the south-west (Pyrénées-Atlantiques, Landes) and the northwestern quarter (Loire-Atlantique, Sarthe, Mayenne, Ille-et-Vilaine) of France (Figure 15.1). Was viticulture not practiced in these regions or are we simply seeing the existence of research bias? It should be remembered that in France, as elsewhere, our current state of knowledge is based on publications of regional syntheses that are still far from exhaustive. Therefore, the apparent data vacuum in some regions could simply be explained by a lack of publication or research on the subject. The archaeological visibility of viniculture is also relatively good for this region due to the diversity of winemaking techniques used, climatic, and cultural reasons. For example, while vinification and storage were carried out in southern Gaul using *dolia* in cellars, with their comparatively

[10]Bernigaud, Bondeau and Guiot 2021.

easily identifiable remains, in temperate Gaul these operations are hypothesized to have occurred in wooden barrels, which are rarely preserved. Viticulture generally, therefore, remains more difficult to archaeologically identify in northern regions than in the south.

While the model bears promise for those regions where data is scarce or absent, these results must only be considered to have heuristic, potential predictive value, and don't prove anything until validated by more concrete data. They are designed to stimulate not only archaeological research in the field, but also within reports of unpublished excavations (e.g. preserved in the Regional Archaeology Services).

Reconstructing favourable areas for viticulture and potential yields from climate-vegetation models: LPJmL and the Viticulture Index

Lund-Potsdam-Jena-managed-Land (LPJmL) agro-ecosystem modelling: reconstructing potential wine yields as a function of climate

LPJmL agro-ecosystem modelling was initially designed by German and Swedish environmental scientists to model current carbon and water cycles on a global scale, as well as yields of the main crops based on daily climate data (e.g. temperature, precipitation, cloud cover) and environmental data (e.g. porosity and soil water reserve). The process-based dynamic global vegetation and hydrology model, LPJmL, calculates carbon and the corresponding water fluxes with a daily time step and a spatial resolution of 0.5 x 0.5 (latitude/longitude). Potential natural vegetation and the main processes controlling its dynamics are calculated from inputs of climate data (e.g. temperature, precipitation and cloud cover), atmospheric CO_2 and soil texture.[11] More recently, LPJmL has been adapted to study past situations, using palaeoclimatic data to calculate potential yields for cereals during Protohistory in southern France.[12]

On the methodological basis of this pioneering work, we conducted first trials with LPJmL to measure the impact of Roman Climate Optimum (RCO) on wine yields.[13] Potential yields were calculated for a rainfed crop and an irrigated crop from values of temperature, precipitation and cloud cover. These palaeoclimatic data were reconstructed from modelling based on the current climate of France (Aire Limitée Adaptation Dynamique Développement International [ALADIN] climate model), modified with anomaly values extracted from a palaeoclimatic reconstruction carried out on a European scale.[14]

---

[11]Bondeau et al. 2007; Fader et al. 2015.
[12]Contreras et al. 2018, 2019.
[13]McCormick et al. 2012; Bernigaud, Bondeau and Guiot 2021.
[14]Guiot and Kaniewski 2015.

Calculations made for the coldest century (sixth century BCE) and the hottest century (first century BCE) of the last three millennia indicate large differences in value. For the sixth century BCE, the potential yields of the vine are between 0–20 hl/ha, while they reach up to 30 hl/ha in the first century BCE in rainfed cultivation (i.e. one-third more) (Figure 15.1). In irrigated cultivation, they increase from 30–45 hl/ha in the coastal plains to 45–60 hl/ha, which represents a growth of one-third to one-half depending on the sector. In both cases, the potential yields decrease with altitude to reach zero values in the Massif Central, the Alps and the Pyrenees. They still appear the highest in the Languedoc and Roussillon plains, the middle and lower Rhone valley, as well as on the Var coast (Figure 15.2). These areas of higher potential, only based on biophysical criteria, coincide well with the results of our predictive model built with other criteria (Figure 15.1). However, some differences should be noted: the high potential that emerges from the predictive model in the Alpes-Maritimes appears for example much more moderate with LPJmL (Figure 15.2).

In general, according to the first results obtained with LPJmL the potential yields of the vine could have increased very sharply – from one-third to double – between the coldest century (sixth century BCE) and the hottest century (first century BCE) of the last three millennium. For *Gallia Narbonensis*, where viticulture reached its peak during the High Empire, archaeological investigations show that this development occurred from the first century BCE.[15] If this development appears to follow the Roman conquest of

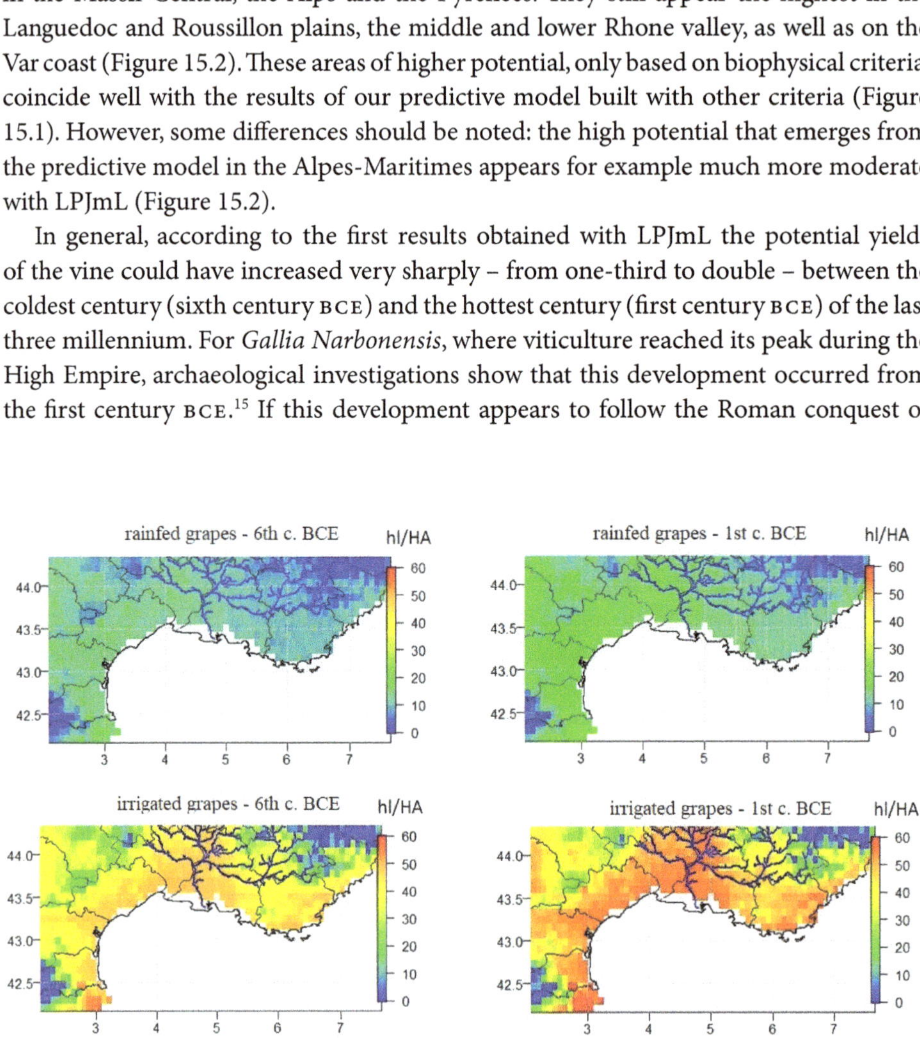

Figure 15.2 Maps of potential grapevine yields calculated by LPJmL for irrigated and rainfed cultivation in southern Gaul according to paleoclimatic data. After Bernigaud et al. 2021.

[15]Carrato 2017b.

southern Gaul (in the 120s BCE), the results of our models now allow us to visualize that climate, together with socio-historical and political events, could have played a decisive role for the growth of commercial viticulture.

The Viticulture Index (VI)

The Viticulture Index (VI) is another recently developed and innovative method for defining areas favourable to the development of the grapevine according to bioclimate.[16] This index (whose values vary between 0 and 1) is calculated from threshold annual values of precipitation, temperature, evapotranspiration and values of net primary production.[17] The climate data used are produced from an ensemble of Earth System Models (ESMs), coupled to a global vegetation model (Biome 4) constrained by palaeoclimatic reconstruction carried out from pollen data.[18]

The VI was calculated for different periods of the Holocene at the scale of the Mediterranean and western Europe to assess the impact of climate change on the biogeography of the vine. The results obtained with the climatic characteristics of the first century BCE show, for example, the highest probabilities in the Mediterranean islands (Balearic Islands, Corsica, Sardinia, Sicily, Elba, Crete, Cyprus, Rhodes), on the continental coasts of the Tyrrhenian, Adriatic, Ionian, Aegean seas, as well as on the Levantine, Tunisian, Algerian and Moroccan shores. The values also appear high on the Atlantic coast (Portuguese, Spanish and French coasts). On the other hand, they are low or zero on the African coasts of the eastern Mediterranean (Libyan and Egyptian coasts) and continental areas, with the notable exception of the Iberian Peninsula and a large part of the French/Gaul territory which show both potential in the large river valleys (Tagus, Ebro, Douro, Rhone, Loire) (Figure 15.3).

To measure more precisely the predictive potential of this index on the geography of vineyards, it would be useful to compare it with an exhaustive mapping of archaeological and historical data at the scale of the Roman Empire. While this long work of data collection remains to be undertaken, it can already be roughly seen that viticulture is attested by archaeological evidence and texts in many areas where the model indicates that climatic conditions allowed the development of the grapevine. Considering the

---

[16]We propose the following index: $VI = (1 - I_{NPPtrop}) \cdot I_{NPP} \cdot I_{Pann} \cdot I_{MTWA} \cdot I_{MTCO} \cdot I\alpha$
Where each of the factors denoted I_x follows this function:
$I_x = 0$ *for* $X < X_{min}$,
$I_x = 1$ *for* $X > X_{mix}$,
$I_x =$ [eqn_297] *for* $X_{min} < X < X_{max}$
and NPP_{trop}, the net primary production of tropical plants with interval $[X_{min}, X_{max}]$ equal to [0, 10 kg m^{-2}]; NPP, the total ecosystem net primary production with interval equal to [500, 1000 kg m^{-2}]; P_{ann}, the total annual precipitation, with interval equal to [400, 800 mm]; MTWA, the mean temperature of the warmest month with interval equal to [18, 23°C]; MTCO, the mean temperature of the coldest month with interval [3, 12°C]; α, the actual to potential evapotranspiration ratio with interval equal to [30, 60%] (see Guiot et al., forthcoming).
[17]Guiot et al., forthcoming.
[18]Kaplan, Prentice and Buchman 2002; Kay et al. 2014; Guiot and Kaniewski 2015.

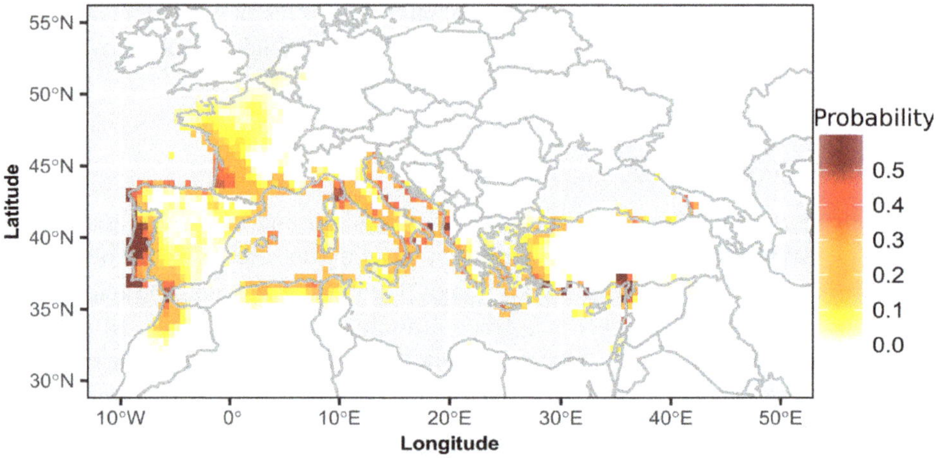

Figure 15.3 Map of climatic potential for viticulture according to the Viticulture Index in the Mediterranean and western Europe for the first century BCE. After Guiot et al. 2023.

range of vinicultural structural, ceramic and archaeobotanical evidence, this can be seen for the regions of *Hispania Tarraconensis* and *Baetica*,[19] Gaul (*Narbonensis*, *Aquitania*, *Belgica*) (Figure 15.1), Italy and the Adriatic region,[20] and the Ionian, Aegean and Levantine coast.[21] The viticulture of most Mediterranean islands in Roman periods is also relatively well attested by historical and archaeological data.[22]

To our knowledge, however, no, or few data are known for some regions where this index displays high values; for example, in the case of the northwest coast of Spain, the Ebro Valley, and northern Morocco, Algeria and Tunisia. Additionally, while ancient viticulture has been identified in Lusitania (Portugal),[23] the generally moderate importance of the production perceived by archaeological investigations does not correlate to the very high climate probabilities of the Index (Figure 15.4). Conversely, it should also be noted that the probabilities appear rather modest in the eastern Mediterranean, where important wine regions were undoubtedly located in Egypt and Israel.[24]

For regions with good probability but where data is scarce or absent efforts may be made to verify on a case-by-case basis whether this deficit could be linked to research bias. In the case of northern Morocco, *Mauretania Tingitana* remains often perceived as

[19]E.g. Brun 1997; Martín i Oliveras, Martín-Arroyo Sánchez and Revilla Calvo 2017.
[20]Van Limbergen 2011, 2019; Dodd 2022a.
[21]E.g. Komar 2016; Dodd 2020a.
[22]See contributions in Van Limbergen et al., forthcoming.
[23]Brun 1997.
[24]cf. Ayalon, Frankel and Kloner 2009; Dzierzbicka 2018; Fuks et al. 2020.

an olive oil producing region, above that of wine. Recent excavations, however, conducted on the tell of Rirha in the region of Sidi Slimane near the ancient city of *Volubilis*, have uncovered a press with a highly suggestive vinicultural attribution thanks to the identification of large quantities of charred grape seeds, as well as biochemical analyses.[25] This discovery, therefore, provides an insight into wine production in this Roman province where our VI indicates good climatic potential. Conversely, the identification of areas where the Index gives zero values, but where Roman-era viticulture clearly occurred (e.g. Egypt, Israel), should make it possible to critically improve aspects of calibration. These discrepancies may also be impacted by local conditions favourable to the grapevine but unable to be detected due to the model's analytical resolution (*c.* 50 kilometres) or anthropic adaptations to climatic conditions (e.g. irrigation technologies).

Calculating the potential benefits of winemaking: the agent-based model ROMCLIM

Arising from the United States in the 1990s in the field of archaeology, ABM makes it possible to carry out dynamic simulations through the programming of agents that interact with each other and with their environment, thus producing phenomena that we try to reproduce, or unexpected 'emerging' phenomena.[26] The ROMCLIM agent-based model was developed to simulate at the scale of southern Gaul the impact of global warming in the Roman period (RCO) and the cooling of Late Antiquity (LALIA) on agricultural production and its potential income.[27] This model, in which the agents are virtual farms, has the particularity of reproducing in a simplified way the more complex functioning of the agro-ecosystem modelling LPJmL (above).

The wineries we have programmed operate with a staff of sixteen people over an area of 100 *iugera* (25 hectares), corresponding to the main characteristics given by the agronomist Cato (*Agr.* 11) in the second century BCE in Italy. For each century of the chronological period considered (sixth century BCE to the seventh century CE), the annual production (averaged by century) is calculated by multiplying the area exploited by the value of the potential wine yield, provided by an emulation function of LPJmL. The market value of the annual wine production is then calculated by multiplying its quantity by a market selling price. This price is chosen by the user in an interval (0.57–240 *sesterces*/L) provided by historical sources.[28] The profits generated (in *sesterces*)

[25] Carrato et al. 2020.
[26] E.g. Lake 2020; Romanowska, Wren and Crabtree 2021.
[27] McCormick et al. 2012; Büntgen et al. 2016; Bernigaud et al., forthcoming. The NETLOGO code is available at: https://github.com/Bernigaud2021/ROMCLIM/blob/NETLOGO/ROMCLIM2021.nlogo
[28] In the first century CE, Columella (*Rust.* 3.3.2) estimated that 40 urns (or 1 *culleus*) of wine sold for at least 300 *sesterces* (0.57 *sesterces*/L). We therefore chose this value as the minimum price. The maximum is the price of the old Falernian, established at the beginning of the fourth century CE by the edict of Diocletian at 34 *denarii* per *sextarius* (240 *sesterces*/L).

by each of these wineries are obtained by subtracting from this market value operating and transport costs to the nearest capital city.[29]

According to multiple simulations, the yields and potential incomes of the grapevine appear particularly sensitive to climate change.[30] Overall, the results indicate a sharp increase in yields and potential benefits as early as the third century BCE, at the beginning of the RCO, before a decrease during Late Antiquity (fourth to seventh century CE) due to LALIA. The model's visualization shows the areas where the profitability of the wineries is higher or lower depending on the climate, but also their distance to the capitals of cities (proportional to transport costs).

In general, simulations carried out with the climatic characteristics of the High Empire show that the Roman colonies located in the Rhone valley (Orange, Avignon, Glanum) as well as Gard area (Nîmes, Uzès) and pre-Alpine foothills (Vaison, Carpentras, Cavaillon) form the heart of the most profitable area, which extends to the west into the Languedoc (Béziers, Narbonne) and Roussillon (Perpignan) plains (Figure 15.4). This strong

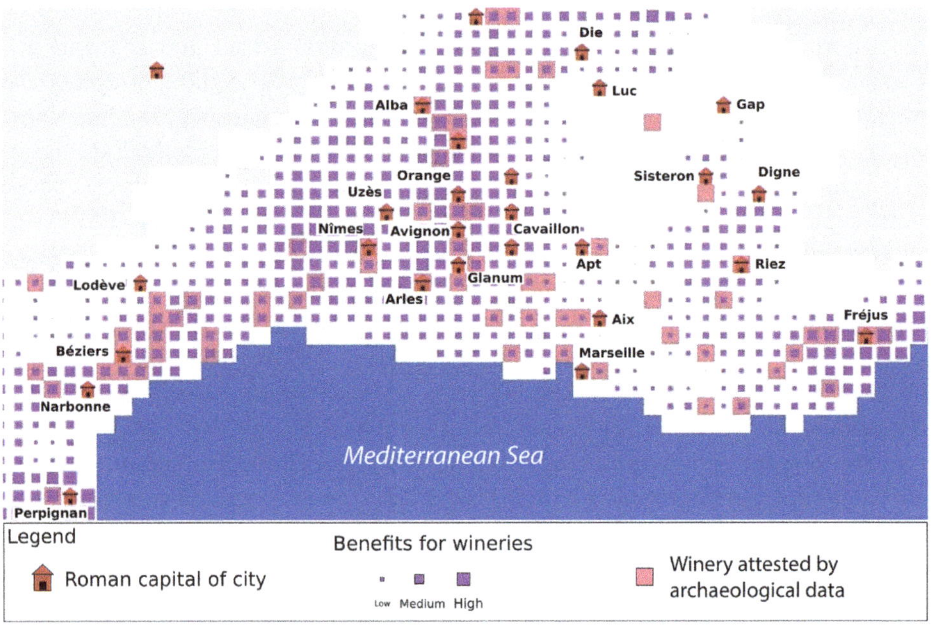

Figure 15.4 Map of potential profit obtained for southern Gaul with climatic data of the first century CE (1900 BP). Each winery is represented by a purple square whose size is proportional to the profits made, and the cells where viticulture is attested by archaeological data are indicated in pink. Courtesy N. Bernigaud.

[29]The annual cost for the maintenance of staff is based on 140 *sesterces*/individual/year or 2,240 *sesterces*/year for the sixteen people on the property. Transport costs are defined by the user within a range whose maximum given by Diocletian's Edict is 20 *denarii*/mile/1,200 kg (0.145 *sesterces*/km/kg).

[30]For more details regarding sensitivity analysis and the distribution of outcomes, see Bernigaud et al., forthcoming.

potential also emerges in the southeast of *Narbonensis* (Fréjus), as far as the Alps (Digne, Riez). On the other hand, it appears unsurprisingly close to zero at high altitude (Pyrenees, Hautes-Alpes), but also in a region east of Aix-en-Provence and Marseille (Figure 15.4).

It should be noted that archaeological surveys conducted in recent decades found viticultural evidence in these areas of highest potential, including the middle and lower Rhone valley, plains of Languedoc and Roussillon, and Var.[31] It is interesting to note that little or no data are known in some of the regions favourable according to the model; for example, in the case of the Cevennes hinterland (west of Nîmes and Uzès), the city of Antibes (east of Fréjus), the alpine area between Riez and Digne, or the Perpignan area (southwest). Here, the results have a heuristic value that raises new questions and encourages further research: are they simply an effect of the model or do they indicate a past reality? Again, this encourages efforts to verify whether the scarcity or absence of archaeological data is related to a lack of research or publications. The positive signs of potential displayed, however, could also be an artefact of the model, unless it has not been exploited for historical reasons that remain to be clarified.

Conclusion and future perspectives

The different types of modelling presented in this chapter offer potential for analysis that opens new perspectives for research on ancient viticulture by their ability to integrate climate, to which the vine appears particularly sensitive. These preliminary exploratory models can be improved, hybridized and adapted to the scale of the Roman Empire to reconstruct the geography of wine regions, but also to analyse the causes of their development and decline. After the booming period of the High Empire, it is thought that the vineyards of the Italian Peninsula, Iberia and Gaul were affected in the early third century CE by a relative downturn, coinciding with the development of viticulture in North Africa and then in the eastern Mediterranean (e.g. Cilicia, Cyprus, Gaza, Egypt) where it enjoyed great prosperity during Late Antiquity from the sixth to seventh centuries CE.

The first results of the models we have developed support the hypothesis of a potentially positive effect of Roman-era global warming on the yields and profitability of commercial viticulture, which could, on the other hand, have been significantly affected in Gaul by the subsequent cooling of Late Antiquity. However, the rise of the eastern Mediterranean wine regions at the end of antiquity suggests on the contrary favourable climatic conditions. To try to resolve this apparent paradox, several avenues can be considered. In the warmer regions of the Mediterranean, climate cooling may have favoured the development of viticulture, to the point of making it possible in the Negev desert.[32] But it must also be considered that the climates of the western and eastern

[31] Buffat et al. 2001; Brun 2001; Jung et al. 2013.
[32] cf. Fuks et al. 2020.

Mediterranean sit in contrast.[33] The effect of climate change was therefore not necessarily the same.

On the methodological basis of the exploratory examples presented in this chapter, other models could be developed to identify more broadly the different wine regions of the Roman Empire, the chronology of their spatio-temporal evolution, as well as various factors involved in these changes. The design of a predictive model will benefit from the one we have developed for Gaul (above), but also from those recently proposed for the Laetanian vineyards of northeast Spain and for the Adriatic region.[34] By combining these different methods and considering an increased number of historical-geographical, environmental and palaeoclimatic criteria, potential maps produced for different periods (High Empire, Lower Empire/Late Antiquity) would make it possible to identify regions where viticulture is still underrepresented or even unknown. Dynamic simulations carried out with a hybridized ABM, integrating the results of predictive and agro-ecosystem models, would finally offer the possibility of analysing not only the potential role of climate change in the reorganization of the Empire's wine regions, but also that of other historical, socio-economic, demographic and environmental factors which remain under-integrated in these first models.

---

[33]Guiot and Kaniewski 2015.

[34]Stubert et al. 2020; McLean in this volume.

CHAPTER 16
MODELLING VITICULTURE IN THE ADRIATIC REGION: A QUANTIFICATION OF AGRICULTURAL SUITABILITY

Andrew McLean[1]

Abstract

This chapter analyses how different factors, including slope, aspect and proximity to infrastructure, affect the distribution of Roman agricultural sites in the Adriatic region based on a proxy for agricultural suitability. The chapter directly compares two study regions – the Adriatic and South Etruria – and provides a quantitative approach for understanding agricultural suitability that can be applied across the ancient world. The analysis demonstrates that preferences for slope are fairly consistent across the two regions, but that this aspect appears to have been more important in South Etruria than the Adriatic, where the changeable Bora and Mistral winds made a preference for direct southerly slopes less apparent. Additionally, slope and aspect are shown to be potential indicators to aid in the identification of wine production rather than oil production at Roman sites.

Introduction

At its core, the ancient economy was an agricultural one. The cultivation of crops, particularly grains, olives and vines, was the cornerstone of the economy and permeated all levels of society.[2] Understanding the cultivation of such crops, therefore, provides invaluable insight into the wider society and economic system of the Roman world. While the so-called Mediterranean triad was important everywhere in the Empire, the Roman world was not a homogeneous landscape where any crop could be grown the same way with the same result. The diverse geography and ecology of the landscape resulted in different levels of agricultural suitability across the Roman world.[3] While this concept of varying suitability has often been discussed in studies of the ancient economy,

---

[1]Please contact the author for additional information, including access to full datasets and appendices associated with this work.
[2]Purcell 1985: 2; Brun 2003.
[3]See especially Horden and Purcell 2000.

these are usually qualitative in nature, though quantitative studies are becoming more common.[4] Additionally, there has been a tendency to rely on the problematic ancient agronomic texts as a basis for analyses; instead, it can be more useful to use these sources as a point of comparison for the outputs of analyses. Indeed, by applying more scientific and quantitative approaches, we can expand upon and inform these qualitative studies. By quantifying agricultural suitability, we can more meaningfully compare regional and chronological trends, and, importantly, begin to identify viticulture and wine production in a more scientifically informed manner, including for areas where intensive archaeological investigation and analysis has not yet been possible.

This chapter presents a methodology for quantifying agricultural suitability in the Roman world. Applying this quantification to the Roman Adriatic allows for the creation of a model of viticulture in the region. The data produced through this approach is invaluable for informing and building upon more qualitative analysis. To demonstrate this, I first outline some previous approaches to ancient agricultural suitability and provide a brief overview of the study region. Following this, the materials, methods and results are detailed, covering the data used to quantify agricultural suitability, and the Adriatic data used to model viticulture. All this helps to create a framework for approaching agricultural suitability in the Roman world, and highlights some of the specific trends presented by viticulture and wine production.

Agricultural suitability

Multiple studies have sought to determine the agricultural suitability of different areas of the Roman world.[5] These often rely on the ancient sources, and their descriptions of various parts of the Empire. Indeed, ancient agronomists make frequent references to what conditions they believe to be most suitable for the cultivation of particular crops, including grapevines. Columella (*Rust.* 3.1.3) praises the vine 'not only for the sweetness of its fruit' but for the hardiness of the plant, claiming that it can be cultivated in most regions and climates, other than the excessively hot or cold, whether on flat plain or hillside. He expands on this, claiming that for grapevines, warm climates are better than cold, gentle slopes are better than flat or steep terrain, and drier conditions better than waterlogged (*Rust.* 3.1.8–10). At the centre of much of these discussions is the economic viability of vineyards in specific terrain. Depending on whether quantity or quality of wine is the main concern, Columella (*Rust.* 3.2.6) suggests gently sloping plains on a northern slope or slightly steeper hillsides on a southern slope respectively. Indeed, Cato (*Agr.* 1.3) suggests that a southern slope near transport routes or a settlement

[4]For some influential, but essentially qualitative studies, see Rossiter 1981, 1998; Brun 2003, 2004b; Goodchild 2009; Bowman and Wilson 2013; Brun et al. 2020. For some more quantitative studies, see Stubert et al. 2020; Trapero Fernández 2016; Angourakis et al. 2018.

[5]Van Joolen 2003; Thomas and Wilson 1994; Goodchild 2009, 2019.

is the best location for any profit-motivated agricultural venture. Thus, even brief analysis of the ancient sources illuminates specific factors as vital considerations to ensure a productive, and, importantly, profitable, vineyard. These factors include slope, aspect and proximity to market or transport infrastructure.[6] In a study of the productive landscape of central Italy, slope, aspect and proximity to roads and urban centres were all used in order to understand the landscape, and a strong correlation between archaeological sites and the conditions outlined by the agronomists identified.[7] Cato's recommendations were then used to generate a 'suitability map', though the generation of this map was 'essentially qualitative'.[8] Such approaches are extremely useful for visualization and comparison. However, the reliance on interpreting ancient sources results in a qualitative analysis, as highlighted by Goodchild. The ancient sources are vital for understanding viticulture but can be more usefully utilized as data against which to compare quantitative analysis, rather than the basis of the analysis itself.

Beyond relying on ancient textual sources for measures of viticultural suitability, there has been a more recent trend to quantify these factors. This is often aimed at predictive modelling, using various factors to predict the most likely locations for the archaeological remains of wine-producing villas and rural sites. Quantitative approaches have been applied to predict the location of vinicultural sites in a coastal Catalan area.[9] These models draw primarily on factors of accessibility and potential connectivity, analysing proximity to rivers, ports, roads, the steepness of the terrain and its impact on movement, akin to factors alluded by Cato (*Agr.* 1.3).[10] More recent work has expanded upon earlier approaches, adding additional criteria to the model, including climate, yield and soil properties, such as drainage and chemistry.[11] Others have utilized more complicated modelling to understand climate change in an ancient context and its impact on agricultural yields.[12] Following developments in the use of agent-based modelling in archaeology, such approaches have also been used to simulate different forms of ancient agricultural practice, with interesting methodological results from studies in the Dutch *limes* zone specifically.[13] It is this background against which the current methodology is framed and theoretical basis drawn, applying computational analysis in the pursuit of quantifying agricultural suitability and in order to model Adriatic viticulture.

[6]For some example discussions of these factors, see Trapero Fernández 2016: 54–8; Stubert et al. 2020.
[7]Goodchild 2009: 774–7.
[8]Goodchild 2009: 779–80.
[9]Stubert et al. 2020.
[10]See also McLean and Rubio-Campillo 2022 for a quantitative approach to potential connectivity.
[11]Goodchild 2019: 89.
[12]Bernigaud et al. 2021. Also see Bernigaud et al. in this volume.
[13]Joyce 2019; Verhagen et al. 2021.

The Adriatic region

The Adriatic region represents a particularly interesting case study to apply such a methodology. From the Strait of Otranto to the gulf of Trieste, the region separates the eastern and western Mediterranean basin. The opposing coasts not only cover several modern countries and economic blocs but were split across multiple provincial boundaries during the Roman period.[14] Analysing such a diverse region allows both widespread patterns of agricultural suitability to be compared as well as those within provincial boundaries. Furthermore, being spread across numerous modern countries means that research has experienced diverse archaeological traditions and accessibility to data.[15] While this can complicate research, it also means that region-wide analysis draws on different archaeological biases, akin to a big data approach where different biases can act to cancel one another.[16]

While this case study is applied to the Adriatic, the methodology is not restricted to this region. Anywhere with accessible elevation data and some form of archaeological site distribution can be analysed using this approach. Indeed, it is hoped that the diversity of geography and the archaeological availability of data will highlight the wide spatial applicability of this methodology.

Materials and methods

The methodology for quantifying agricultural suitability and modelling Adriatic viticulture draws directly on this history of work and utilizes probability to determine the impact of various factors on site distribution.[17] Two main processes were undertaken: first, quantification of agricultural suitability; and second, applying this to the Adriatic case study and analysing the different categories of rural site, allowing for a model of viticulture to be proposed (see Figure 16.1 for an overview of this process).

Agricultural suitability

Rather than draw on interpretations of the ancient sources, actual data for central Italian site distribution are used to generate a proxy for agricultural suitability.[18] These data are far more complete than what is available for most other parts of the Roman world, given a long history of intensive and relatively well published survey in the area, though this is becoming less acute as more surveys are conducted across Italy and elsewhere in the

[14]For some wide-ranging historical perspectives on the region, see de Vivo 2003: 161–62; Hodges 2010: 107; Reill 2012.

[15]See, for example, historic issues with archaeological survey in Croatia, specifically Matijašić 1993: 257.

[16]For an important discussion of big data in archaeology, see Van Valkenburgh and Dufton 2020.

[17]Especially the data outlined in Goodchild 2009. For some discussion of statistics and probability in archaeology, see Buxeda i Garrigós et al. 2003; Baxter 2008; Li et al. 2014; Rubio-Campillo 2016; Rubio-Campillo et al. 2017.

[18]Goodchild 2009: 774–9.

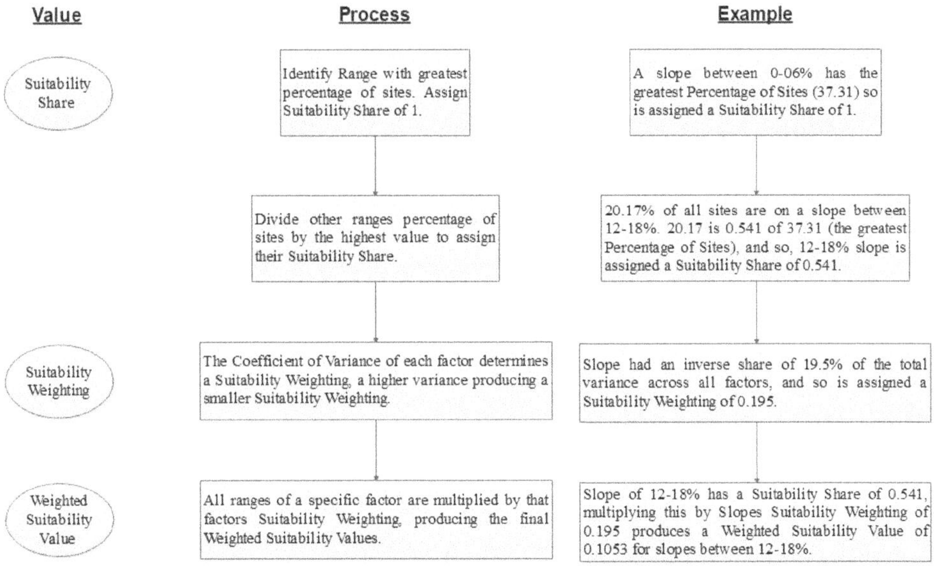

Figure 16.1 Workflow of the methodology for building the weighted suitability value, with examples. Courtesy A. McLean.

Empire.[19] As such, the data is more likely to represent real trends in Roman rural site distribution as opposed to archaeological biases. Over 1,600 sites with relevant slope, aspect and distance data allow us to quantify agricultural suitability in a practical manner. The trends observed in these samples can then be compared to different case studies, such as the Adriatic or viticulture.

The percentage of sites that fall within specific ranges of slope, aspect and distances from relevant features is provided by a previous study.[20] As it is the slope and aspect data that are important, rather than exact geographic locations, the data used in this study are extrapolated from that provided by Goodchild, rather than seeking access to the original point data. While limited information on individual sites is offered, it is the percentages that are important for the current model (found in tables 16.2 to 16.6, below). This forms the basis for generating the numerical value for agricultural suitability, with a higher percentage of sites within a specific range resulting in a higher suitability value. The range with the highest percentage of sites for each factor is given a value of one, and each of the remaining ranges an equivalent value based on the relative percentage of sites. For example, 37.31 per cent of sites are on a slope of between 0 and 6 per cent, representing the single largest percentage of sites for any range of slope. As such, a slope of between 0 and 6 per cent is given a suitability share of one. Regarding the 32.5 per cent of sites on a slope between 6 and 12 per cent, the suitability value is calculated by dividing the

[19]For example, Ward-Perkins 1962; Harrison et al. 2004; Pasquinucci and Menchelli 2017.
[20]Goodchild 2009.

percentage (32.5) by the largest percentage (37.31), giving a suitability share of 0.871. This is repeated for all factors and ranges, giving suitability shares for all ranges in each of the five factors. While it is impossible to completely overcome the bias of using known archaeological sites, the choice of a relatively well researched and documented study region helps to keep these biases to a minimum and ensure that this proxy for suitability can offer useful comparative data.

Some factors have a greater impact on site distribution than others, and so a weighting can be added to the suitability share. To determine this weighting, the coefficient of the variance of each factor was measured. More complex statistical analyses such as regression analysis can provide useful alternatives to variance, but in this situation the required dependent variable would be geographic location, which is difficult to quantify meaningfully. As such, variance was determined to be the most straightforward and useful measure to determine a weighting value. Using these values, across every factor, the coefficient of the variance was calculated by dividing the standard deviation by the mean.[21] In the case of Aspect, the R package 'circular' was used, to calculate accurate means and standard deviations for circular values.[22] These coefficients of variance were then used to generate proportional weighting values, with a smaller variance resulting in a larger weighting value (Table 16.1). These weighting values are then applied to the suitability share by multiplying the two together, and so a higher weighting value gives the factor more impact (see tables 16.2–16.6). These weighted suitability values are based on real data with no qualitative interpretation. While the dataset these are drawn from is relatively robust, the same process can be readily applied to other datasets to generate similar suitability values. It should be emphasized that these values are ultimately proxies, and differences between regional values likely represent real differences in preference of land exploitation, whether due to local ecological, economic or other conditions.

Table 16.1 Factor suitability weighting

Factor	*Suitability Weighting*
Slope	0.195
Aspect	0.248
City Distance	0.208
Road Distance	0.167
River Distance	0.183

[21] Van Pool and Leonard 2011: 54–7.
[22] Agostinelli and Lund 2017.

Modelling Adriatic viticulture

To apply the quantification of agricultural suitability to the Adriatic setting, we need to gather the same data for the sites in this study region. The model draws on five main factors: slope, aspect, and the distances from urban centres, roads and major rivers. Acquiring the relevant data for slope and aspect is relatively straightforward. Digital Elevation Models (DEMs) of 25 metres resolution are available for much of Europe from the Copernicus Project.[23] Using the QGIS gdal functions 'Slope' and 'Aspect', this DEM can be transformed into a Slope and Aspect raster. The distance data requires additional steps but is similarly straightforward and computationally inexpensive. Shapefiles of urban centres, roads and major rivers are transformed into raster data (using the 'Rasterize' function), with each cell in the rasters having a value of, for example, 1 where a relevant feature is present, and a NULL/no data value wherever a feature is not present. These three rasters can then be converted into distance using the 'Proximity' function. This assigns a value to every cell in the output raster to the corresponding distance, in user-defined units, from the relevant features. All of these can then be transformed into a single suitability raster. For example, if a cell in the Slope raster has a value of 7.3 per cent, it falls within the slope range of 6–12 per cent and so is assigned a suitability value of 0.1697 (Table 16.2). This is repeated for all five rasters, with new suitability rasters based on each of the factors generated. These can then be summed together, using the R package 'terra', to produce a total suitability raster (Figure 16.2). These values are cumulative, so the weighted values for every factor in a cell are summed for the overall suitability value.

The data for the sites here were drawn primarily from the work of Van Limbergen and the Ancient Archaeological Sites of Croatia website (IARH).[24] While these data are not exhaustive, the 405 sites, split into four main categories, offer a robust enough dataset from which to draw conclusions (Figure 16.3). Categories include: no evidence for wine or oil production; only evidence for wine production; only evidence for oil production; or evidence for wine and oil production (or it is impossible to discern one from the other). These categorizations are largely based on existing databases, established through archaeological investigation or residue analysis, and while they can rarely be certain, they offer a good basis against which to test this methodology.[25] However, if *dolia* are present at the site, it is assumed wine production likely occurred, as there is no archaeological evidence for these having been used for oil storage. Indeed, *dolia* can be cumbersome containers to fill, empty and clean (cf. Cato, *Agr.* 154), and fermenting wine for longer periods of time limits the need to frequently transfer liquid to or from *dolia*.

---

[23]Copernicus Land Monitoring Service website: https://land.copernicus.eu/imagery-in-situ/eu-dem/eu-dem-v1.1

[24]IARH, Baza antičkih arheoloških lokaliteta Republike Hrvatske Instituta za arheologiju: http://baza.iarh.hr/public/locality/map

[25]See for example, Karaman 1930: 211; Brun 2004b: 34–5; Glicksman 2007: 43; Van Limbergen 2016: 173; Pecci et al. 2013; and others within this volume.

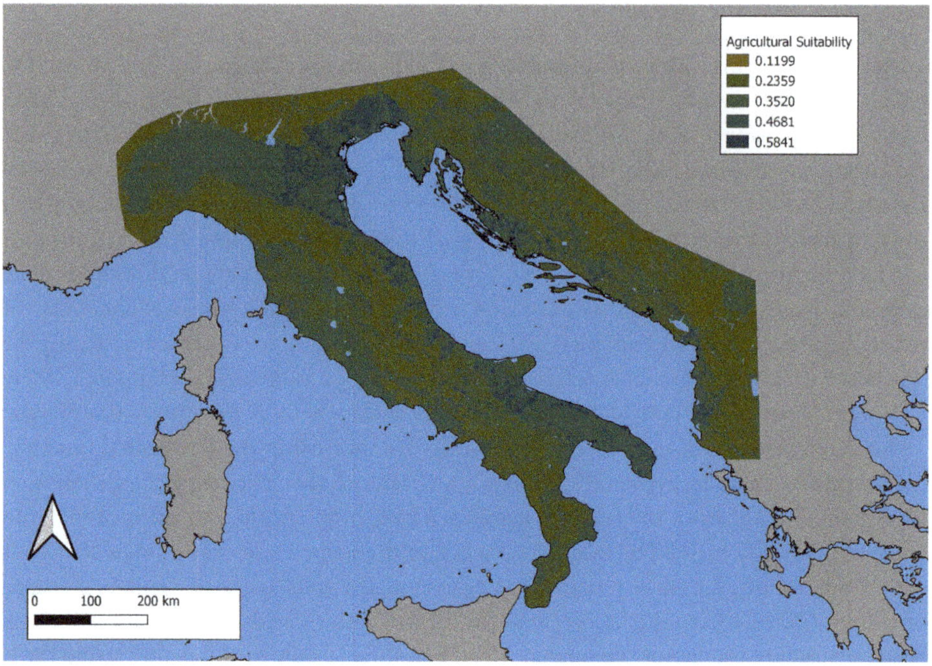

Figure 16.2 Map with cost surface of the weighted Agricultural Suitability Value, including infrastructure. Courtesy A. McLean.

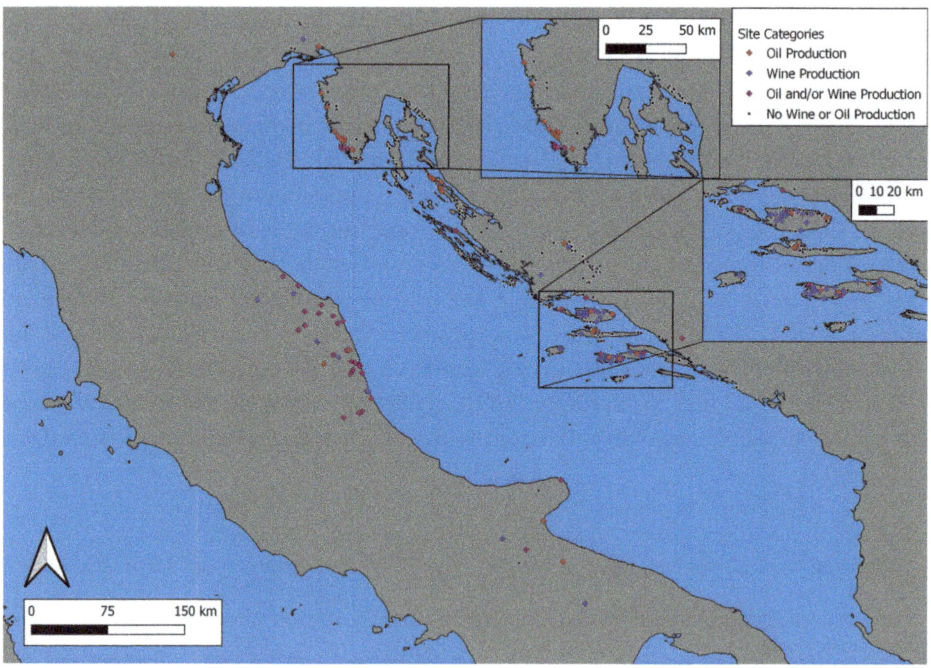

Figure 16.3 Map showing an outline of the Adriatic site dataset broken down into site categories based on likely production type. Courtesy A. McLean.

While certainly not a rule, they are also a less practical vessel for storing oil, which does not necessitate such long-term storage. There remains debate around this matter, and sites with *dolia* are here considered as only *likely* producing wine.[26]

The location of urban centres was determined using Hanson's Oxford Roman Economy Project database (2016) and expanded using the work of Wilkes and de Ligt.[27] While the data of the former are extremely useful, it should be noted that the database has been criticized for inaccuracies and even *lacunae* regarding smaller sites.[28] Nevertheless, it provides a solid base from which to comprehend general trends. For roads and rivers, the shapefiles from the Ancient World Mapping Centre website and the European Environment Agency's (EEA) Catchments and Rivers Network System were used.[29] In order to extract the relevant slope, aspect and distance for each site, a buffer of 2 kilometre radius was drawn around each site, and the mean value for each factor calculated using the terra and rgdal R packages and the buffer function in QGIS. This radius was chosen as many of the locations of sites in the database are accurate only within 2 kilometres, and, while the locations of urban centres, rivers and roads are unlikely to be exact, they should be expected to be significantly more precise than 2 kilometres, meaning this is a reasonable buffer to use for the analysis. Furthermore, such a radius allows for a reasonable average of the territory around a site to be included, allowing for any changes to elevation or slope due to minor erosion or excavation at the exact site itself to be negated.

Results

This methodology produces a great deal of comparative data which can be used to better understand agricultural suitability and Adriatic viticulture in a quantitative and easily replicable manner. In this section, the weighted suitability ratings for the factors are outlined before the resultant suitability raster map is briefly analysed. The differences between the distributions of different types of sites are then visualized, highlighting the specific trends for sites with evidence for wine production and viticulture. Following this, these results and their wider significance are discussed.

Weighted suitability

The five factors (Slope, Aspect and distances from major rivers, roads and urban centres) have differing levels of variance; thus, the weighting values change the overall suitability meaningfully (Figure 16.4A). Distance from roads had the highest coefficient of variance,

---

[26]For recent discussions of differentiating wine and oil production, see Dodd 2022a, especially pp. 457–8; Feige 2021: 35; 2022.
[27]Cf. Wilkes 1969; de Ligt 2012.
[28]Donev and Willet 2017; Pfunter 2017.
[29]EEA 2012; Datasets – MERCURY-MINERVA-SIMREC: https://projectmercury.eu/datasets/

and so has the smallest weighting value, at 0.16. Indeed, this factor is the only one that had a coefficient of variance more than one standard deviation above the mean of all five factors' variances. The distance from urban centres, major rivers, and the slope of the terrain each had relatively similar variances, and thus similar weighting values. Aspect was a clear outlier, being more than one standard deviation below the mean variance, suggesting it was of greater significance in determining site distribution than other factors; thus aspect can be thought of as particularly important for agricultural suitability in this model.

Regarding Slope, flatter terrain was clearly preferable, with almost 90 per cent of sites situated on a slope of less than 18 per cent (Figure 16.4B; Table 16.2). However, the suitability between the flatter two ranges, 0–6 per cent and 6–12 per cent is very small, suggesting that anywhere on these gentle slopes would have been highly suitable for agriculture. Steeper slopes, on the other hand, become rapidly less suitable. Aspect shows a similarly convincing trend, with a clear preference for southerly slopes at the expense, primarily, of northerly slopes (Figure 16.4C; Table 16.3). Most sites are around 2 kilometres from an urban centre, with a steep drop off beyond this distance and, interestingly, relatively few sites closer than 1 kilometre (Figure 16.4D; Table 16.4), though this is likely as a result of the precision of the sites being less than 1 kilometre in some cases. The distances from major roads and rivers are similar, though generally sites are closer to these transport routes than to urban centres (Figures 16.4E and F;

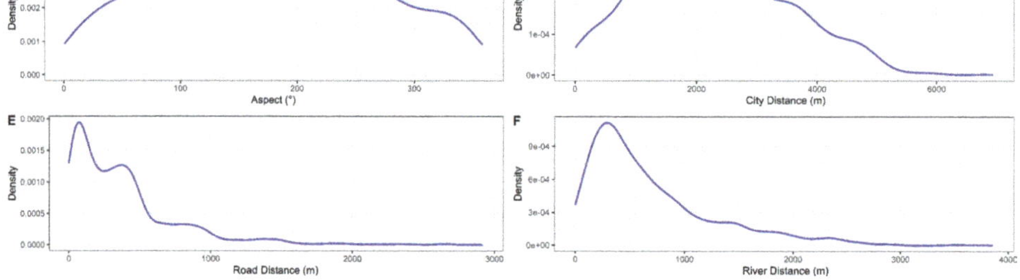

Figure 16.4 Graphs detailing the density of sites in South Etruria based on specific criteria: (A) breakdown of the Coefficient of Variance for the five criteria in South Etrurian sites (the horizontal lines represent one standard deviation above and below the mean value of Coefficient of Variance across all five criteria); (B) density of sites that are on a slope of a specified percentage; (C) density of sites that are on a slope with a specified aspect value in degrees; (D) density of sites lying a specified distance from Roman urban centres; (E) density of sites lying a specified distance from major Roman roads; (F) density of sites lying a specified distance from major rivers. Courtesy A. McLean.

Tables 16.5 and 16.6). All of this highlights the relative importance of these factors in determining the viability of a rural site in South Etruria, and, while distribution patterns varied and sites beyond these general ranges were by no means infeasible, quantification of these factors allows for their relative importance to be easily compared and for wider regional comparisons to be made.[30]

Table 16.2 Slope site percentages and suitability values

Slope (%)	*Percentage of Sites*	*Suitability Share*	*Weighted Suitability Value*
0–6	37.31	1.000	0.1949
6–12	32.5	0.871	0.1697
12–18	20.17	0.541	0.1053
18–24	7.88	0.211	0.0412
24–30	1.56	0.042	0.0081
30–36	0.46	0.012	0.0024
36–42	0	0.000	0.0000
42–48	0.08	0.002	0.0004
48–51	0.04	0.001	0.0002

Table 16.3 Aspect site percentages and suitability values

Aspect	*Percentage of Sites*	*Suitability Share*	*Weighted Suitability Value*
N	7	0.438	0.108
NE	10	0.625	0.155
E	14	0.875	0.217
SE	15	0.938	0.232
S	16	1.000	0.248
SW	15	0.938	0.232
W	12	0.750	0.186
NW	8	0.500	0.124
Flat	3	0.188	0.046

[30]For more detailed discussion of these trends in South Etruria, see Goodchild 2009, 2013.

Table 16.4 City distance site percentages and suitability values

City Distance (m)	*Percentage of Sites*	*Suitability Share*	*Weighted Suitability Value*
1000	13.9	0.470	0.097
2000	29.6	1.000	0.208
3000	29.45	0.995	0.207
4000	16.875	0.570	0.118
5000	8.975	0.303	0.063
6000	0.975	0.033	0.007
7000	0.225	0.008	0.002

Table 16.5 Road distance site percentages and suitability values

Road Distance (m)	*Percentage of Sites*	*Suitability Share*	*Weighted Suitability Value*
100	27.975	0.5730	0.0957
500	48.825	1.0000	0.1671
1000	16.5	0.3379	0.0565
1500	4.4	0.0901	0.0151
2000	1.275	0.0261	0.0044
>2000	1.025	0.0210	0.0035

Table 16.6 River distance site percentages and suitability values

River Distance (m)	*Percentage of Sites*	*Suitability Share*	*Weighted Suitability Value*
100	5.65	0.427	0.078
200	10.25	0.775	0.141
300	11.75	0.888	0.162
400	13.225	1.000	0.183
500	7.45	0.563	0.103
600	8.975	0.679	0.124
700	5.85	0.442	0.081
800	5.325	0.403	0.073
900	4.7	0.355	0.065
1000	4.3	0.325	0.059
1500	11.65	0.881	0.161
2000	6.15	0.465	0.085
2500	3.35	0.253	0.046
3000	0.975	0.074	0.013
>3000	0.4	0.030	0.006

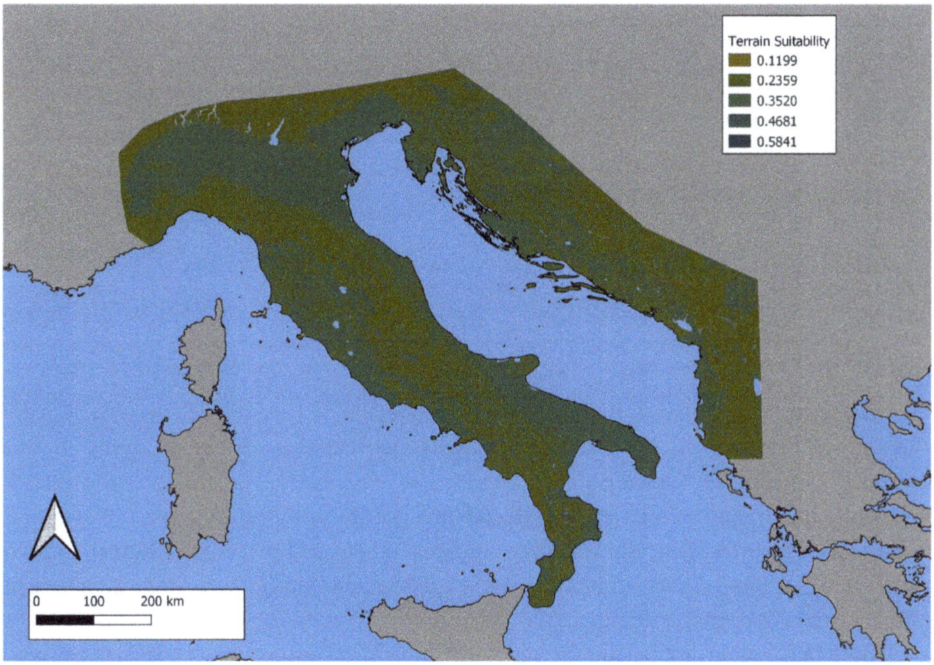

Figure 16.5 Cost surface of the weighted Agricultural Suitability Value based solely on terrain-slope and aspect and excluding infrastructure. Courtesy A. McLean.

It is helpful to consider the landscape of suitability in the Adriatic region before analysing the rural sites in relation to this. Looking at the slope and aspect alone (Figure 16.5), some areas immediately stand out. In particular, the Salento peninsula in southern Italy and the Po Plain in the north, appear to be highly suitable based on this model. The western coast of Istria and the Adriatic coasts of Albania and Montenegro also have high suitability values. Adding the distance from cities, roads and rivers into the suitability model changes this picture to an extent (Figure 16.2). The Salento and especially the Po Plain remain highly suitable, and the western coast of Istria and Albania and Montenegro are relatively less suitable considering these factors. On the other hand, the central Italian Adriatic coast and the Zadar Plateau appear more significant, with relatively dense urban populations and networks of roads and rivers improving the suitability value of the areas.

Adriatic sites

The mean suitability values within the 2 kilometre buffer around each of the sites are consistently above the overall mean suitability value of the region as a whole, including areas beyond the site buffers (Table 16.7). While the values may not appear vastly greater than the overall mean, for all subsets of sites, excluding wine-producing, the p-value of single sample t-tests using a *mu* value of 0.322 (the overall mean) was significantly below

Table 16.7 Site subcategories mean suitability and p-values against overall mean suitability (0.322)

Type of Site	*Site Mean Suitability*	*P-value*
All	0.3678924	2.20E-16
Wine	0.3445644	8.29E-02
Oil	0.4002311	6.04E-05
Wine and/or Oil	0.4128138	1.46E-06

0.01, indicating that the null hypothesis can be rejected. It appears that site distribution did indeed favour locations that have a higher agricultural suitability rating in this model. When considering wine-producing sites, the mean is still greater than the overall, and while the p-value is relatively small (0.08) it is only significant within a 90 per cent confidence interval. This immediately appears to suggest that viticulture followed different trends, and perhaps indicates more limited wine production and viticulture in South Etruria than in the Adriatic.[31]

Most apparent in analysing the site distribution is the apparent lack of sites on the Salento peninsula and the Po Plain, which have some of the highest suitability values anywhere in the region (Figure 16.3). This may be explained, in large part, due to archaeological biases and incomplete data. Indeed, the model appears to suggest that we are missing evidence for agricultural production, particularly wine, in these areas when considering the ancient sources.[32] The central Italian Adriatic, the western coast of Istria and the southern Croatian islands all show dense concentrations of sites. Both Istria and the central Italian Adriatic are highly suitable in the model, due largely to the terrain in Istria, and the proximity of urban centres and rivers in the central Italian Adriatic.[33] The Croatian islands represent a slightly different scenario. There is limited evidence for major road networks or navigable river systems on these islands, and only Issa, on Vis, has evidence for an urban settlement.[34] While the terrain is suitable for agriculture, the lack of infrastructure leads to a relatively limited suitability value overall, and thus an alternative to the factors considered here must be proposed to explain the viticulture and centuriation on these islands, for example proximity to the sea. This approach has already highlighted potential data gaps, the exploitation of highly suitable land, as well as revealing the fact that factors beyond slope, aspect and distances from cities, roads and rivers must have played an important role in some areas of the Adriatic.

[31] Goodchild 2009: 770 in fact highlights the area as a cereal producer.

[32] For some discussion of this, see Buchi 1996: 373; Busana 2003; Paci 2009; Van Limbergen 2011: 72–3.

[33] For more detailed analysis of some of these sites, see Matijašić 1993, 2007, 2012; Cipriano and Mazzochin 2002; Glicksman 2005; Busana et al. 2009; Cipriano 2009; Van Limbergen 2011, 2016, 2018, 2019; Auriemma and Degrassi 2015; Van Limbergen et al. 2017; Šprem 2021.

[34] Gaffney et al. 1997; Stančič 1999; Gaffney and Kirigin 2006.

Discussion

Comparing the distribution of Adriatic sites to the distribution of sites in South Etruria, provides further interesting comparisons, and allows for a model of Adriatic viticulture to be proposed. In terms of slope, both datasets are similar, with a clear preference for gentle slopes (Figure 16.6A). Breaking this down into categories, sites with evidence for wine production tend to favour slightly steeper slopes on average (Figure 16.6B), and those with evidence for oil production more gentle slopes. Furthermore, those sites with evidence for either or both wine and oil production have two distinct peaks for slope, corresponding roughly to the peaks of wine and oil production. Indeed, these peaks are present, though less clear, when considering all sites without categorization (Figure 16.6A). This begins to suggest that preference for slope is relatively uniform across regional divides, but that wine and oil production show different trends. This is perhaps because grapevines generally require more sunlight and heat than olive trees, which is facilitated by more direct sunlight and air flow on steeper slopes.[35] If the ancient agronomists are to be taken at face value, which is of course problematic, this may also reflect a preference for high quality wine over high quantity – Columella (*Rust.* 3.2.5) recommends slightly steeper slopes for more flavoursome wines.[36]

The fact that these peaks are present without categorization of the sites and are prominent following categorization leads to two important conclusions. First, that distinguishing between these two groups of sites is sensible. Such differences in slope between these groups clearly suggests that there was indeed something different about them. Importantly, these distinctions were initially made based purely on likely wine or oil production but are emphatically reinforced by the quantitative analysis that did not take wine or oil production into account. This indicates that the methodological distinction between wine and oil used is sound. Second, if we follow this conclusion, it is possible to extrapolate this to future uncategorized data. If the data has two distinct peaks in slope, we might consider that the lower slope peak represents likely oil production, with the higher wine production. Of course, it is impossible to be certain about this, and other methods should be used in conjunction with any such predictive modelling, such as archaeological and scientific methods discussed elsewhere in this volume. This quantitative approach simply adds an additional element that is not reliant on intensive archaeological investigation or analysis.

Aspect – considered the most important factor by the model – is slightly different between the Adriatic and South Etrurian sites. While a clear preference for southerly slopes is evident in Goodchild's previous data, this is far less pronounced for the Adriatic sites, with easterly and westerly slopes generally preferred, though northerly slopes are similarly uncommon (Figure 16.6C). Sites with evidence for wine production have a more pronounced preference for easterly and westerly slopes (Figure 16.6D). Oil-producing sites, on the other hand, are dominated by easterly slopes, again suggesting a different trend between wine and oil production, and one which might be explained by

---

[35]See Trapero Fernández 2016 for a viticulture study in what is typically thought of as an olive growing region.
[36]Goodchild 2009.

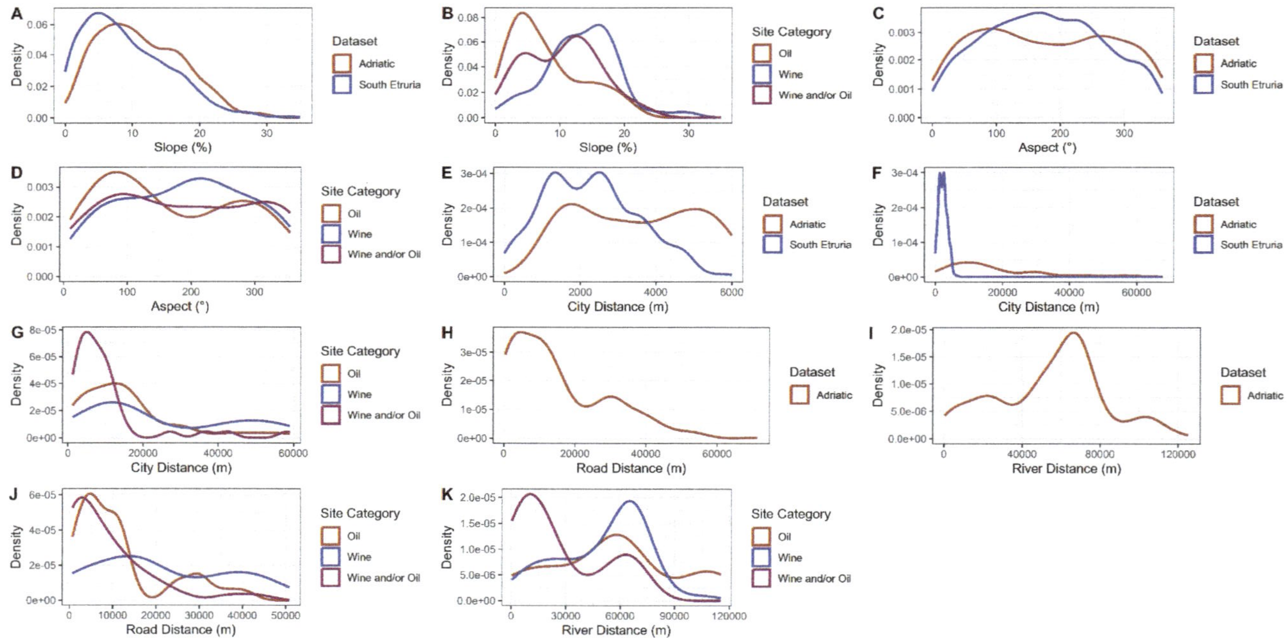

Figure 16.6 Graphs comparing South Etrurian and Adriatic sites, and the different categories of Adriatic sites: (A) density of sites that are on a slope of a specified percentage; (B) density of Adriatic sites that are on a slope of a specified percentage separated into likely production categories; (C) density of sites that are on a slope with a specified aspect value in degrees; (D) density of Adriatic sites that are on a slope with a specified aspect value in degrees separated into likely production categories; (E) density of sites lying a specified distance from Roman urban centres (limited to sites within 6 kilometres of an urban centre); (F) density of all sites lying a specified distance from Roman urban centres; (G) density of Adriatic sites lying a specified distance from Roman urban centres separated into likely production categories; (H) density of Adriatic sites lying a specified distance from major Roman roads (cf. sites from South Etruria in Figure 4E); (I) density of Adriatic sites lying a specified distance from major rivers (cf. sites from South Etruria in Figure 4F); (J) density of Adriatic sites lying a specified distance from major Roman roads separated into likely production categories; (K) density of Adriatic sites lying a specified distance from major rivers separated into likely production categories. Courtesy A. McLean.

a preference for slope. Fundamentally, aspect does not appear to have been the important factor in determining the suitability of land in the Adriatic in the same way that it was for South Etruria.[37] The ancient sources suggest that aspect preference is often an attempt to limit the impact of wind (cf. Columella throughout Book 3). Importantly, the dominant wind of the Adriatic is the northeasterly Bora, with the northwesterly Mistral blowing daily throughout the summer.[38] As such, a slope facing directly south may not be preferable in the same way as it was in central Italy. Depending on the season, a southeasterly or southwesterly slope would provide the greatest cover from these winds in the Adriatic region, and so a more varied distribution of sites appears in the Adriatic, albeit with exposed northerly slopes still unfavourable.[39]

The criteria of distance from cities, roads and rivers reveals a greater regional disparity. The distance from cities within 6 kilometres has peaks around 2 kilometres for both the Adriatic and South Etrurian sites (Figure 16.6E); however, there is a second peak for Adriatic sites around 5 kilometres, while very few South Etrurian sites are more than 2 kilometres from an urban centre. Furthermore, some Adriatic sites are as distant as 60 kilometres from an urban centre (Figure 16.6F). This likely reflects the higher level of urbanization in South Etruria, as well as the prominence of the Croatian islands in the Adriatic, which are generally further from urban centres. Indeed, sites with evidence for wine production are generally further from urban centres, with peaks at around 15 kilometres and 50 kilometres (Figure 16.6G). Oil-producing sites, on the other hand, are generally closer to urban centres, with a clear majority of sites being around 15 kilometres from an urban centre. Again, if the ancient texts are believed, this might indicate that very few of these sites were exporting fresh grapes as well as wine, with Columella (*Rust.* 3.2.1) suggesting that for a profit to be made on fresh grapes they must be grown very close to a city. Grapes across the Adriatic were perhaps primarily destined for wine presses and the wider export market. Virtually all the South Etrurian sites are within 2 kilometres of a road, while there are Adriatic sites as far as 60 kilometres from a road, though these are relatively few (Figures 16.4E and 16.4H). The Croatian islands again are likely the source of this, with no known major roads on these islands. The situation for river distances is much the same, with most sites being some 70 kilometres from a major river, though this may also reflect diversity in the definition of 'major river' (Figures 16.4F and 16.4I). This apparent discrepancy between distances to urban markets and transport infrastructure can probably be explained by the presence of the Adriatic itself, particularly if this is true for other coastal regions. Maritime transport of wine was vital to the Roman economy, particularly in coastal areas like the Adriatic.[40] Indeed, sites with evidence for wine production are relatively evenly spread out across the region in terms of the road network, with a slight peak at 15 kilometres (Figure 16.6J). Sites with evidence for oil production cluster more closely to the road network. This is much the same for distance to rivers, with a steep peak for wine-producing sites at around 70 kilometres

[37]Cf. Goodchild 2009, 2016.
[38]Bignami et al. 2007; Pullen et al. 2007; Boldrin et al. 2009.
[39]See Goodchild 2009: 776 for a possible alternative to this, focussing on soil temperature.
[40]Rice 2016; Komar 2021.

(Figure 16.6K). This may be due to the process of fermentation in wine production and storage for a longer time before being sent to market. The turnover and frequency of transporting the commodity would presumably be higher for oil, possibly explaining this greater proximity to a transport network. The importance of maritime transport appears to be a major regional difference in determining agricultural suitability and site distribution. Sites in the Adriatic region cluster far less closely to terrestrial and fluvial transport infrastructure and urban centres. This is particularly the case for viticulture and wine production, where it can be argued that access to maritime transport was of far greater importance than other forms of transport.

Conclusion

The model for agricultural suitability provides a quantitative approach to understand agricultural production across the Roman world. It allows for direct regional comparisons as well as trends unique to viticulture that are difficult to identify without such an approach. Preferences for slope appear fairly consistent across the two regions considered in this study; however, aspect appears to have been more important in South Etruria than the Adriatic, where the changeable Bora and Mistral winds made a preference for direct southerly slopes less apparent. The importance of maritime transport is also emphatically shown in the Adriatic, where proximity to roads and rivers is far less prominent in determining site distribution. Analysis of port and harbour infrastructure would be particularly useful to better understand this but can be a complex undertaking and is thus beyond the scope of this chapter. Additional factors that could not be considered in depth here include soil type and climate, which undoubtedly impact site distribution and suitability. Furthermore, slope appears to be a possible indicator to differentiate viniculture and oleiculture. Sites with evidence for wine production are generally on steeper slopes than those with evidence for oil production, with the two peaks observed for those sites where wine and oil production cannot be distinguished between perhaps representing sites producing wine and oil. Closer analysis of individual sites would allow for a better understanding of those that have remained ambiguous through archaeological investigation and qualitative approaches alone.

While this suitability model drew upon sites from South Etruria, the analysis clearly shows that there was not a standard measure for suitability across the Roman world. This reassuring conclusion allows us to pick up often subtle differences and use them to better understand economic nuances – in particular, whether a site was better positioned for viniculture or oleiculture. Wider comparison with larger and more varied datasets will allow for more robust suitability models to be produced and for regional nuances to be better highlighted. Even with the limitations of this early model, it has been shown that the suitability value is a strong indicator for identifying areas of Roman agricultural activity and presents additional possibilities to differentiate viticulture from other forms of cultivation. Ultimately, this chapter provides an outline for an approach for quantifying agricultural suitability, a method which can help to demystify the complexities in modelling wine production and viticulture in the Roman world.

CHAPTER 17

LINKING EXPERIMENTAL ARCHAEOLOGY AND WINEMAKING: FROM THE DIG TO THE WINERY

Mkrtich Harutyunyan, Manuel Malfeito Ferreira and Mario Indelicato

Abstract

This chapter discusses how experimental archaeological research methodologies enable us to deepen the study of viticulture and oenology in the Roman world. It describes a general application of experimental archaeology to the study of human nutrition followed by an exploration of several case studies in experimental wine production specific to the Roman era. This clearly highlights the beneficial role experimental archaeology can play in illuminating aspects of and nuance in production otherwise undetectable through traditional methods.

Introduction

The discovery of fire between 500,000 and 300,000 years ago, after a journey of experimentation that probably lasted over 50,000 years, made humanity capable of cooking foods provided by nature.[1] Since then, all food operations – including cooking, seasoning, grinding, slicing and slaughtering – have had the primary function of making food more digestible, edible and tasteful.[2] Alongside these operations, humans increasingly refined their agricultural techniques to produce and collect foods that were increasingly suitable for human consumption and more adapted to tastes and dietary habits. In the Mediterranean basin, this trend towards crop specialization led to the generation of the so-called Mediterranean triad of cereals, grapevines and olive trees. The development of these foods formed the foundation of proto-urban civilization and the traditional economy and culinary basis of Greece and Rome, continuing in various forms in the Mediterranean area until the present. A discussion of wine in the Roman

[1]On the underlying reasons for cooking food, see the chapter dedicated to nutrition in the seminal book by Morris 2017: 201–15.
[2]Flandrin and Montanari 1997: 9.

world, therefore, needs to consider the production of this commodity within the broader context of the history of human nutrition.

An archaeology of human nutrition

Various forms of human nutrition over the centuries have left traces in archaeological sites around the world in different ways. Archaeologists have studied these and a history of nutrition has developed as early as the end of the nineteenth century, where Boudreau asserted that 'for the exact science of human facts, the menu of a meal is more instructive than a war facts narration, a cookbook more than a diplomatic acts collection, a statistic of alimony more than a court intrigues report'.[3] For a long time, however, the study of ancient nutrition was relegated to the margins of 'great' history.

However, even though historians now habitually refer to studies on food habits and practices, there is yet no common use of the term 'nutrition archaeology',[4] and the definition itself continues to cause confusion.[5] In fact, just like the history of nutrition, nutrition archaeology is not a single discipline but a broad collection of studies in which different disciplinary fields converge, often distinct in both methods and materials. Its general object of study is nutrition, but it includes all those areas of research and investigation that produce information in some way related to nutrition. The various disciplines thus become research tools by which to study and deepen the characteristics of ancient societal food habits. This implies that it should investigate all aspects of food production as an anthropological expression – that is, cultural, economic, social and ritual – and all related material traces in archaeological deposits.

Other operations connected with nutrition, however, are more complex to investigate. For example, the traces of agro-pastoral activities are difficult to investigate because they were varied and diverse across and between territories.[6] Even more challenging to detect and evanescent in the archaeological record are testimonies of hunter-gatherer activities.[7]

Nutrition archaeology, still far from being commonly accepted and discussed, has many characteristics that can make it, however, 'a container in which large masses of data can converge and above all the results obtained from research and quantification'.[8]

---

[3]Boudreau 1894: 5.

[4]In Italian also known as *archeologia dell'alimentazione* which is usually translated as 'archaeology of diet' or 'archaeology of food'. Here, it is preferable to opt for 'nutrition' as this is a more direct translation of *alimentazione*, expressing more holistically the broad meaning of human nutrition that transcends merely food or diet.

[5]Mariottini and Salvadori 2012: 69.

[6]Giannichedda 2006: 13–14.

[7]Giannichedda 2006: 13.

[8]Mariottini and Salvadori 2012: 72.

The contribution of experimental archaeology

As mentioned, the history of nutrition draws its data from a diverse set of disciplines, including, for example, ethnology, history and sociology. Likewise, there are numerous disciplines that have contributed to the development of the archaeological study, *stricto sensu,* of human nutrition throughout history, including archaeobotany,[9] archaeozoology,[10] archaeometry[11] and ethnography.[12] Alongside these archaeological disciplines, experimental archaeology plays an important role in the reconstruction, interpretation and understanding of past food and lifeways. Through this discipline, and by following rigorous scientific methods, it becomes possible to deepen our knowledge on aspects of ancient nutrition and plan experimental fieldwork. It also allows us to better understand the function and methods of an object's use, organization of practices and techniques (e.g. related to agriculture, hunting, breeding, cooking), but also the subsistence and development of ancient communities. This methodology has been defined as 'a sub-field of archaeological research which employs a number of different methods, techniques, analyses, and approaches within the context of a controllable imitative experiment to replicate past phenomena (from objects to systems) in order to generate and test hypotheses to provide or enhance analogies for archaeological interpretation'.[13] Experimental archaeology is, therefore, configured as a scientific instrument that uses the replication of 'past phenomena' to generate verifiable analogies useful for archaeological interpretation. By offering the opportunity to develop theories based on information otherwise not available, replicated through multiple trials and repeated experiments,[14] it is a tool capable of 'providing the practice that supports the theory',[15] through a single method of investigation that allows archaeologists to examine opinions on past human behaviour (Figure 17.1).[16] The main tool of experimental archaeology is the 'experiment', interpreted in its Galilean and positivist meaning,[17] which is the final activity of a process started with the collection of information, the observation of the context of the reference phenomenon, and the evaluation of the

[9]See, for example, Willcox 1999; Lodwick and Rowan 2022 and the contributions of Dodd and Van Limbergen, Marchesini et al., Basso and Dobreva, and Carmenati et al. in this volume.

[10]Including the reconstruction of the human capacity to eat other animals and the techniques used, for example, to prepare and use animal tissues for food: see, for example, Crovetto et al. 1994.

[11]Including the use of chemical-physical analyses to identify and analyse organic residues: see, for example, Notarstefano 2012 and the contributions of McGovern, Pecci and Rousse et al. in this volume.

[12]By studying human behaviours useful to reconstruct and understand analogous ancient behaviours: in relation to nutrition, see, for example, Giannitrapani, Simone and Tinè 1990.

[13]Mathieu 2002: 1.

[14]Marsh and Ferguson 2010: 4.

[15]Vidale 2000: 281. Also see Coles for an identification of the characteristics common to the multiplicity of possible experiments in archaeology. All the experiments 'begin with reconstruction, all proceed with experiments of use, all follow a series of stages: problem → idea → procedure → result → evaluation' (Coles 1981: 8–9).

[16]Godino, Lebole and Di Gangi 2020: 8.

[17]Galilean in the sense of experimentally reproducible. Positivist with 'the hypothesis' experimental confirmation formulated on the basis of repeated observation of a given phenomenon' (Giardino 2012: 13).

Figure 17.1 The experimental process of pitching wine jars (*right*); mosaic scene of the pitching of the vases from the calendar of agricultural works of St Romain en Gal (third century CE) (*left*). Courtesy M. Indelicato.

hypotheses. This phase must respect rules, methodologies and protocols to establish and make the scientific process.

In the case of food production, numerous experiments have been carried out globally over time: these experiments were linked directly to the production of food and related to various fields, from agricultural production to the reconstruction of ancient recipes. The first attempt at an overview of food-related experiments, carried out in the late 1970s, distinguished in various categories several archaeological experiments related to food production: deforestation and cultivation, ploughing, harvesting and food storage, and preparing and consuming food.[18]

The experimental production of Roman wine

Classical agricultural and other written sources

Ancient texts with technical information relating to viniculture include those from the philosopher and naturalist Theophrastus (371–287 BCE), as well as several Roman agricultural manuals: Cato (*c.* 234–149 BCE), Varro (*c.* 116–27 BCE), Columella (*c.* first century CE), Pliny the Elder (*c.* 23/24–79 CE), Palladius (late fourth to early fifth centuries

[18]See Coles 1981: 13–48. For a review of experimental archaeology studies applied to food production in the ancient world, see Indelicato 2020a.

CE), and an anonymous treatise of the tenth century CE entitled the *Geoponika*.[19] It is suspected that many of these Roman authors depended to a great extent on translations of an agricultural treatise from Mago (date unknown, but perhaps *c.* third to second centuries BCE) – a Carthaginian writer whose compendium was lost after the fall of Carthage in the Third Punic War (149–146 BCE). The early treatises on medicine by physicians, including Hippocrates (*c.* 460–375 BCE) and Galen (*c.* 129–216 CE), are relevant to understand wine's sensory characteristics.[20] Other ancient texts, from Horace's *Odes* (65–8 BCE), Homer's *Iliad* and *Odyssey* (most likely eighth century BCE),[21] or Virgil's *Aeneid* and *Georgics* (70–19 BCE)[22] together with the Holy Books of the Bible, Four Gospels, Torah, Talmud and Quran, shed additional light on the colour of the grapes, grape-growing techniques and types of wine.[23]

Wine as an object of experimental archaeology

Wine is a drink of both social use and high symbolic value, closely related to the practice of *συμπόσιον* (*symposium*), typical of the Greek world.[24] Similarly, in Roman times, feasting and lying together as part of the so-called *convivium* was also widespread.[25] Despite its symbolic use, wine remained, however, a food and source of nutrition, albeit a particular one. From this point of view, the methods of its production were closely linked to technique and materiality, and they can thus be studied using practical experiments.

The production of wine itself can be described as between agricultural and food production. In fact, like many other productions, before being able to prepare wine it is necessary to plant, cultivate and raise the vine and, subsequently, collect the grapes and transform them first into must and then, through fermentation, into wine. However, its particularity lies in the fact that from the planting of the grapevines to the production of wine, the cycle is unique and is often – though not always – performed by the same owner. For example, those who cultivate wheat are not always the ones who produce the bread; those who grow legumes are not necessarily the ones to prepare, for example, soups.

Thus, grapes constitute the starting point from which one produces a 'good' wine. In fact, the better the quality of the grapes, the better the potential quality of the wine. This principle was already known in the Roman era, with Columella (cf. *Rust.* Book 3) describing the existence of a viticulture founded on solid foundations of personal

[19]Dodd 2020a: 17–21.
[20]Jouanna 2012: 179.
[21]Donahue 2015: 278–9.
[22]Donahue 2015: 285.
[23]Donahue 2015: 112.
[24]The origins of the Greek *symposium* can be traced back to the royal banquet and the Mesopotamian practice of 'drinking in common' (cf. Lissarague 1989; Roaf 1992; Musti 2005; Nencini 2009).
[25]Dunbabin 2003: 13; Hudson 2010; Donahue 2017.

empirical experience[26] throughout the Mediterranean basin and firmly concerned with the quality of grapevines, essential to obtain a good product and to make a profit.[27] This agricultural practice constituted the main economic activity of the ancient world (until at least the seventeenth century), and was central in Graeco-Roman art where it represented the virtuous integration between practical knowledge (τέχνη) and nature (φύσις).[28] Through agriculture, man interacted with nature and earned his livelihood. This importance of agriculture is evidenced by the flourish of agronomic literature in the classical age.[29]

Roman winemakers understood the basic process of fermentation but were ignorant to the biochemical mechanisms underlying the transformation of must into wine.[30] This is suggested in the text of Columella, who, despite leaving an exceptional compendium of agronomic sciences, hardly discusses the actual transformation of grapes into wine. This is true for most agronomists, with little description about what we might term 'scientific' winemaking operations and thus many open questions regarding the oenological aspects of Roman wine. What kind of data can archaeology provide to recreate the work in the vineyard and in the winery? And what can experimental archaeology in particular add to the limited and often opaque literary sources?

We can create archaeological experiments that begin to answer these questions, and which, alongside critical re-reading of written sources, can help to reconstruct the entire vinicultural cycle in the Roman world (the experiments described below apply most to the period between the second century BCE and the first century CE). In addition to the Roman texts, it is possible to utilize ethnographic data which can help, for example, to interpret, clarify and decode procedural descriptions. This also allows us to reconstruct the tools and instruments used in Roman viticultural practices (Figure 17.2).

Modern replication of Roman winemaking techniques

In addition to the literary and archaeological study of Roman viticultural techniques, it is also possible to use experimental archaeology to increase our understanding of vinicultural activities that occurred in the ancient winery. Thanks to the data obtained from archaeological research, it is possible to reconstruct not only the places – the *calcatoria* and *torcularia* – where the grape juice extraction took place but also all various practices

[26]Columella (*Rust.* 3.333) reminds us that he owns various farms in central Italy, especially near Caere, ad Ardea, Carseoli, ed Alba (Kolendo 1980: 21; Marcone 1997: 27).
[27]Saltini 1979: 36–7.
[28]Lelli 2010: VII.
[29]Marcone 1997: 205. Among the innumerable written sources available to us, the category of authors called *scriptores rei rusticae* includes, since the Renaissance, the work of Cato, Varro, Columella and Palladius (fourth and fifth centuries CE). To these must be added the *Georgics* of Virgil and some books of the *Historia Naturalis* of Pliny the Elder (first century CE), in particular Book 14. The work of Mago the Carthaginian in 28 books, which underpinned much of this, was an accumulation of agricultural knowledge of Aegean/eastern Mediterranean derivation (see above and Marcone 1997: 19).
[30]Peña Cervantes 2010: 31–3; Thurmond 2017: 174–8.

Figure 17.2 The experimental Roman vineyard planted in 2013 on the eastern slopes of Mount Etna (Sicily, Italy). After Indelicato 2017: 171, Figure 133.

and the operations. As described by McGovern in Chapter 2, pioneers in this field were Brun and Tchernia who in the 1990s at Mas des Tourelles embarked upon one of the first experiments in Roman winemaking by combining archaeological knowledge with the experience of winemakers.[31] By applying the guidelines of the Latin agronomists and local archaeological evidence, the first reconstruction of a Gallo-Roman wine cellar with *torcularium* was attempted (Figure 17.3). This experiment also included the reproduction of numerous Roman oenological practices, such as covering *dolia* with pitch, fermentation in *dolia*, and the creation of various types of wine according to Roman recipes.[32]

Researchers at the Musée Gallo-Romain St. Romain en Gal have collected more than 300 vines of fifteen different varieties, some of which are among the oldest in France, planting a vineyard cultivated according to six different methods documented in Roman times.[33] Every two years they hold a *Vinalia*, a Roman festival in honour of wine, during which local wine is experimentally reproduced alongside the recreation of ceramics and the coating of wine vessels.[34] This panorama of experimental archaeology projects also extends to Catalonia, where the archaeological park of *Cella Vinaria* was built near the

[31]Tchernia and Brun 1999. See McGovern in this volume.
[32]Tchernia and Brun 1999: 106 and 111–13.
[33]For this museum and its activities, see: https://exarc.net/members/venues/musee-gallo-romain-fr and https://musee-site.rhone.fr/archeologie-experimentale/
[34]Caillaud 2018.

Figure 17.3 Reconstruction of Cato's *torcularium* at Mas des Tourelles. Courtesy Mas des Tourelles, photograph by C. Durand.

site of Vallmora.[35] Here were revealed the structures of a Roman wine production centre active from the first century BCE to the fifth century CE. The experimental archaeology project includes both the restoration of the structural remains and the functional study and *in situ* reconstruction of two large Roman lever-operated wine presses.[36] The project also includes the experimental development of a Roman vineyard in which to reproduce and study different grapevine planting and cultivation techniques.

Similar experiments have also been undertaken in Italy. A number of urban vineyards have been replanted in Pompeii following the study of the gardens, of which the most famous is a vineyard near the amphitheatre in Insula II.5 ('Foro Boario').[37] Here, autochthonous grapevines were planted with the same techniques preceding the great eruption (e.g. high planting density, rows supported by wooden poles).[38] A Roman harvest open to the public was carried out annually at this reconstructed vineyard for a number of years and the (modern) wine made from its grapes is marketed all over the world.

[35] Martín i Oliveras 2013.

[36] Martín i Oliveras and Bayès 2009.

[37] Jashemski 1979; Carbonneau and Rotunno 2000; Dodd 2017, 2022a.

[38] Support poles were planted exactly on their imprints located during the excavation (Jashemski 1979: 204–5). Also see the contribution of Seiler in this volume.

From 2013–17, a research project was launched in Sicily with the aim of improving our knowledge of wine production in the Roman world.[39] This project, inspired largely by the text of Columella, consisted initially of a first phase using Roman planting techniques in a small vineyard on the slopes of Mount Etna to experimentally investigate ancient viticultural practices and to identify, reconstruct and use numerous agricultural tools (Figure 17.2).[40] In a second phase, Roman oenological practices (the set of operations that transformed grapes into wine) were studied. Starting from an analysis of the remains of winery structures excavated across Italy and Europe, it is possible to understand the various stages of processing, including fermentation, decantation and transport.[41] Additionally, analyses conducted on ceramic materials from production sites make it possible to reproduce the coating of wine vessels and evaluate their effectiveness and impact on the production and taste of wine.[42] Combining all this data makes it possible to draft a proper experimental winemaking protocol.[43] The latter can guide researchers in setting up pilot (or 'first generation') experiments through which to acquire data for the construction and execution of 'second generation' experiments to produce wines with commercial potential.

As an example of this approach, Harutyunyan and colleagues replicated several protocols described by Columella concerning the production of sweet wines from dehydrated/withered Muscat of Alexandria grapes.[44] Juices were obtained by directly pressing the grapes or after a rehydration step using saltwater or white wine. Pressed pomace (i.e. the processed skins, seeds and stems) were diluted with water (5 g/L tartaric acid) to obtain a beverage named 'piquette', corresponding to the low-quality wine-based beverage described by Columella as *lora*. The wines had an alcoholic content ranging from 14.8 per cent to 17.0 per cent (*v/v*), while the *piquette* analysis showed ethanol levels ranging from 10.2 per cent to 16.0 per cent (*v/v*). These wines were further tasted and ranked by consumers and the highest scores were obtained by the samples that were aged at 45°C for four months to mimic the Roman roof storage conditions.[45]

Conclusions

From the planting of grape vines to the production of wine, from the use of basic tools or instruments to the study of wine preparation techniques, experimental archaeology is

[39]Born from the collaboration between the University of Catania and the Italian National Research Council (CNR).
[40]Indelicato, Malfitana and Cacciaguerra 2017.
[41]E.g. Brun 2004b; Dodd 2022a; Feige 2022.
[42]Indelicato 2017. Also see the contribution of Pecci in this volume.
[43]Indelicato 2020b.
[44]Harutyunyan et al. 2022: 1.
[45]Harutyunyan et al. 2023: 14.

building to become a fundamental field-research tool. One that allows us and others to obtain 'living' data on material culture(s) of ancient societies. Experimental archaeology has been an important tool for the reconstruction of the ancient ways of food storage and preparation in pots, cooking techniques, bread baking, traditional beverage-making recipes and ceramics firing technology. Overall, the technical interpretation and replication of ancient winemaking protocols will provide solid historical support for the current rebirth of ancient production methods, particularly those using ceramic vessels.

BIBLIOGRAPHY

Acconcia, V. and A. Piergrossi (2021), 'L'archeologia del vino nella penisola italiana e nelle grandi isole del Tirreno tra il neolitico e la romanizzazione: tematiche, dati e approcci possibili', *Bollettino di Archeologia online*, 12 (1–2): 183–230.

Accorsi C. A., M. Bandini Mazzanti, L. Forlani, N. Giordani, M. Marchesini, S. Marvelli and G. Bosi (1998), 'Archaeobotany of the Cognento hiding well (Modena; Northern Italy; 34 m a.s.l.; 44°40'N 10°55'E; Late Roman – Modern Age)', in *1st International Congress on 'Science and Technology for the Safeguard of Cultural Heritage in the Mediterranean Basin', November 27–December 2 1995*, 1537–54, Catania.

Agostinelli, C. and U. Lund, (2017), 'R package "circular": Circular Statistics (version 0.4-93). Available online: https://cran.r-project.org/web/packages/circular/citation.html (accessed 21 February 2022).

Ahmed, M. A. M. (2019), *Rural Settlement and Economic Activity: Olive Oil, Wine and Amphorae Production on the Tarhuna Plateau During the Roman Period*, Society for Libyan Studies Open Access Monograph 1, Oxford: Oxbow Books.

Aimar, F., P. Gullino and M. Devecchi (2021), 'Towards Reconstructing Rural Landscapes: A Case Study of Italian Mongardino.' *Journal of Rural Studies* 88: 446–61.

Akerman, J. R., ed. (2009), *The Imperial Map: Cartography and the Mastery of Empire*, Chicago: University of Chicago Press.

Akkemik, Ü., H. Caner, G. A. Conyers, M. J. Dillon, N. Karlioğlu, N. K. Rauh and L. O. Theller (2012), 'The Archaeology of Deforestation in South Coastal Turkey', *International Journal of Sustainable Development and World Ecology* 1 (1): 1–11.

Alba, V., C. Bergamini, M. Gasparro, F. Mazzone, S. Roccotelli, D. Antonacci and A. R. Caputo, eds (2015), *Il Progetto Basivin SUD: Recupero e valorizzazione delle principali varietà locali e dei vitigni autoctoni minori in Basilicata*, Bari: Mario Adda Editore.

Albert, R. M., R. Shahak-Gross, D. Cabanes, A. Gilboa, S. Lev-Yadun, M. Portillo et al. (2008), 'Phytolith-rich Layers from the Late Bronze and Iron Ages at Tel Dor (Israel): Mode of Formation and Archaeological Significance.' *Journal of Archaeological Science* 35 (1): 57–75.

Al-Gaadi, K. (2012), 'Employing Electromagnetic Induction Techniques for the Assessment of Soil Compaction', *Am. J. Agric. Biol. Sci.* 4: 425–34.

Allegrezza, V. (2017), 'Torcular e gli olivi nel Lazio settentrionale: nuove evidenze', *Rivista di Storia dell'Agricoltura*, 57 (1): 3–24.

Allevato E., M. Buonincontri, A. Pecci, M. A. Cau, M. Vairo, M. Yoneda, G. F. De Simone, C. Angelelli, S. Matsuyama, K. Takeuchi and G. Di Pasquale (2012), 'Persistence of the Cultural Landscape in Campania (Southern Italy) Before the AD 472 Vesuvius Eruption: Archaeoenvironmental Data', *Journal of Archaeological Science* 39 (2): 399–406.

Amouretti, M.-C. and J.-P. Brun, eds (1993), *La production du vin et de l'huile en Méditerranée*, BCH Supp. XXVI, Athens: École Française d'Athènes.

Andersen, S. T. (1970), 'The Relative Pollen Productivity and Representation of North European Trees, and Correction Factors for Tree Pollen Spectra', *Dan Geologiske Undersogelse II* 96: 1–99.

Andreev, A.A., P. E. Tarasov, F. A. Romanenko, and L. D. Sulerzhitsky (1997), 'Younger Dryas pollen records from Sverdrup Island (Kara Sea)', *Quat. Int.*, 41/2, 135–9.

Andrieu-Ponel, V., P. Ponel, A. J. T. Jull, J.-L. de Bealieu, H. Bruneton and P. Leveau (2000), 'Towards the Reconstruction of Holocene Vegetation History of Lowert Provence: Two New Pollen Profiles from Marais des Baux', *Vegetation History and Archaeobotany* 9: 71–84.

Angourakis, A., V. Martínez Ferreras, A. Torrano and J. M. Gurt Esparraguera (2018), 'Presenting Multivariate Statistical Protocols in R Using Roman Wine Amphorae Productions in Catalonia, Spain', *Journal of Archaeological Science* 93: 150–65.

Antonini, E. and M. Marchesini (2017), 'La piantata e la viticoltura: persistenza nel paesaggio moderno', in L. Malnati, S. Pellegrini, F. Piccinini and C. Stefani (eds), *Mutina splendidissima: la città romana e la sua eredità*, Catalogo della Mostra, Modena 2017–18, 290–3, Rome: De Luca Editori d'Arte.

Areshian G. E., B. Gasparyan, P. S. Avetisyan and R. Pinhasi (2012), 'The Chalcolithic of the Near East and South-Eastern Europe: Discoveries and New Perspectives from the Cave Complex Areni-1, Armenia', *Antiquity* 86: 115–30.

Armstrong, K. L. and T. Kalayci (2015), 'Images of the Past: Magnetic Prospection in Archaeology', in A. Sarris (ed.), *Best Practices of Geoinformatic Technologies for the Mapping of Archaeolandscapes*, 1–12, Oxford: Archaeopress.

Arobba D., R. Caramiello and G. P. Martino (1999), 'Analisi paleobotaniche di resine dal relitto navale romano del Golfo Dianese', *Rivista di Studi Liguri* 63–4: 339–55.

Arobba, D., R. Caramiello and G. P. Martino (2005), 'Il contributo delle analisi archeobotaniche per la storia del paesaggio agrario della città romana di Albintimilium (Ventimiglia)', *Rivista di Studi Liguri* 70: 283–306.

Arobba, D., F. Bulgarelli and R. Caramiello (2010), 'Preliminary Study on Carpological Remains of Food Resources from a Roman Well at Vada Sabatia', in C. Delhon, I. Théry-Parisot and S. Thiébault (eds), *Des Hommes et des Plantes: Exploitation du milieu et gestion des ressources végétales de la préhistoire à nos jours, XXX rencontres internationals d'archéologie et d'histoire d'Antibes*, 123–7. Antibes: Editions APDCA.

Arobba, D., F. Bulgarelli, C. Siniscalco and R. Caramiello (2013), 'Roman Landscape and Agriculture on the Ligurian Coast through Macro and Microremains from a Vada Sabatia Well (Vado Ligure, Italy)', *Environmental Archaeology* 18 (2): 114–31.

Arobba, D., F. Bulgarelli, F. Camin, R. Caramiello, R. Larcher and L. Martinelli (2014), 'Paleobotanical, Chemical and Physical Investigation of the Content of an Ancient Wine Amphora from the Northern Tyrrhenian Sea in Italy', *Journal of Archaeological Science* 45: 226–33.

Arriaga-Varela, E. J. and T. Takahashi (2019), 'Automatic Georeferencing of Heterogeneous Historic and Illustrated Maps', in *Abstracts of the International Cartographic Association 1, 29th International Cartographic Conference (ICC 2019), 15–20 July 2019, Tokyo, Japan*, 1–2.

Arroyo García, R. A. and E. Revilla (2013), 'The Current Status of Wild Grapevine Populations (Vitis vinifera ssp sylvestris) in the Mediterranean Basin', in D. Poljuha and B. Sladonja (eds), *The Mediterranean Genetic Code: Grapevine and Olive*, 51–72. London: Intech Open.

Aspinall, A., C. Gaffney and A. Schmidt (2008), *Magnetometry for Archaeologists*, Lanham: AltaMira.

Atherden, M., J. Hall and J. C. Wright (1993), 'A Pollen Diagram from the Northeast Peloponnese, Greece: Implications for Vegetation History and Archaeology', *The Holocene* 3 (4): 351–6.

Atkins, R. and I. Meadows (forthcoming), *Neolithic Pits, Pit Alignments and Iron Age to Roman Settlements at Wollaston Quarry Northamptonshire*, Museum of London Publications.

Auriemma, R. and V. Degrassi (2015), 'Flussi di circolazione e redistribuzione in Adriatico tra tarda Repubblica e Impero: anfore da contesti terrestrie subacquei', in *AdriAtlas et l'histoire de l'espace adriatique du VIe s. a.C. au VIIIe s. p.C.: actes du colloque international de Rome (4–6 novembre 2013)*, 453–78, Bordeaux.

Aversano, R., B. Basile, M. P. Buonincontri, F. Carucci, D. Carputo, L. Frusciante and G. Di Pasquale (2017), 'Dating the Beginning of the Roman Viticultural Model in the Western Mediterranean: The Case Study of Chianti (Central Italy).' *PLoS ONE* 12 (11): e0186298. https://doi.org/10.1371/journal.pone.0186298.

Ayalon, E., R. Frankel and A. Kloner, eds (2009), *Oil and Wine Presses in Israel from the Hellenistic, Roman and Byzantine Periods*, Oxford: BAR Publishing.

Aydinoğlu, Ü. and A. K. Şenol, eds (2010). *Olive Oil and Wine Production in Anatolia During Antiquity: Symposium Proceedings 06–08 November 2008, Mersin, Turkey*, Istanbul: Research Centre of Cilician Archaeology, KAAM.

Bacchetta, A. (2003), *Edilizia Rurale Romana. Materiali e tecniche costruttive nella Pianura Padana (II sec. a.C. – IV sec. d.C.)*, Firenze: All'Insegna del Giglio.

Bacilieri, R., L. Bouby, I. Figueiral, C. Shaal, J.-F. Terral, C. Breton, S. Picq, A. Weber and A. Schlumbaum (2017), 'Potential of Combining Morphometry and Ancient DNA Information to Investigate Grapevine Domestication', *Vegetation History and Archae P obotany*, 26 (3): 345–56.

Bacilieri, R., T. Lacombe, L. Le Cunff, M. Di Vecchi-Staraz, J.-P. Genna and B. Péros (2013), 'Genetic Structure in Cultivated Grapevines Is Linked to Geography and Human Selection', *BMC Plant Biology*, 13: 25.

Badiali, F. (2009), 'Dagli acini d'uva alla ricostruzione del paesaggio: l'esempio dell'ecomuseo della collina e del vino di castello di Serravalle (Bologna)', *Bollettino Associazione Italiana Cartografia*, 136–7: 81–92.

Badiali, F. (2012), *Metodologia e ricerche sperimentali sull'evoluzione del paesaggio in contesti 'naturali' (Montese e Castello di Serravalle, Appennino emiliano) e urbani (Oradea, Romania)*, Scuola di Dottorato in ESS: Environment Resources and Cultural Heritage, XXIV (2009–12), Università degli studi di Modena e Reggio Emilia.

Bakels, C. (2020), 'Pollen and archaeology', in A. G. Henry (ed.), *Handbook for the Analysis of Micro-particles in Archaeological Samples*, 203–24, Cham: Springer.

Baker, A., G. Mariethoz, L. Comas-Bru, A. Hartmann, S. Frisia, A. Borsato, P. C. Treble and A. Asrat (2021), 'The Properties of Annually Laminated Stalagmites: A Global Synthesis', *Reviews of Geophysics*, 59 (2): e2020RG000722.

Balmelle, C. and J.-P. Brun (2005), 'La vigne et le vin dans la mosaïque romaine et byzantine', in H. Morlier (ed.), *La Mosaïque gréco-romaine IX*, 899–921, Rome.

Bandinelli, R. and G. Di Collalto (1997), 'La presenza del vitigno Ansonica in Toscana', in *Ansonica: La riscoperta di un antico vitigno*, Atti del Convegno Nazionale, Isola del Giglio 1996, 23–35, Firenze: Regione Toscana.

Bandini Mazzanti, M., C. A. Accorsi, R. Curina, A. Cattini and M. Marchesini (1995), 'Carpological Remains from a Pit Fill at the Roman Villa (1st–4th Century AD) of Casteldebole (Bologna–Emilia Romagna, North Italy)', *Giornale Botanico Italiano*, 129 (2): 222.

Baratta, G. (1994a), 'Circa Alpes ligneis vasis condunt circulisque cingunt', *Archeologia Classica*, 46: 233–60.

Baratta, G. (1994b), 'Bolli su botti', in *Epigrafia della produzione e della distribuzione: Actes de la VIIe Rencontre franco-italienne sur l'épigraphie du monde romain (CEFR 193)*, 555–65, Rome.

Baratta, G. (1997), Le botti: dati e questioni', in *Techniques et économie antiques et médiévales: le temps de l'innovation, Colloque international C.N.R.S. Aix-en-Provence 21–23 Mai 1996*, 109–12, Paris.

Baratta, G. (2001), 'Un'alternativa all'anfora: la botte', in J. Maldonado Rosso (ed.), *Actas del I simposio de la Asociación Internacional de Historia y Civilización de la Vid y del Vino*, Vol. I, 149–55, El Puerto de Santa Maria.

Baratta, G. (2005), 'La cupa nell'ambito femminile: dalla caupona al loculus?', in F. Cenerini and A. Buonopane (eds), *Donna e vita cittadina nella documentazione epigrafica: Epigrafia e antichità* 23, 95–108, Faenza.

Barisashvili, G. (2017), *Making Wine in Qvevri*, Tbilisi: Biological Farming Association.

Barker, P. (1998), *Techniques of Archaeological Excavation*, 3rd edn, London and New York: Routledge.

Barnabei, F. (1901), *La villa pompeiana di P. Fannio Sinistore scoperta presso Boscoreale*, Rome: Tipografia della R. Accademia dei Lincei.
Barzanò, A. (2018), 'La crisi del III secolo', in C. Bearzot and A. Barzanò (eds), *Istituzioni di storia antica: Dalla preistoria all'espansione araba,* 279–90, Rome.
Basso, P., G. Falezza, M. Migliavacca and F. Saggioro (forthcoming), *Veronensium mensa: Food and Wine in Ancient Verona*, Roma: Quasar.
Baxter, M. J. (2008), 'Mathematics, Statistics and Archaeometry: The Past 50 Years or So', *Archaeometry* 50 (6): 968–82.
Beaurredon, J. (1887), *La Culture de la Vigne dans l'Antiquité*, Imprimerie Dax: Labèque .
Bellelli, V. (2018), 'Ischia, le anfore etrusche di Nocera e il vino amineo', *La Parola del Passato* 73 (2): 359–429.
Bellin, D. and G. Bolognesi (forthcoming), 'Analisi morfometriche e genetiche sui vinaccioli', in P. Basso, G. Falezza, M. Migliavacca and F. Saggioro (eds), *Veronensium mensa: Food and Wine in Ancient Verona*, Roma: Quasar.
Benčić G. (2006), 'L'origine di Torre: dalla torre all'abitato', in D. L. Ratković (ed.), *Torre, Fratta, Abrega: Patrimonio culturale*, 275–98, Poreč.
Benoit, M. F. (1954), 'Amphores et céramique de l'épave de Marseille', *Gallia* 12 (1): 35–51.
Benoit, M. F. (1958), 'Typologie et épigraphie amphoriques: Les marques de Sestius', *Rivista di Studi Liguri* 23: 247–86.
Benoit, M. F. (1961), *Fouilles sous-marines: L'épave du Grand Congloué à Marseille*, Gallia Supp. 14, Paris.
Berger, J. F. and C. Jung (1996), 'Fonction, Evolution et "Taphonomie" des parcellaires en moyen vallée du Rhône: Un exemple d'approche integrée en archéomorphologie et en géoarcheologie', in G. Chouquer (ed.), *Les formes du paysage: Vol. 2. Archéologie des parcellaires*, 95–112, Paris: Éditions Errance.
Bernigaud, N., A. Bondeau, J. Guiot, F. Bertoncello, M.-J. Ouriachi, L. Bouby, P. Leveau, L. Bernard and D. Isoardi (forthcoming), 'The Impact of Climate Change on the Agriculture and the Economy of Southern Gaul: New Perspectives of Agent-Based Modeling', *PLoS ONE.*
Bernigaud, N., A. Bondeau and J. Guiot (2021), 'Understanding the Development of Viticulture in Roman Gaul during and after the Roman Climate Optimum: The Contribution of Spatial Analysis and Agro-Ecosystem Modelling', *Journal of Archaeological Science: Reports* 38: 103099.
Bernal Casasola, D. (2015), 'What Contents Do We Characterize in Roman Amphorae? Methodological and Archaeological Thoughts on a Trending Topic', in C. Oliveira, R. Morais and A. Morillo (eds), *Archaeoanalytics: Chromatography and DNA analysis in archaeology. International Symposium*, 61–83, Porto: University of Porto.
Bertolazzi, R. and V. Guidorizzi (2017), 'Regio X – Venetia et Histria. Arilica – Sirmio (I.G.M. 47, II, NE; 48, III, NE, NO, IV, SO)', in M. L. Lazzarini (ed.), *Supplementa Italica: Nuova Serie*, Vol. 29, 309–72, Rome: Quasar.
Bevan, A. (2014), 'Mediterranean Containerization', *Current Anthropology* 55 (4): 387–418.
Bezeczky T. (1998), *The Laecanius Amphora Stamps and the Villas of Brijuni*, Wien: Österreichische Akademie der Wissenschaften, Philosophisch-historische Klasse, Denkschriften, 261 Band.
Bielefeld, D. (1997), *Stadtrömische Eroten-Sarkophage*, ASR V, Teil 2, Fasz. 2.
Biggs, M. (1999), 'Putting the State on the Map: Cartography, Territory and European State Formation', *Comparative Studies in Society and History* 41 (2): 374–405.
Bignami, F., R. Sciarra, S. Carniel and R. Santoleri (2007), 'Variability of Adriatic Sea Coastal Turbid Waters from SeaWiFS Imagery', *Journal of Geophysical Research: Oceans* 112 (C3).
Billiard, R. (1913), *La Vigne dans l'Antiquité*, Lyon: Librairie H. Lardanchet.
Birley, A. R. (1991), 'Vindolanda: New Writing-tablets 1986–9', in V. Maxfield and M. J. Dobson (eds), *Roman Frontier Studies 1989: Proceedings of the XVth International Congress of Roman Frontier Studies (Exeter)*, 16–20, Exeter.

Bitella, G., R. Rossi, A. Loperte, A. Satriani, V. Lapenna, M. Perniola and M. Amato (2015), 'Geophysical Techniques for Plant, Soil and Root Research Related to Sustainability', in A. Vastola (ed), *The Sustainability of Agro-Food and Natural Resource Systems in the Mediterranean Basin*, 353–72, New York: Springer.

Boissinot, P. (1998), 'Les vignobles antiques du midi de la France', Poster at *Geoarchaeology of the Landscapes of Classical Antiquity Conference*, Ghent, 23–4 October.

Boissinot, P. (2009), 'Les vignobles des environs de Mégara Hyblaea et les traces de la viticulture italienne durant l'Antiquité', *MEFRA* 121 (1): 83–132.

Boissinot, P. and J.-E. Brochier (1998), 'Pour une archéologie du champ', in G. Chouquer (ed.), *Les formes des paysages,* 35–56, Paris: Éditions Errance.

Boldrin, A., S. Carniel, M. Giani, M. Marin, F. Bernardi Aubry, A. Campanelli, F. Grilli and A. Russo (2009), 'Effects of Bora Wind on Physical and Biogeochemical Properties of Stratified Waters in the Northern Adriatic', *Journal of Geophysical Research* 114 (C8).

Bolla, M. (2008), 'Le Iscrizioni dal Teatro Romano di Verona', in P. Basso, A. Buonopane, A. Cavarzere and S. Pesavento Mattioli (eds), *Est enim ille flos Italie: Vita Economica e Sociale nella Cisalpina Romana*, 77–101, Verona: Qui Edit.

Bondeau, A., P. C. Smith, Z. Zaehle, S. Schaphoff, W. Lucht, W. Cramer, D. Gerten, H. Lotze-Campen, C. Müller, M. Reichstein and B. Smith (2007), 'Modelling the Role of Agriculture for the 20th Century Global Terrestrial Carbon Balance', *Global Change Biology*, 13: 67–706.

Bonhomme, V., S. Picq, C. Gaucherel and J. Claude (2014), 'Momocs: Outline Analysis Using R', *Journal of Statistical Software*, 56: 1–24.

Bonhomme, V., S. Picq, S. Ivorra, A. Evin, T. Pastor, R. Bacilieri, T. Lacombe, I. Figuerail, J.-F. Terral and L. Bouby (2020), 'Eco-evo-devo Implications and Archaeobiological Perspectives of Trait Covariance in Fruits of Wild and Domesticated Grapevines', *PLoS ONE* 15 (11): e0239863. https://doi.org/10.1371/journal.pone.0239863.

Bonhomme, V., J.-F. Terral, V. Zech-Matterne, S. Ivorra, T. Lacombe, G. Deborde, P. Kuchler, B. Limier, T. Pastor, P. Rollet and L. Bouby (2021), 'Seed Morphology Uncovers 1500 Years of Vine Agrobiodiversity before the Advent of the Champagne Wine', *Scientific Reports* 11: 1–14.

Borgna E., S. Corazza, M. Marchesini, A. Pecci and G. Petrucci (2022), 'Fire Installations at the Bronze Age Site of Ca' Baredi Near Aquileia: An Interdisciplinary Insight into Subsistence and Social Practices', *Journal of Archaeological Science: Reports*, 46: 103648.

Borgongino, M. (1999), 'Le colture extraurbane', in A. Ciarallo and E. de Carolis (eds), *Homo faber: Natura, scienza e tecnica nell'antica Pompei*, 89–91, Milano: Electa.

Bosi, G., A. M. Mercuri, M. B. Mazzanti, A. Florenzano, M. C. Montecchi, P. Torri, D. Labate and R. Rinaldi (2015a), 'The Evolution of Roman Urban Environments through the Archaeobotanical Remains in Modena – Northern Italy', *Journal of Archaeological Science*, 53: 19–31.

Bosi, G., A. Benatti, R. Rinaldi, D. Dallai, C. Santini, M. Carbognani, M. Tomaselli and M. B. Mazzanti (2015b), 'The Memory of Water: Archaeobotanical Evidence of Wetland Plants from Modena (Emilia Romagna, Northern Italy) and Palaeoecological Remarks', *Plant Biosystems*, 149: 144–53.

Bosi, G., M. B. Mazzanti, M. C. Montecchi, P. Torri and R. Rinaldi (2017), 'The Life of a Roman Colony in Northern Italy: Ethnobotanical Information from Archaeobotanical Analysis', *Quaternary International*, 460: 135–56.

Bosi, G. and M. Marchesini (2017), 'Reperti archeobotanici di vite a Mutina', in L. Malnati, S. Pellegrini, F. Piccinini and C. Stefani (eds), *Mutina splendidissima: La città romana e la sua eredità*, 288–9, Rome: De Luca Editori d'Arte.

Bosi, G., E. Castiglioni, R. Rinaldi, M. Mazzanti, M. Marchesini and M. Rottoli (2020), 'Archaeobotanical Evidences of Food Plants in Northern Italy during the Roman Age', *Vegetation History and Archaeobotany*, 29 (6): 681–97.

Bosi, G., D. Labate, R. Rinaldi, M. C. Montecchi, M. Mazzanti, P. Torri, F. M. Riso and A. M. Mercuri (2019), 'A Survey of the Late Roman Period (3rd–6th century AD): Pollen, NPPs and Seeds/Fruits for Reconstructing Environmental and Cultural Changes after the Floods in Northern Italy', *Quaternary International*, 499: 3–23.

Bottema, S. (1999), 'Landscape Archaeology and Reconstruction of the Mediterranean Environment based on Palynology', in P. Leveau, F. Trément, K. Walsh and G. Barker (eds), *Environmental Reconstruction in Mediterranean Landscape Archaeology*, The Archaeology of Mediterranean Landscapes 2, 9–16, Oxford: Oxbow Books.

Bottema, S., G. Entjes-Nieborg and W. Van Zeist, eds (1990), *Man's Role in the Shaping of the Eastern Mediterranean Landscape*, Rotterdam: A. A. Balkema.

Bottema, S. and A. Sarpaki (2003), 'Environmental Change in Crete: A 9000-year Record of Holocene Vegetation History and the Effect of the Santorini Eruption', *The Holocene* 13 (5): 733–49.

Bottema, S. and H. Woldring (1990), 'Anthropogenic Indicators in the Pollen Record of the Eastern Mediterranean', in S. Bottema, G. Entjes-Nieborg and W. Van Zeist (eds), *Man's Role in the Shaping of the Mediterranean Landscape*, 231–64, Rotterdam: Balkema.

Botto M. (2013), 'The Phoenicians and the Spread of Wine in the Central West Mediterranean', in S. Celestino Pérez and J. Blánquez Pérez (eds), *Patrimonio Cultural de la Vid y el Vino*, 103–31, Madrid: UAM.

Bouby, L., I. Figueiral, A. Bouchette, N. Rovira, S. Ivorra, T. Lacombe, T. Pastor, S. Picq, P. Marinval and J.-F. Terral (2013), 'Bioarchaeological Insights into the Process of Domestication of Grapevine (Vitis vinifera L.) during Roman Times in Southern France', *PLoS ONE*. https://doi.org/10.1371/journal.pone.0063195

Bouby, L. and P. Marinval (2001), 'La vigne et le début de la viticulture en France: apports de l'archéobotanique', *Gallia*, 58: 13–28.

Bouby, L., P. Marinval and J. Terral (2014), 'From Secondary to Speculative Production? The Protohistory History of Viticulture in Southern France', in A. Chevalier, E. Marinova and L. Pena-Chocarro (eds), *Plants and People: Choices and Diversity through Time*, 175–210, Oxford: Oxbow.

Bouby, L., N. Wales, M. Jalabadze, N. Rusishvili, V. Bonhomme, J. Ramos-Madrigal, A. Evin, S. Ivorra, T. Lacombe, C. Pagnoux, E. Boaretto, M. T. P. Gilbert, R. Bacilieri, D. Lordkipanidze and D. Maghradze (2021), 'Tracking the History of Grapevine Cultivation in Georgia by Combining Geometric Morphometrics and Ancient DNA', *Vegetation History and Archaeobotany*, 30 (1): 63–76.

Boudreau L. (1894), *Études d'histoire générale: Histoire de l'alimentation*, Paris.

Boulay, T. (2018), 'Tastes of Wine: Sensorial Wine Analysis in Ancient Greece', in K. C. Rudolph (ed.), *Taste and the Ancient Senses*, 197–211, New York: Routledge.

Bouvier, M. (2000), 'Recherches sur les gouts de vins antiques', *Pallas*, 53: 115–34.

Bowman, A. and A. Wilson, eds (2009), *Quantifying the Roman Economy: Methods and Problems*, Oxford: OUP.

Bowman, A. K. and A. Wilson, eds (2013), *The Roman Agricultural Economy: Organization, Investment, and Production*, 1st edn, Oxford: OUP.

Braconi, P. (2012), 'In vineis arbustisque: Il concetto di vigneto in età romana', in A. Ciacci, P. Rendini and A. Zifferero (eds), *Archeologia della Vite e del Vino in Toscana e nel Lazio: Dalle tecniche dell'indagine archeologica alle prospettive della biologia molecolare*, 291–306, Borgo San Lorenzo: Edizioni All'Insegna del Giglio.

Brandaglia, M. (1997), 'Le origini e la viticoltura dell'Ansonica dell'Isola del Giglio', in *Ansonica: La riscoperta di un antico vitigno*, Atti del Convegno Nazionale, Isola del Giglio 1996, 36–60, Firenze: Regione Toscana.

Brandaglia, M. (2001), *Il vitigno Ansonica: Dalla ricerca delle antiche colture allo stato attuale della viticoltura sull'Isola del Giglio*, Grosseto: Provincia di Grosseto.

Braudel, F. (1949), *La Méditerranée et le monde méditerranéen à l'époque de Philippe II*, Paris: Armand Colin.

Breitenbach, S. F. M., K. Rehfeld, B. Goswami, J. U. L. Baldini, H. E. Ridley, D. J. Kennett, K. M. Prufer, V. V. Aquino, Y. Asmeron, V. J. Polyak, H. Cheng, J. Kurths and N. Marwan (2012), 'Constructing Proxy-Record Age Models (COPRA)', *Climate of the Past*, 8 (5): 1765–79.

Brewer, S., J. Guiot and D. Barboni (2013), 'Pollen Methods and Studies: Use of Pollen as Climate Proxies', in S. A. Elias and C. J. Mock (eds), *Encyclopedia of Quaternary Science*, 2nd edn, 805–15, Elsevier.

Briggs L., S. Demesticha, S. Katzev, H. W. Swiny, O. E. Craig and L. Drieu (2021), 'There's More to a Vessel than Meets the Eye: Organic Residue Analysis of "Wine" Containers from Shipwrecks and Settlements of Ancient Cyprus (4th–1st century BCE)', *Archaeometry*, 64 (3): 779–97.

Broadhead, W. (2007), 'Colonization, Land Distribution, and Veteran Settlement', in P. Erdkamp (ed.), *A Companion to the Roman Army*, 148–63, Oxford & Carlton: Blackwell Publishing Ltd. Malden.

Broecker, W. S., E. A. Olson and P. C. Orr (1960), 'Radiocarbon Measurements and Annual Rings in Cave Formations', *Nature*, 185: 93–4.

Broise, H. and V. Jolivet (1995), 'Bonification agraire et viticulture antiques autour du site de Musarna (Viterbe)', in L. Quilici and S. Quilici Gigli (eds), *Interventi di Bonifica Agraria nell'Italia Romana*, Atlante tematico di topografia antica 4, 107–16, Rome: L'Erma di Bretschneider.

Broise, H. and V. Jolivet (1996), 'Musarna (Viterbe): le site étrusco-romain', *MEFRA*, 108 (1): 455–60.

Broodbank, C. (2013), *The Making of the Middle Sea: A History of the Mediterranean from the Beginning to the Emergence of the Classical World*, London: Thames and Hudson.

Brown, A. G. and I. Meadows (2000), 'Roman Vineyards in Britain: Finds from the Nene Valley and New Research', *Antiquity*, 74: 491–2.

Brown, A. G., I. Meadows, S. D. Turner and D. Mattingley (2001), 'Roman Vineyards in Britain: Stratigraphic and Palynological Data from Wollaston in the Nene Valley, England', *Antiquity*, 75: 745–57.

Brown, A. G., K. Walsh, D. Fallu, S. Cuchiarro, P. Tarolli, P. Zhao, B. Pears, K. van Oost, L. Snape, A. Lang, R. M. Albert, I. G. Alsos and C. Waddington (2021a), 'Ending the Cinderella Status of Terraces and Lynchets in Europe: The Geomorphology of Agricultural Terraces and Implications for Ecosystem Services and Climate Adaptation', *Geomorphology*, 379: 107579.

Brown, A. G., K. Walsh, D. Fallu, S. Cuchiarro and P. Tarolli (2021b), 'European Agricultural Terraces and Lynchets: From Archaeological Theory to Heritage Management', *World Archaeology*, 52: 566–88.

Brown, A. G., T. Fonville, M. van Hardenbroek, G. Cavers, A. Crone, F. McCormick, E. Murray, H. Mackay, N. Whitehouse, A. Henderson, P. Barratt, K. Head, I. G. Alsos and D. Pirrie (2022), 'New Integrated Molecular Approaches for Understanding Lake Settlements in NW Europe', *Antiquity* 96 (389): 1179–99.

Brown, J. (2010), 'A Possible Roman Vineyard on Land off Tavistock Avenue, Ampthill, Bedfordshire', *Northamptonshire Archaeology Report* 10/132.

Brun, J.-P. (1986), *L'Oléiculture Antique en Provence: Les Huileries du department du Var*, Paris: CNRS.

Brun J.-P. (1993), 'La discrimination entre les installations oléicoles et vinicoles', in M.-C. Amouretti and J. P. Brun (eds), *La production du vin et de l'huile en Méditerranée*, BCH Supp. XXVI, 511–38, Athens and Paris: École Française d'Athènes.

Brun, J.-P. (1997), 'Production de l'huile et du vin en Lusitanie romaine', *Conimbriga*, 36: 45–72.

Brun, J.-P. (2001), 'La viticulture antique en Provence', in J.-P. Brun and F. Laubenheimer (eds), *La viticulture en Gaule*, 69–89, Paris: CNRS.

Brun, J.-P. (2003), *Le vin et l'huile dans la Méditerranée antique: Viticulture, oleiculture et proceeds de transformation*, Paris: Éditions Errance.
Brun, J.-P. (2004a), *Archéologie du vin et de l'huile de la préhistoire à l'époque hellénistique*, Paris: Editions Errance.
Brun, J.-P. (2004b), *Archéologie du vin et de l'huile dans l'Empire romain*, Paris: Editions Errance.
Brun, J.-P. (2005), *Archéologie du vin et de l'huile en Gaule romaine*, Paris: Editions Errance.
Brun, J.-P., N. Garnier and G. Olcese, eds (2020), *A. Making Wine in Western-Mediterranean. B. Production and the Trade of Amphorae: Some new data from Italy – International Congress of Classical Archaeology – Panel 3.5*, Proceedings of the 19th International Congress of Classical Archaeology (Cologne-Bonn 22–6 May 2018), Heidelberg: Propylaeum.
Brun, J.-P. and K. J. Gilles (2001), 'La viticultura Antique en Rhénanie', *La viticulture en Gaule, Gallia*, 58: 165–79.
Brun, J.-P. and F. Laubenheimer, eds (2001), *La viticulture en Gaule*, Paris: CNRS.
Brun, J.-P. and Cl. Pommerpuy (1992), 'La nécropole de Bucy-le-Long (Picardie), Le fond du Petit Marais', *Association Française pour l'Étude de l'Âge du Fer (AFEAF): Bulletin Intérieur*, 10: 17–18.
Brun, J.-P., M. Poux and M.-L. Hervé-Monteil (2011), *La vigne et le vin dans les Trois Gaules*, Paris: CNRS.
Brun, J.-P., M. Poux and A. Tchernia, eds (2004), *Le Vin: Nectar des Dieux, Génie des Homme*, Gollion.
Buffat, L., C. Pellecuer, S. Mauné and H. Pomarèdes (2001), 'La viticulture antique en Languedoc-Roussillon', in J.-P. Brun and F. Laubenheimer (eds), *La viticulture en Gaule*, 91–111, Paris: CNRS.
Bucelli, P. and D. I. Costantini (2009), 'Wine Grape and Vine Zoning', in E. A. C. Costantini (ed.), *Manual of Methods for Soil and Land Evaluation*, 353–400, Enfield: Science Publishers.
Buchi, E. (1996), 'La vitivinicultura cisalpina in età romana', in G. Forni and A. Scienza (eds), *2500 anni di cultura della vite nell'ambito alpino e cisalpino*, 373–89, Trento: Istituto Trentino del Vino.
Buchi, E. (1987), 'Porta Leoni e la Fondazione di Verona Romana', *Museum Patavinum*, 5: 13–45.
Büntgen, Ü., V. S. Myglan, F. Charpentier Ljunqvist, M. McCormick, N. Di Cosmo, M. Sigl, J. Jungclaus, S. Wagner, P. J. Krusic, J. Esper, J. O. Kaplan, M. A. C. de Van, J. Lutherbacher, L. Wacker, W. Tegel, A. V. Kirdyanov (2016), 'Cooling and Societal Change during the Late Antique Little Ice Age from 536 to around 660 AD', *Nature Geoscience*, 9: 231–36.
Burzynski-Chang E. A., E. J. Brown, N. Reshef and G. L. Sacks (2019), 'Malate Content in Wild *Vitis* Spp. Demonstrates a Range of Behaviors during Berry Maturation', *American Journal of Enology and Viticulture*. https://doi.10.5344/ajev.2019.19015
Busana, M. S. (2002). *Architetture rurali nella Venetia romana*, Rome: L'Erma di Bretschneider.
Busana, M. S. (2003), 'La produzione vinaria dalle fonti archeologiche nella Valpolicella di età romana', *Annuario Storico della Valpolicella* 19: 117–32.
Busana, M. S. (forthcoming), 'Wine Production in the Roman West: The Role of Artificial Heating', in D. Van Limbergen, E. Dodd and M. S. Busana (eds), *Vine-Growing and Winemaking in the Roman World*, Leuven: Peeters.
Busana, M. S., C. D'Inca and S. Forti (2009), 'Olio e pesce in epoca romana nell'alto e medio Adriatico', in S. Pesavento Mattioli and M.-B. Carre (eds), *Olio e Pesce in Epoca Romana: Produzione e Commercio nelle Regioni dell'Alto Adriatico: Atti del Convegno (Padova 16 febbraio 2007)*, 35–79, Rome: Quasar.
Butorac, L., K. Hančević, K. Lukšić, Z. S. Kvorc, M. Leko, E. Maul and G. Zdunić (2018), 'Assessment of Wild Grapevine (*Vitis vinifera* ssp. *sylvestris*) Chlorotypes and Accompanying Woody Species in the Eastern Adriatic Region' *PLoS ONE*, 13 (6): e0199495.
Buxeda i Garrigós, J., M. A. Cau Ontiveros and V. Kilikoglou (2003), 'Chemical Variability in Clays and Pottery from a Traditional Cooking Pot Production Village: Testing Assumptions in Pereruela', *Archaeometry*, 45: 1–17.

Cabanes, D. (2020), 'Phytolith Analysis in Paleoecology and Archaeology', in A. G. Henry (ed.), *Handbook for the Analysis of Micro-Particles in Archaeological Samples*, 255–88, Cham: Springer.

Caillaud, C. (2018), 'Vinalia, fête du vin et de la cuisine antiques', *Territoires du vin*, 8: https://preo.u-bourgogne.fr/territoiresduvin/index.php?id=1339#toc

Calamita, G., A. Perrone, L. Brocca, B. Onorati and S. Manfreda (2015), 'Field Test of a Multi-Frequency Electromagnetic Induction Sensor for Soil Moisture Monitoring in Southern Italy Test Sites', *J. Hydrol*, 529 (1): 316–29.

Callender, M. H. (1965), *Roman Amphorae*, London: Oxford University Press.

Calzolari, M. (2008), *Città dell'Aemilia: Mutina: le fonti letterarie di Modena romana*, Modena: G.s.B.M.

Campana, S. (2016), 'Sensing Ruralscapes: Third-Wave Archaeological Survey in the Mediteranean Area', in M. Forte and S. Campana (eds), *Digital Methods and Remote Sensing in Archaeology: Quantitative Methods in the Humanities and Social Sciences*, 113–45, Cham: Springer.

Campanile, T. (1922), 'Negrar di Valpolicella: Avanzi di una villa romana con magnifici mosaici', *Notizie degli Scavi di Antichità*: 347–61.

Campbell, J. B. (1995), 'Sharing Out Land: Two Passages in the "Corpus Agrimensorum Romanorum"', *The Classical Quarterly*, 45: 540–6.

Campbell, J. B. (1996), 'Shaping the Rural Environment: Surveyors in Ancient Rome', *Journal of Roman Studies*, 86: 74–99.

Campbell, J. B. (2000), *The Writings of the Roman Land Surveyors: Introduction, Text, Translation and Commentary*, Journal of Roman Studies Monograph 9, London: Society for the Promotion of Roman Studies.

Campbell, J. B. (2006), 'Surveyors, Topography, and Definition of Landholding, in Ancient Rome', in D. Conso, A. Gonzales, J. Y. Guillaumin (eds), *Les vocabulaires techniques des arpenteurs romains: Actes du Colloque International (Besançon 2002)*, 540–46, Besançon: Presses universitaires de Franche-Comté, Institut des Sciences et Techniques de l'Antiquité.

Carandini, A. (1989), 'L'economia italica fra Repubblica e Medio Impero considerata dal punto di vista di una merce: il vino', in *Amphores Romaines et Histoire Économique: Dix ans de recherche*, 505–21, Rome.

Carandini, A. and A. Ricci, eds (1985), *Settefinestre una villa schiavistica nell'Etruria romana*, Modena: Panini.

Carbonneau, A. and R. Rotunno (2000), 'Reconstitution du vignoble de Pompéi', *Pallas*, 53: 135–40.

Cardarelli, A., G. Bosi, R. Rinaldi, M. Ucchesu and G. Bacchetta (2015), 'Vino o non vino? Nuovi dati sui vinaccioli della Terramara di Montale (Modena) tra la fine della media età del Bronzo e il Bronzo Recente', in *Preistoria del cibo: 50ma Riunione Scientifica dell'Istituto Italiano di preistoria e Protostoria*, 1–9, Rome: Istituto Italiano di Preistoria.

Cardell, M. F., A. Amengual and R. Romero (2019), 'Future Effects of Climate Change on the Suitability of Wine Grape Production across Europe', *Regional Environmental Change*, 19 (3): 1–12.

Carmenati, R., G. Fiorentino and R. Perna (2019), 'The Roman Villa of Villamagna (Urbisaglia (Macerata), Italy): Pilot Archaeobotany Analysis', in G. Fiorentino (ed), *Book of Abstracts of 18th International Workgroup for Palaeoethnobotany Conference* (Lecce, 3rd–8th June 2019), 169, Lecce.

Carrato, C. (2017a), *Le dolium en Gaule Narbonnaise (Ier s. av. J.-C. – IIIé s. ap. J.-C.): Contribution à l'histoire socio-économique de la Méditerranée nord-occidentale*, Ausonius Éditions, Collection Mémoires 46, Bordeaux.

Carrato, C. (2017b), 'Dynamique de la production vinicole en Gaule Narbonnaise à l'époque romaine à partir des capacités de stockage (Ier s. av. J.-C. – Ve s. ap. J.-C.)', in J.-P. Brun, N. Garnier and G. Olcese (eds), *A. Making Wine in Western-Mediterranean. B. Production and the Trade of Amphorae: Some New Data from Italy: Panel 3.5*, 159–66, Heidelberg: Propylaeum.

Carrato, C., V. M. Ferreras, J.-M. Dautriam and M. Bois (2019), The Biggest Opus doliare Production in Narbonese Gaul Revealed by Archaeometry (first to second centuries A.D.), *ArcheoSciences, revue d'Archéométrie*, 43 (1): 69–82.

Carrato, C., A. Ichkhakh, M. Kbiri Alaoui, E. Rocca, M.-P. Ruas, M. Alilou, V. Mathieu, S. Sanz-Laliberté, N. Garnier and J.-B. Pineau (2020), 'Recent Discovery of an Urban Winery in Rirha (Sidi Slimane, Morocco), 2nd–3rd Century CE', in J.-P. Brun, N. Garnier and G. Olcese (eds), *A. Making Wine in Western-Mediterranean. B. Production and the Trade of Amphorae: Some New Data from Italy: Panel 3.5*, 167–75, Heidelberg: Propylaeum.

Carre M.-B., V. Kovačić and F. Tassaux (2011), *L'Istrie et la mer: La côte du Parentin dans l'Antiquité,* Bordeaux: Ausonius Mémoires 25.

Carre, M.-B. and S. Pesavento Mattioli (2003), 'Anfore e Commerci nell'Adriatico', in F. Lenzi (ed.), *L'archeologia dell'Adriatico dalla Preistoria al Medioevo*, 268–85, Firenze: All'insegna del Giglio.

Carre, M.-B., S. Pesavento Mattioli and P. Monsieur (2014), 'Transport Amphorae Lamboglia 2 and Dressel 6A: Italy and/or Dalmatia? Some Clarifications', *Journal of Roman Archaeology*, 27: 417–28.

Carreras, C. (2014), 'Nuevas tendencias y datos sobre la demografía romana en la Península Ibérica', *Boletín del Seminario de Estudios de Arte y Arqueología*, 80: 53–82.

Carroll, M. (2022), 'Viticulture, Opus Doliare, and the Patrimonium Caesaris at the Roman Imperial Estate at Vagnari (Puglia)', *Journal of Roman Archaeology*, 35 (1): 1–26.

Carruthers, W. J. and K. L. Hunter Dowse (2019), 'A Review of Macroscopic Plant Remains from the Midland Counties', *Research Report Series*, 47/2019, Historic England.

Cartaxo-Pinto, S., G. H. C. Paulo, B. R. Jackes, V. Gonçalves-Esteves and C. B. F. Mendonça (2022), 'A Palynological Perspective on the Tribe Viteae of the Grape Family (Vitaceae)', *Grana,* 61: 27–44.

Carver, E. (2001), *The Visibility of Imported Wine and Its Associated Accoutrements in Later Iron Age Britain,* British Archaeological Reports.

Castagnoli, F. (1984), 'Sulle più antiche divisioni agrarie romane', *Accademia Nazionale Lincei: Rendiconti classe scienze morali, storiche e filologiche*, 39: 7–17.

Castagnoli, F. (1993), *Topografia antica: Un metodo di studio*, Vol. 2, Rome: Istituto Poligrafico e Zecca dello Stato.

Castelletti, L., E. Castiglioni, M. Cottini and M. Rottoli (1996), 'Analisi paleobotaniche', in P. Melli (ed.), *La città ritrovata: Archeologia urbana a Genova 1984–1994*, 263–9, Genova: Tormena Editore.

Castelletti, L., E. Castiglioni, M. Cottini and A. Di Vora (1998), 'Analisi morfometrica dei vinaccioli di vite (Vitis vinifera L.) provenienti da scavi archeologici', *Atti UISPP XIII*, 11–24, Forlì.

Castiglione d'Orcia 2011 = Biagini, B., ed. (2011), *Origini della viticoltura*, Atti del Convegno, Castiglione d'Orcia 2010, Podere Forte: Tipografia Castiglione d'Orcia.

Castiglioni, E. and M. Rottoli (1994), 'Resti vegetali: carboni, semi e frutti. Ricostruzione dell'ambiente naturale e coltivato', in E. Cavada (ed.), *Archeologia a Mezzacorona: Documenti per la storia del popolamento rustico di età romana nell'area atesina*, 205–31, Trento.

Castrioto-Scanderberg, A., G. E. Hagberg, A. Cerasa, G. Committeri, G. Galati, F. Patria, S. Pitzalis, C. Caltagirone and R. Frackowiak (2005), 'The Appreciation of Wine by Sommeliers: A Functional Magnetic Resonance Study of Sensory Integration', *Neuroimage*, 25: 570–8.

Cau, M. Á., V. Martínez, A. Pecci, C. Mas and L. Fantuzzi (2018), 'Archaeometric Analysis for Provenance and Content of Roman Amphorae from the Site of Sa Mesquida (Mallorca, Spain)', *Mediterranean Archaeology and Archaeometry*, 18: 2–105.

Cavalieri Manasse, G. (1998), 'Banchi d'Anfore Romane a Verona: Nota Topografica', in S. Pesavento Mattioli (ed.), *Bonifiche e Drenaggi con Anfore in Epoca Romana: Aspetti Tecnici e Topografici*, 185–96, Modena: Franco Cosimo Panini.

Chabal, L. (2005), 'Charbons de bois et bois gorgé d'eau des puits antiques: des jardins de Lattara aux forêts du delta du Lez', *Lattara*, 18: 221–34.

Cheng, H., H. Zhang, C. Spötl, J. Baker, A. Sinha, H. Li, M. Bartolomé, A. Moreno, G. Kathayat, J. Zhao, X. Dong, Y. Li, Y. Ning, X. Jia, B. Zong, Y. A. Brahim, C. Pérez-Mejías, Y. Cai, V. F. Novello, F. W. Cruz, J. P. Severinghaus, Z. An and R. L. Edwards (2020), 'Timing and Structure of the Younger Dryas Event and its Underlying Climate Dynamics', *Proceedings of the National Academy of Sciences*, 117 (38): 23408–17.

Cheung, C. (2021a), 'Precious Pots: Making and Repairing Dolia', in H. Hochscheid and B. Russell (eds), *The Value of Making: Theory and Practice in Ancient Craft Production*, 171–88, Turnhout: Brepols.

Cheung, C. (2021b), 'Managing Food Storage in the Roman Empire', *Quaternary International*, 597: 63–75.

Chevalier, M., B. A. S. Davis, O. Heiri, H. Seppä, B. M. Chase, K. Gajewski, T. Lacourse, R. J. Telford, W. Finsinger, J. Guiot, N. Kühl, S. Y. Maezumi, J. R. Tipton, V. A. Carter, T. Brussel, L. N. Phelps, A. Dawson, M. Zanon, F. Vallé, C. Nolan, A. Mauri, A. de Vernal, K. Izumi, L. Holmström, J. Marsicek, S. Goring, P. S. Sommer, M. Chaput and D. Kupriyanov (2020), 'Pollen-based Climate Reconstruction Techniques for Late Quaternary Studies', *Earth-Science Reviews* 210: 103384.

Chouquer, G. (1983), 'La genèse des paysages du Centre-Est de la Gaule: Polymorphisme et production d'une identité rurale', *Dialogues d'histoire ancienne*, 9: 113–39.

Chouquer, G., M. Clavel-Lévêque, F. Favory and J. P. Vallat (1987), *Structures agraires en Italie Centro-méridionale: Cadastres et paysage ruraux*, Rome: Collection de l'École Française de Rome.

Chouquer, G. and F. Favory (1979), 'Contribution a la recherche des cadastres antiques', *Annales Littéraires de l'Université de Besançon: Centre de Recherches Spécialisées d'Histoire Ancienne. Equipe de Recherche Associée au CNRS*, Vol. 31, Paris.

Chouquer, G., and F. Favory (1991), *Les paysages de l'Antiquité: Terres et cadastres de l'Occident romain*, Paris: Éditions Errance Paris.

Chouquer, G. and F. Favory (2001), *L'arpentage Romain: histoire des textes, droit, techniques*, Paris: Éditions Errance, Paris.

Ciacci, A. and M. Giannace, eds (2012), *Senarum Vinea: Il paesaggio urbano di Siena. Forme di recupero e valorizzazione dei vitigni storici*, Siena: Nuova Immagine Editrice.

Ciacci, A., P. Rendini and A. Zifferero, eds (2012), *Archeologia della Vite e del Vino in Toscana e nel Lazio: Dalle tecniche dell'indagine archeologica alle prospettive della biologia molecolare*, Borgo San Lorenzo: Edizioni All'Insegna del Giglio.

Ciacci, A. and A. Zifferero, eds (2005), *VINUM: Un progetto per il riconoscimento della vite silvestre nel paesaggio archeologico della Toscana e del Lazio settentrionale*, Siena: Ci.Vin.

Ciacci, A. and A. Zifferero (2007a), 'Il "Progetto VINUM": prime considerazioni conclusive', in A. Ciacci, P. Rendini and A. Zifferero (eds), *Archeologia della Vite e del Vino in Etruria: Atti del Convegno Internazionale di Studi, Scansano 2005*, 249–72, Siena: Ci.Vin.

Ciacci, A. and A. Zifferero (2007b), 'Progetto VINUM: prospettive di ricerca in area populoniese', in L. Botarelli, M. Coccoluto and M. C. Mileti (eds), *Materiali per Populonia 6*, 397–419, Pisa: Edizioni ETS.

Cianferoni, R., Z. Ciuffoletti and L. Rombai (2002), *Storia dell'Agricoltura Italiana. L'Età Contemporanea: Dalle 'rivoluzioni agronomiche' alle trasformazioni del Novecento*, Florence: Edizioni Polistampa.

Ciarallo, A. and E. de Carolis, eds (1999), *Homo faber: Natura, scienza e tecnica nell'antica Pompeii*, Milan: Electa.

Cipriano, S. (2009), 'Le Anfore Olearie Dressel 6b', in S. Pesavento Mattioli and M.-B. Carre (eds), *Olio e Pesce in Epoca Romana: Produzione e Commercio nelle Regioni dell'Alto Adriatico: Atti del Convegno (Padova 16 febbraio 2007)*, 178, Rome.

Cipriano, S. and S. Mazzochin (2002), 'Analisi di alcune serie bollate di anfore Dressel 6B (AP. PVLCRI, FLAV.FONTAN/FONTANI, L.IVNI.PAETINI, L.TRE.OPTATI)', *AquilNost*, 73: 305–40.

Ciurletti, G. (1996), 'La viticoltura nel Trentino Alto Adige fra preistoria ed età romana: Contributi archeologici', in G. Forni and A. Scienza (eds), *2500 anni di cultura della vite nell'ambito Alpino e Cisalpino*, 441–74, Trento: Istituto Trentino del Vino.

Clark, A. (1996), *Seeing Beneath the Soil: prospecting methods in archaeology*, London: Routledge.

Clarke, J. R., S. De Caro and A. Lagi (2015), 'Oplontis e le sue ville', in P. G. Guzzo and G. Tagliamonte (eds), *Città vesuviane: Antichità e fortuna. Il suburbio e l'agro di Pompei, Ercolano, Oplontis e Stabiae*, 142–55, Roma: Treccani.

Clarke, J. and I. Meadows (2014), 'Archaeological evaluation and open area at Appleby Lodge, Wellingborough, Northamptonshire', *MOLA Northampton Report*, 14/126.

Clavel-Lévêque, M. (1983), *Cadastres et espace rural: Approches et réalités antiques*, Paris: Éditions du CNRS.

Clavel-Lévêque, M., D. Conso and A. Gonzales (1996), *Hygin l'arpenteur: L'établissement des limites*, Naples: Jovene.

Clavel-Lévêque, M. and A. Orejas, eds (2002), *Atlas historique des cadastres d'Europe*, Vol. II, Luxemburg.

Clavel-Lévêque, M. and A. Vignot, eds (1998), *Atlas historique des cadastres d'Europe*, Vol. I, Bruxelles.

Coles, J. (1981), *Archeologia Sperimentale*, Milano: Longanesi.

Columbu, A., V. Chiarini, C. Spötl, S. Benazzi, J. Hellstrom, H. Cheng and J. D. Waele (2020), 'Speleothem Record Attests to Stable Environmental Conditions during Neanderthal-Modern Human Turnover in Southern Italy', *Nature Ecology & Evolution*, 4 (9): 1188–95.

Comas-Bru, L., K. Rehfeld, C. Roesch, S. Amirnezhad-Mozhdehi, S. P. Harrison, K. Atsawawaranunt, S. M. Ahmad, Y. A. Brahim, A. Baker, M. Bosomworth, S. F. M. Breitenbach, Y. Burstyn, A. Columbu, M. Deininger, A. Demény, B. Dixon, J. Fohlmeister, I. G. Hatvani, J. Hu, N. Kaushal, Z. Kern, I. Labuhn, F. A. Lechleitner, A. Lorrey, B. Martrat, V. F. Novello, J. Oster, C. Pérez-Mejías, D. Scholz, N. Scroxton, N. Sinha, B. M. Ward, S. Warken, H. Zhang and SISAL Working Group members (2020), 'SISALv2: A Comprehensive Speleothem Isotope Database with Multiple Age–Depth Models', *Earth Syst. Sci. Data*, 12 (4): 2579–2606.

Conca Messina, S. A., S. Le Bras, P. Tedeschi and M. Vaquero Piñeiro (2019), *A History of Wine in Europe, 19th to 20th Centuries,* Vol. I, Cham: Palgrave Macmillan.

Conca Messina, S. A., S. Le Bras, P. Tedeschi and M. Vaquero Piñeiro (2020), *A History of Wine in Europe, 19th to 20th Centuries,* Vol. II, Cham: Palgrave Macmillan

Condamin, J. F., F. Formenti, M. O. Metais, M. Michel and P. Blond (1976), 'The Application of Gas Chromatography to the Tracing of Oil in Ancient Amphorae', *Archaeometry*, 18 (2): 195–201.

Conso, D., A. Gonzales and J. Y. Guillaumin, eds (2005), *Les vocabulaires techniques des arpenteurs romains: Actes du Colloque International (Besançon 2002)*, Besançon: Presses universitaires de Franche-Comté, Institut des Sciences et Techniques de l'Antiquité.

Contreras, D. A., A. Bondeau, J. Guiot, A. Kirman, E. Hiriart, L. Bernard, R. Suarez and M. Fader (2019), 'From Paleoclimate Variables to Prehistoric Agriculture: Using a Process-Based Agro-Ecosystem Model to Simulate the Impacts of Holocene Climate Change on Potential Agricultural Productivity in Provence, France', *Quaternary International*, 501 (Part B): 303–16.

Contreras, D. A., E. Hiriart, A. Bondeau, A. Kirman, J. Guiot, L. Bernard, R. Suarez, S. Van Der Leeuw (2018), 'Regional Paleoclimates and Local Consequences: Integrating GIS Analysis of Diachronic Settlement Patterns and Process-Based Agroecosystem Modeling of Potential Agricultural Productivity in Provence (France)', *PLoS ONE*, 13 (12): e0207622.

Conyers, L. (2012), *Ground-Penetrating Radar for Archaeology*, Plymouth: AltaMira Press.

Conyers, L. B., E. G. Ernenwein, M. Grealy and K. M. Lowe (2008), 'Electromagnetic Conductivity Mapping for Site Prediction in Meandering River Floodplains', *Archaeological Prospection*, 15: 81–91.

Corsi, C. and F. Vermeulen (2007), 'Digital vs. Historical Cartography for Studying Archaeological Sites: A Case Study from Corsica', in *Proceedings of the 11th Workshop 'Archäologie und Computer: Kulturelles Erbe und Neue Technologien', Wien 18–20 Oktober 2006*, 1–19, Wien: Stadtarchäologie.
Cosgrove, D. (1992), 'Mapping New Worlds: Culture and Cartography in Sixteenth-Century Venice', *Imago Mundi*, 44: 65–89.
Costantini, L. and L. Costantini Biasini (1999), 'La viticoltura della Grecia alla Magna Grecia: La documentazione archeobotanica', in O. Failla and G. Forni (eds), *Alle radici della civiltà del vino in Sicilia*, 169–92, Menfi: Cantine Settesoli.
Costantini, E. A. C., P. Bucelli and S. Priori (2012), 'Quaternary Landscape History Determines the Soil Functional Characters of Terroir', *Quaternary International*, 265: 63–73.
Costantini, E. A. C., R. Lorenzetti and G. Malorgio (2016), 'A Multivariate Approach for the Study of Environmental Drivers of Wine Economic Structure', *Land Use Policy*, 57: 53–63.
Cottica, D., M. Marchesini, S. Marvelli, M. Novello and P. Ventura (2017), 'Per uno studio integrato di uomo e ambiente ad Aquileia: alcune riflessioni a partire da recenti indagini archeologiche', *Rivista di Archeologia*, 41: 99–123, Tav. XIV–XVI.
Craig, O. E. (2021), 'Prehistoric Fermentation, Delayed-Return Economies, and the Adoption of Pottery Technology', *Current Anthropology*, 62, Suppl. 24. https://doi.org/10.1086/716610
Craig, O., H. Saul and C. Spiteri (2020), 'Residue Analysis', in M. Richards and K. Britton (eds), *Archaeological Science: An Introduction*, 70–98, Cambridge: CUP.
Cremonini, S., D. Labate and R. Curina (2013), 'The Late-Antiquity Environmental Crisis in Emilia Region (Po River Plain, Northern Italy): Geoarchaeological Evidence and Paleoclimatic Considerations', *Quaternary International*, 316: 162–78.
Crovetto C., M. Ferrari, C. Peretto, L. Longo and F. Vianello (1994), 'The Carinated Denticulates from the Paleolithic Site of Isernia La Pineta (Molise, Central Italy): Tools or Flaking Waste? The Results of The 1993 Lithic Experiments', *Human Evolution*, 9 (3): 175–207.
Curtel, G. (1903), *La Vigne et la Vin chez les Romains*, Paris: Masson et Cie.
Curtis, R. I. (2001), *Ancient Food Technology*, Leiden: Brill.
Curtis, R. I. (2008), 'Food Processing and Preparation', in J. P. Oléson (ed.), *The Oxford Handbook of Engineering and Technology in the Classical World*, 369–92, Oxford: Oxford University Press.
Dalmasso, G. (1941), *La Viticoltura ai Tempi dell'Impero Romano*, Rome: Libreria Editrice Aseq.
Dark, P. (2000), *The Environment of Britain in the First Millennium AD*, Bristol: Bristol Classical Press.
Day, J. (1932), 'Agriculture in the Life of Pompeii', *YaleClSt*, 3: 165–208.
Davis, M. B. (1963), 'On the Theory of Pollen Analysis', *AmJSci*, 261: 897–912.
De Caro, S. (1994), *La villa rustica in località Villa Regina a Boscoreale*, Rome: Bretschneider.
De Carolis, E., M. R. Senatore and A. Ciarallo (2012), 'Rinvenimento di resti di viti e di un palo di sostegno al di sotto dei depositi dell'eruzione del 79 d.C.', *RStPomp*, 23: 140–2.
de Grummond, N. T. (2018), 'Grape Pips from Etruscan and Roman Cetamura del Chianti: On Stratigraphy, Literary Sources and Pruning Hooks', *Etruscan Studies*, 21: 27–57.
de Ligt, L. (2012), *Peasants, Citizens and Soldiers: Studies in the Demographic History of Roman Italy 225 BC–AD 100*, Cambridge: Cambridge University Press.
De Lorenzis, G., F. Mercati, C. Bergamini, M. F. Cardone, A. Lupini, A. Mauceri, A. R. Caputo, L. Abbate, M. G. Barbagallo, D. Antonacci, F. Sunseri and L. Brancadoro (2019), 'SNP Genotyping Elucidates the Genetic Diversity of *Magna Graecia* Grapevine Germplasm and Its Historical Origin and Dissemination', *BMC Plant Biology*, 19 (7): 1–15.
De Lorenzis, G., F. Mercati, C. Bergamini, M. F. Cardone, M. G. Barbagallo, F. Sunseri, L. Brancadoro, A. Scienza and O. Failla (2020), 'Genomic Tools to Reconstruct the Grapevine Domestication and Evolution in the Western Mediterranean Basin' in J.-P. Brun, N. Garnier and G. Olcese (eds), *A. Making Wine in Western-Mediterranean. B. Production and the Trade of Amphorae: Some New Data from Italy*, 23–5, Heidelberg: Propylaeum.

De Neef, W. (2020), 'A Crossroads in the Central Potenza Valley: Non-Invasive Research into Settlement (Dis-)Continuity at Monte Franco (Pollenza, Marche, Italy)', in F. Boschi, E. Giorgi and F. Vermeulen (eds), *Picenum and the Ager Gallicus at the Dawn of the Roman Conquest: Landscape Archaeology and Material Culture*, 131–44, Oxford: Archaeopress.

De Sena, E. (2005), 'An Assessment of Wine and Oil Production in Rome's Hinterland: Ceramic, Literary, Art Historical and Modern Evidence', in B. Frizell and A. Klynne (eds), *Roman Villas Around the Urbs: Interaction with Landscape and Environment. Proceedings of a Conference at the Swedish Institute in Rome, September 17–18, 2004*, 135–49, Rome: Swedish Institute in Rome.

De Simone, G. F. (2017), 'The Agricultural Economy of Pompeii: Surplus and Dependence', in M. Flohr and A. Wilson (eds), *The Economy of Pompeii*, 23–51, Oxford: Oxford University Press.

De Smedt, Ph., P. Garwood, H. Chapman, K. Deforce, J. De Grave, D. Hanssens and D. Vandenberghe (2021), 'Identifying Elusive Prehistoric Land Use by Integrating Electromagnetic and Invasive Survey Approaches', *ArcheoSciences*, 45 (1): 231–3.

De Smedt, P., T. Saey, A. Lehouck, B. Stichelbaut, E. Meerschman, M. Monirul Islam, E. Van De Vijver and M. Van Meirvenne (2013), 'Exploring the Potential of Multi-receiver EMI Survey for Geoarchaeological Prospection: A 90 ha Dataset', *Geoderma*, 199: 30–6.

De' Spagnolis, M. (1989), 'Il ritrovamento di località Tre Ponti di Scafati e la via extraurbana Pompei-Sarno', *RStPomp*, 3: 41–52.

De' Spagnolis, M. (1991–2), 'Ville rustiche a Scafati', *RStPomp*, 5: 67–88.

De' Spagnolis, M. (1995), 'Osservazioni sulle fasi edilizie di alcune ville rustiche di Scafati, suburbio orientale di Pompei, seppellite dalla eruzione del 79 d.C', in *Archäologie und Seismologie: La regione vesuviana dal 62 al 79 d.C. Problemi archeologici e sismologici. Colloquium, Boscoreale 26–27 November 1993*, 93–102, München.

De' Spagnolis, M. (2002), *La villa N. Popidi Narcissi Maioris in Scafati suburbio orientale di Pompei*, Studia Archaeologica 119, Roma: L'Erma di Bretschneider.

De Stefani, S. (1887), 'Negrar di Valpolicella', *Notizie degli Scavi di Antichità*: 431–2.

de Vivo, F. (2003), 'Historical Justifications of Venetian Power in the Adriatic', *Journal of the History of Ideas*, 64 (2): 159–76.

De Voguë, M. (1865–77), *Syrie Centrale: Architecture Civile et Religieuse du Ier au VIIe Siècle*, Paris: J. Baudry Libraire-Èditeur.

Dechandol, H., M.-P. Feuillet, T. Odiot (1983), 'Donzère (Drôme): le grand domaine viticole du Molard', *Histoire et Archéologie, les dossiers*, 78: 167–207.

Deiana, R., G. Leucci and R. Martorana, eds (2018), *New Perspectives on Geophysics for Archaeology*, Special Issue: Surveys in Geophysics 39.6.

Del Lungo, S. (2016), 'Centro terziario di domesticazione: la topografia antica e la genetica in Enotria, dalle Siriche alla multivarietà viticola della Lucania', in V. Alba, C. Bergamini, M. Gasparro, F. Mazzone, S. Roccotelli, D. Antonacci and A. R. Caputo (eds), *Basivin_SUD: La ricerca del germoplasma viticolo in Basilicata*, 37–86, Bari: Mario Adda Editore.

Del Lungo, S. (2017), 'Third Center of Domestication: Interaction between Genetics and Archaeological Sciences in Lands Devoted to Biodiversity of Grapevine Varieties', *Journal of Horticulture*, 4 (1): 194.

Del Mastro, B., P. Munzi, J.-P. Brun, H. Duday and N. Garnier (2021), 'Vino per gli Opikoi: L'Esempio delle Tombe Preelleniche di Cuma', in D. Frère, B. del Mastro, P. Munzi and C. Pouzadoux (eds), *Manger, Boire, se Parfumer pour l'Éternité: Rituels Alimentaires et Odorants en Italie et en Gaule du ixe siècle avant au ier siècle après J.- C*, 165–89, Napoli: Centre Jean Bérard.

Del Mastro, B., P. Munzi, J.-P. Brun and N. Garnier (2017), 'Wine Before the Greeks: The Contribution of the Organic Chemistry Analyses', in *3rd IMEKO International Conference on Metrology for Archaeology and Cultural Heritage, Lecce (October 23–25, 2017)*, 1–6, Lecce: HAL.

Del Zan, F. (2004), 'Grecia', in F. Del Zan, O. Failla and A. Scienza (eds), *La vite e l'uomo: Dal rompicapo delle origini al salvataggio delle reliquie. Evidenze storico-ampelografiche per ripercorrere il viaggio della vite da Oriente alle soglie dell'Occidente*, 531–658, Gorizia: ERSA.

Della Corte, M. (1921a), 'Altra villa rustica, scavata dal Sig. Cav. Carlo Rossi-Filangieri nel Fondo di Raffaele Brancaccio, nella stessa Contrada Civita-Giuliana (Comune di Boscoreale), nei mesi da Gennaio a Marzo 1904', *NSc*, 18: 423–6.

Della Corte, M. (1921b), 'Villa rustica, Esplorata dal Sig. Cav. Carlo Rossi-Filangieri nel Fondo di Antonio Prisco, in Contrada Civita-Giuliana (Comune Di Boscoreale), nei mesi da Febbraio a Luglio dell'anno 1903', *NSc*, 18: 415–23.

Della Corte, M. (1923), 'Villa rustica esplorata dal Sig. Giacomo Matrone in un Fondo di sua proprietà, posto immediatamente ad occidente della via vicinale che attraversa, in direzione Nord-Sud, la Contrada Messigno, in comune di Gragnano', *NSc*, 20: 271–4.

Delpino, F. (2007), 'Viticoltura, produzione e consumo del vino nell'Etruria protostorica', in A. Ciacci, P. Rendini and A. Zifferero (eds), *Archeologia della vite e del vino in Etruria*, 133–46, Siena: Ci.Vin.

Delpino, F. (2012), 'Viticoltura, produzione e consumo del vino nell'Etruria protostorica', in A. Ciacci, P. Rendini, and A. Zifferero (eds), *Archeologia della vite e del vino in Toscana e nel Lazio: Dalle techniche dell'indagine archeologica alle prospettive della biologia moleculare*, 189–99, Borgo San Lorenzo: All'Insegna del Giglio.

Demesticha, S. (2015), *Per Terram, Per Mare: Seaborne Trade and the Distribution of Roman Amphorae in the Mediterranean*, Uppsala: Äströms Förlag.

Depalmas, A., C. Loi, N. Garnier and A. Pecci (2020), 'Wine in Sardinia: New Archaeological Data and Research Methodology', in J.-P. Brun, N. Garnier and G. Olcese (eds), *A. Making Wine in Western-Mediterranean. B. Production and the Trade of Amphorae: Some new data from Italy: Panel 3.5*, Archaeology and Economy in the Ancient World 9, 61–71, Heidelberg: Propylaeum.

Desbat, A. (1997), 'Le tonneau antique: questions techniques et problème d'origine', in D. Garcia and D. Meeks (eds), *Techniques et économie antiques et médiévales: le temps de l'innovation, colloque international (C.N.R.S.), Aix-e,-Provence 21–23 mai 1996*, 113–20, Paris: Éditions Errance.

Desmarchelier, J. M., J. C. Hellstrom, M. Mcculloch (2006), 'Rapid Trace Element Analysis of Speleothems by ELA-ICP-MS', *Chemical Geology*, 231 (1–2): 102–17.

Di Castri, F., D. W. Godall and R. L. Specht, eds (1981), *Ecosystems of the World II: Mediterranean Type Shrublands*, Amsterdam: Elsevier Scientific Publishing.

Di Filippo Balestrazzi, E. (2004), *Vino e coltura del vino nell'area orientale della Venetia: Dalla vite al vino. Antropologia e storia di una cultura enologica nell'angolo più orientale della Pianura Veneta*, 9, Fondazione 'Antonio Colluto'.

Di Pasquale, G., ed. (2010), *Vinum Nostrum: Arte, Scienza e Miti del Vino nelle Civiltà del Mediterraneo Antico*, Firenze: Giunti Editore.

Di Pasquale, G., A. Migliozzi, E. Allevato and M. Marotta (2012), 'Italian Cultural Landscapes, A Little Known Resource: The Study Case of the Alberata Aversana, A Vineyard 2500 Years Old', in A. Ferrari (ed.), *5th International Congress 'Science and Technology for the Safeguard of Cultural Heritage in the Mediterranean Basin'. Istanbul 22–25 November 2011*, 322–6, Roma: Valmar.

Di Vecchi Staraz, M., R. Bandinelli, M. Boselli, P. This, J.-M. Boursiquot, V. Laucou, T. Lacombe and D. Varès (2007), 'Genetic Structuring and Parentage Analysis for Evolutionary Studies in Grapevine: Kin Group and Origins of the Cultivar Sangiovese Revealed', *Journal of the American Society for Horticultural Science*, 132 (4): 514–24.

Diederich, S. (2005), 'Das römische Agrarhandbuch als Medium der Selbstdarstellung', in T. Fögen (ed.), *Antike Fachtexte: Ancient Technical Texts*, 271–88, Berlin and New York: De Gruyter.

Dilke, O. A. W. (1971), *The Roman Land Surveyors: An Introduction to the Agrimensores*, Newton Abbot: David and Charles.

Ding, W., Q. Xu and E. Tarasov (2017), 'Examining Bias in Pollen-based Quantitative Climate Reconstructions Induced by Human Impact on Vegetation', *Climate of the Past*, 13 (9): 1285–1300.

Dion, R. (1959), *Histoire de la Vigne et du Vin en France*, Paris: Flammarion.

Dodd, E. (2017), 'Pressing Issues: A New Discovery in the Vineyard of Region I.20, Pompeii', *Archeologia Classica*, 68: 577–88.

Dodd, E. (2020a), *Roman and Late Antique Wine Production in the Eastern Mediterranean: A Comparative Archaeological Study at Antiochia ad Cragum (Turkey) and Delos (Greece)*, Oxford: Archaeopress.

Dodd, E. (2020b), 'Late Roman Viticulture in Rough Cilicia: An Unusual Wine Press at Antiochia ad Cragum', *Journal of Roman Archaeology*, 33: 467–82.

Dodd, E. (2020c), 'Population Decline and Viticultural Industry: Societal Transformation on Late Antique Delos (Greece)', in D. Van Limbergen, S. Maréchal and W. De Clercq (eds), *The Resilience of the Roman Empire: regional case studies on the relationship between population and food resources*, 109–28, Oxford: BAR.

Dodd, E. (2022a), 'The Archaeology of Wine Production in Roman and Pre-Roman Italy', *American Journal of Archaeology*, 126 (3): 443–80. https://doi.org/10.1086/719697

Dodd, E. (2022b), 'Wine, Oil, and Knowledge Networks across the Graeco-Roman Cyclades: New Data from Paros and Naxos in 2021', *Mediterranean Archaeology*, 34/5: 155–67.

Dodd, E. (2022c), 'Wine, Oil, and Knowledge Networks across the Graeco-Roman Cyclades (Greece)', *Papers of the British School at Rome*, 90: 352–5.

Dodd, E., G. Galli and R. Frontoni (2023), 'The Spectacle of Production: A Roman Imperial Winery at the Villa of the Quintilii, Rome', *Antiquity* 97 (392): 436–53.

Donahue, J. F. (2015), *Food and Drink in Antiquity: Readings from the Graeco-Roman World: A Sourcebook*, London: Bloomsbury.

Donahue, J. F. (2017), *The Roman Community at Table during the Principate (New and Expanded Edition)*, Ann Arbor: University of Michigan Press.

Donev, D. and R. Willett (2017), 'Book Review of An Urban Geography of the Roman World, 100 BC to AD 300', *American Journal of Archaeology*, 121 (4). https://doi.org/10.3764/ajaonline1214.Donev

Dong, Y. et al. (2023), 'Dual Domestications and Origin of Traits in Grapevine Evolution', *Science* 379: 892–901.

Doolittle, J. A. and E. C. Brevik (2014), 'The Use of Electromagnetic Induction Techniques in Soils Studies', *Geoderma*, 223: 33–45.

Doorenbosch, M. (2013), *Ancestral Healths*, Leiden: Sidestone Press.

Dougherty, P. H. (2012), 'Introduction to the Geographical Study of Viticulture and Wine Production', in P. H. Dougherty (ed.), *The Geography of Wine: Regions, Terroir and Techniques*, 3–36, Dordrecht: Springer.

Dressel, H. (1899), *Corpus Inscriptionum Latinorum* XV, Pars 1, Berlin.

Drieu L., P. Orecchioni, C. Capelli, A. Meo, J. Lundy, V. Sacco, L. Arcifa, A. Molinari, M. Carver and O. E. Craig (2021), 'Chemical Evidence for the Persistence of Wine Production and Trade in Early Medieval Islamic Sicily', *The Proceedings of the National Academy of Sciences USA*, 118 (10): e2017983118.

Drieu L., M. Rageot, N. Wales, B. Stern, J. Lundy, M. Zerrer, I. Gaffney, M. Bondetti, C. Spiteri, J. Thomas-Oates and O. E. Craig (2020), 'Is It Possible to Identify Ancient Wine Production Using Biomolecular Approaches?', *STAR: Science & Technology of Archaeological Research*, 6: 16–29.

Drysdale, R. N., C. Spötl, J. C. Hellstrom and D. A. Richards (2012), 'New Advances in the Dating of Speleothems: An Introduction', *Quaternary Geochronology*, 14: 1–4.

Dunbabin, K. M. D. (2003), *The Roman Banquet: Images of Conviviality*, Cambridge: Cambridge University Press.

Duncan-Jones, R. (1974), *The Economy of the Roman Empire: Quantitative Studies,* Cambridge.

Dzierzbicka, D. (2018), *OINOS: Production and import of wine in Graeco-Roman Egypt*, Warsaw: University of Warsaw.

EEA (2012), *EEA Catchments and Rivers Network System ECRINS v1.1: Rationales, Building and Improving for Widening Uses to Water Accounts and WISE Applications*, Luxembourg.

Ehmig, U. (2003), *Der Römischen Amphoren aus Mainz*, Frankfurter Archäologische Schriften 4, Möhnesee: Bibliopolis.

Erdtman, G. (1952), *Pollen Morphology and Plant Taxonomy: Angiosperms (An Introduction to Palynology I)*, Stockholm: Almqvist and Wiksell.

Eterovich, F. H. and C. Spalatin, eds (1964), *Croatia: Land, People, Culture*, Vol. 1, Toronto: University of Toronto Press.

Eterovich, F. H. and C. Spalatin, eds (1970), *Croatia: Land, People, Culture*, Vol. 2, Toronto: University of Toronto Press.

Evershed, R. (1993), 'Biomolecular Archaeology and Lipids', *World Archaeology*, 25 (1): 74–93.

Evershed, R. P. (2008), 'Organic Residue Analysis in Archaeology: The Archaeological Biomarker Revolution', *Archaeometry*, 50 (6): 895–924.

Fabbrini, L. (2003), *I vigneti sperimentali in Toscana*, Firenze: ARSIA.

Fabre, G., M. Mayer and I. Rodà (1984), *Inscriptions Romaines de Catalogne, I (sauf Barcino)*, Paris: Universitat Autònoma de Barcelona.

Fabrini, G. M. (2000), 'L'area del tempio-criptoportico ad Urbs Salvia: Risultati preliminari delle campagne di scavo 1995–1999', *Picus*, 20: 113–58.

Fader, M., W. von Bloh, S. Shi and W. Cramer (2015), 'Modelling Mediterranean agro-Ecosystems by Including Agricultural Trees in the LPJmL model', *Geosci. Model. Dev.*, 8: 3545–61.

Failla, O. (2014), 'Il vitigno: significato, origine e variabilità', in G. Forni, G. Kezich and A. Scienza (eds), *Le frontiere nascoste della cultura del vino*, Atti del Seminario Permanente di Etnografia Alpina, SPEA 14, *Annali di San Michele*, 25: 115–38.

Fairchild, I. J. and A. Baker (2012), *Speleothem Science: From Process to Past Environments*, Chichester: Wiley-Blackwell.

Faraji, M., A. Borsato, S. Frisia, J. C. Hellstrom, A. Lorrey, A. Hartland, A. Greig and D. P. Mattey (2021), 'Accurate Dating of Stalagmites from Low Seasonal Contrast Tropical Pacific Climate using Sr 2D Maps, Fabrics and Annual Hydrological Cycles', *Scientific Reports*, 11: 2178.

Fassbinder, J. W. E. (2016), 'Magnetometry for Archaeology', in A. S. Gilbert (ed.), *Encyclopedia of Geoarchaeology*, 499–513, Dordrecht: Springer.

Favretto, A. (2012), 'Georeferencing Historical Cartography: A Quality-Control Method', *Cartographica: The International Journal for Geographic Information and Geovisualization*, 47 (3): 161–7.

Feier, I., A. Migala, M. Pietruszka and M. Jackowski (2019), 'Roman Wine in Babaricum: Preliminary Studies on Ancient Wine Recreation', *Heritage*, 2: 331–8.

Feige, M. (2021), 'Decorative Features and Social Practices in Spaces for Agricultural Production in Roman Villas', in A. Haug and M. T. Lauritsen (eds), *Principles of Decoration in the Roman World*, 33–52, Berlin: De Gruyter.

Feige, M. (2022), *Landwirtschaftliche Produktionsanlagen römischer Villen im republikanischen und kaiserzeitlichen Italien*, Berlin: De Gruyter.

Ferdière, A. (1988), *Les Campagnes en Gaule Romaine*, Paris: Errance.

Figueiral, I., L. Bouby, L. Buffat, H. Petitot and J.-F. Terral (2010), 'Archaeobotany, Vine Growing and Wine Producing in Roman Southern France: The Site of Gasquinoy (Béziers, Hérault)', *Journal of Archaeological Science*, 37 (1): 139–49.

Finley, M. I. (1975), *The Ancient Economy*, London: Chatto and Windus.

Finné, M., M. Bar-Matthews, K. Holmgren, H. S. Sundqvist, I. Liakopoulos and Q. Zhang (2014), 'Speleothem Evidence for Late Holocene Climate Variability and Floods in Southern Greece', *Quaternary Research*, 81 (2): 213–27.

Fiorentino, G. and F. Solinas (2006), 'Carboni e carporesti dal palazzo di Monastiraki', in A. Kanta and M. Marazzi (eds), *Monastiraki 1, Missione Monastiraki: Campagne 2002/2004*, 123–40, Naples: Suor Orsola Benincasa nell'Università.

Firenze 2006 = *Il Sangiovese vitigno tipico e internazionale: identità e peculiarità. Atti del Secondo Simposio Internazionale (Firenze 2004)*, Firenze: ARSIA.

Firmati, M. (2012), 'I comprensori indagati nel 2005: la valle dell'Albegna', in A. Ciacci, P. Rendini and A. Zifferero, (eds), *Archeologia della Vite e del Vino in Toscana e nel Lazio. Dalle tecniche dell'indagine archeologica alle prospettive della biologia molecolare*, 517–30, Borgo San Lorenzo: All'Insegna del Giglio.

Firmati, M., ed. (2013), *Scansano. Guida al territorio e al Museo Archeologico e della Vite e del Vino*, Siena: Nuova Immagine Editrice.

Firmati, M., A. Zifferero and V. Zorzi. (2021), 'Il Vinum in Maremma', *Archeo: Attualità del Passato*, 439: 98–105.

Firmati, M., A. Zifferero, V. Zorzi and G. Ferrari Melillo (2015), 'Scansano (GR). Progetto ArcheoVino: l'impianto del vigneto sperimentale etrusco-romano', *Notiziario della Soprintendenza per i Beni Archeologici della Toscana*, 11: 497–501.

Fischer-Genz, B. (2016), 'Ancient Wine and Oil Presses from the Bekaa Valley', *Journal of Eastern Mediterranean Archaeology & Heritage Studies*, 4: 57–71.

Flandrin, J. L. and M. Montanari, eds (1997), *Storia dell'alimentazione*, Rome: Laterza.

Forbes, H. (1993), 'Ethnoarchaeology and the Place of the Olive in the Economy of the Southern Argolid, Greece', in M.-C. Amouretti and J.-P. Brun (eds), *La Production du Vin et de l'Huile en Méditerranée*, BCH Supp. XXVI, 213–26, Athens.

Forbes, R. J. (1955), *Studies in Ancient Technology III*, Leiden: Brill.

Formenti, F., A, Hesnard and A. Tchernia (1978), 'Note sur le contenu d'une amphore Lamboglia 2 de l'épave de la Madrague de Giens', *Archaeonautica*, 2: 95–100.

Fornacelli, C., A. Palmieri and S. Rojo Muñoz (2022), 'Le vie del bucchero', *Archeo: Attualità del Passato*, 447: 20–1.

Forni, G. (1996), 'Genesi e diffusione della viti-vinicoltura dal Mediterraneo orientale alla Cisalpina: aspetti ecologici, culturali, linguistici e tecnologici', in G. Forni, A. Scienza (eds), *2500 anni di cultura della vite nell'ambito alpino e cisalpino*, 19–183, Trento: Istituto Trentino del Vino.

Forni, G. (1999), 'La tecnologia e l'economia vitivinicola della Magna Grecia e della Sicilia antica', in O. Failla and G. Forni (eds), *Alle radici della civiltà del vino in Sicilia*, 131–67, Menfi: Cantine Settesoli.

Forni, G. (2012a), 'La matrice euromediterranea della nostra viticoltura: La prospettiva pluridisciplinare', in A. Ciacci, P. Rendini and A. Zifferero (eds), *Archeologia della Vite e del Vino in Toscana e nel Lazio: Dalle tecniche dell'indagine archeologica alle prospettive della biologia molecolare*, 93–118, Borgo San Lorenzo: All'Insegna del Giglio.

Forni, G. (2012b), 'The Origin of "Old World" viticulture', in D. Maghradze, L. Rustioni, J. Turok, A. Scienza and O. Failla (eds), *Caucasus and Northern Black Sea Region Ampelography*, 27–38, Vitis.

Foxhall, L. (2007), *Olive Cultivation in Ancient Greece: Seeking the Ancient Economy*, Oxford: OUP.

Fraga, H. (2019), *Viticulture and Winemaking Under Climate Change*, Basel, Beijing, Wuhan, Barcelona, Belgrade: MDPI.

Frankel, R. (1999), *Wine and Oil Production in Antiquity in Israel and Other Mediterranean Countries*, Sheffield: Sheffield Academic Press.

Frankel R. (2020), 'Olive and Olive Oil in the Eastern Mediterranean: From Antiquity to the Pre-Industrial Age', in I. Anagnostakis and E. Balta (eds), *Olive and Olive Oil in the Eastern Mediterranean: From Antiquity to the Pre-Industrial Age*, 117–44, Athens: Piraeus Bank Group Cultural Foundation (PIOP) and the Institute of Historical Research (IHR) of the National Hellenic Research Foundation (NHRF).

Franzoni, L. (1990), *La Valpolicella nell'Età Romana*, Verona: Centro di Documentazione per la storia della Valpolicella.

Fredlund, G. G. and L. L. Tieszen (1997), 'Calibrating Grass Phytolith Assemblages in Climatic Terms: Application to Late Pleistocene Assemblages from Kansas and Nebraska', *Palaeogeography Palaeoclimatology Palaeoecology*, 136 (1–4): 199–211.
Fuks, D., G. Bar-Oz, Y. Tepper and E. Weiss (2020), 'The Rise and Fall of Viticulture in the Late Antique Negev Highlands Reconstructed from Archaeobotanical and Ceramic Data', *PNAS*, 117 (33): 19780–91.
Fyfe, R. M., J.-L. de Bealieu, H. Binney, R. H. W. Bradshaw, S. Brewer, A. Le Flao, W. Finsinger, M.-J. Gaillard, T. Giesecke, G. Gil-Romera, E. C. Grimm, B. Huntley, P. Kunes, N. Kühl, M. Leydet, A. F. Lotter, P. E. Tarasov and S. Tonkov (2009), 'The European Pollen Database: Past Efforts and Current Activities', *Vegetation History and Archaebotany*, 18: 417–24.
Gaffney, C. and J. Gater (2003), *Revealing the Buried Past: Geophysics for Archaeologists*, Stroud: Tempus.
Gaffney, V. L. and B. Kirigin, eds (2006), *The Adriatic Islands Project: The Archaeological Heritage of Vis, Biševo, Svetac, Palagruža and Šolta*, BAR International Series 1492, Oxford: BAR.
Gaffney, V. L., B. Kirigin, M. Petric and N. Vujnovic, eds (1997), *The Adriatic Islands Project: Contact, Commerce and Colonialism, 6000 BC–AD 600*, BAR International Series 660, Oxford: BAR.
Gaillard, M. J., S. Sugita, M. J. Bunting, R. Middleton, A. Broström, C. Caseldine, T. Giesecke, S. E. V. Hellman, S. Hicks, K. Hjelle, C. Langdon, A.-B. Nielsen, A. Poska, H. von Stedingk, S. Veski and POLLANDCAL members (2008), 'The Use of Modelling and Simulation Approach in Reconstructing Past Landscapes from Fossil Pollen Data: A Review and Results from the POLLANDCAL Network', *Vegetation History and Archaebotany*, 17: 419–43.
Gallo, D., M. Ciminale, H. Becker and N. Masini (2009), 'Remote Sensing Techniques for Reconstructing a Vast Neolithic Settlement in Southern Italy', *JAS*, 36: 43–50.
Galtier, G. (1960), *Le vignoble du Languedoc méditerranéen et du Roussillon*, Montpellier: Thèse Paris.
Garcés-Pastor, S., S. Lavergne, E. Coissac, Y. Lammers, O. W. Wangensteen, J.-P. Theurillat, C. Schwörer, A. G. Brown, W. Tinner, F. Rey, O. Heiri, M. Heer, A. Rutzer, K. Walsh, P. D. Heintzman, T. Goslar and I. G. Alsos (2022), 'High Resolution Ancient Sedimentary DNA Shows That Alpine Plant Diversity is Associated with Human Land Use and Climate Change', *Nat. Comms.*, 13 (6559).
García Armendáriz, J. I. (1995), *Agronomía y tradición clásica. Columela en España*, Sevilla: Universidad de Cádiz y Sevilla.
Garcia, J.-P., S. Chevrier, A. Dufraisse, M. Foucher and R. Steinmann (2010), 'Le vignoble gallo-romain de Gevrey-Chambertin "Au-dessus-de Bergis", Côte-d'Or (Ier–IIe s. ap. J.-C.): Modes de plantation et de conduite de vignes antiques en Bourgogne', *Revue archéologique de l'Est*, 59 (2): 505–37.
García Sanz, O. (1991), 'Ad summam ubertatem vini', *Gerión*, 3: 219–22.
Garnier, N. (2003), 'Structural Analysis of Organic Materials Preserved in Ancient Results: New Results from Chromatography and Mass Spectrometry Analyses', unpublished PhD thesis, Paris: Université Pierre et Marie Curie (Paris VI).
Garnier, N. (2007), 'Analyse de résidus organiques conservés dans des amphores: un état de la question', in M. Bonifay and J. C. Tréglia (eds), *LRCW 2 Late Roman Coarse Wares, Cooking Wares and Amphorae in the Mediterranean: Archaeology and Archaeometry*, BAR International Series 1662, 39–49, Oxford: Archaeopress.
Garnier, N. (2015), 'Methodologies d'analyse chimique organique en archéologie', in C. Oliveira, R. Morais and A. Morillo (eds), *Archaeoanalytics: Chromatography and DNA analysis in archaeology. International Symposium*, 13–39, Porto: University of Porto.
Garnier, N. (forthcoming), 'Elaborazione e consumo di cibi e bevande attraverso il prisma delle analisi organiche dei contenuti invisibili', in P. Basso, G. Falezza, M. Migliavacca and F. Saggioro (eds), *Veronensium mensa: Food and Wine in Ancient Verona*, Rome: Quasar.
Garnier, N., P. Richardin, V. Cheynier and M. Regert (2003), 'Characterization of Thermally Assisted Hydrolosis and Methylation Products of Polyphenols from Modern and

Archaeological Vine Derivatives Using Gas Chromatography-Mass Spectrometry', *Analytica Chimica Acta*, 493 (2): 137–57.

Garnier, N., T. Silvino and D. Bernal Casasola (2011), 'The Identification of the Content of Amphorae: Oils, Salsamenta and Pitch', in *French Society for the Study of Archaeological Ceramics from Gallia, Proceedings of the Congress SFECAG (Arles, June 2011)*, 397–416, Arles.

Garnier, N. and A. Pecci (2021), 'Amphorae and Residue Analysis: Application to the Determination of the Content of Amphorae', in M. Bonifay, D. Bernal, A. Pecci and V. Leich (eds), *Roman Amphora Contents: Reflecting on the Maritime Trade of Foodstuffs in Antiquity*, 113–26, Oxford: Archaeopress.

Garnier, N. and S. M. Valamoti (2016), 'Prehistoric Wine Making at Dikili Tash (Northern Greece): Integrating Residue Analysis and Archaeobotany', *JAS*, 74: 195–206.

Gauthier, E. and M. Joly (2003), 'Nouvelles données sur la culture de la vigne dans l'est de la Gaule (Bourgogne/Franche-Comté)', in F. Favory and A. Vigot (eds), *Actualité de la recherche en histoire et archéologie agraires, Actes du colloque AGER V, Besançon 19–20 septembre 2000*, 191–208, Besançon: Presses Universitaires Franc-Comtoises.

Giannichedda, E. (2006), *Uomini e cose*, Bari: Edipuglia.

Giannitrapani, E., L. Simone and S. Tinè, eds (1990), *Interpretazione funzionale dei 'fondi di capanna' di età preistorica: Atti del seminario di archeologia sperimentale, Milano, 29–30 aprile 1989*, Genova.

Giardino, C. (2012), 'L'Archeologia sperimentale e la preistoria: un contributo alla conoscenza del passato', *Conimbriga*, 51: 5–33.

Giguet-Covex, C., J. Pansu, F. Arnaud, P.-J. Rey, C. Griggo, L. Gielly, I. Domaizon, E. Coissac, F. David, P. Choler, J. Poulenard and P. Taberlet (2014), 'Long Livestock Farming History and Human Landscape Shaping Revealed by Lake Sediment DNA', *Nature Communications*, 5 (3211).

Giguet-Covex, C., G. F. Ficetola, K. Walsh, J. Poulenard, M. Bajard, L. Fouinat, P. Sabatier, L. Gielly, E. Messager, A. L. Develle, F. David, P. Taberlet, E. Brisset, F. Guiter, R. Sinet and F. Arnaud (2019), 'New Insights on Lake Sediment DNA from the Catchment: Importance of Taphonomic and Analytical Issues on the Record Quality', *Scientific Reports*, 9 (14676).

Gilles, K.-J. (1995), *Neuere Forschungen zum roemischen Weinhau an Mosel und Rhein*, Schriftenreihe des Rheinischen Landesmuseums Trier 11, Trier: Rheinisches Landesmuseum.

Ginori, L. (1978), *Cabrei in Toscana: Raccolte di Mappe, Prospetti e Vedute*, Pistoia: Cassa di Risparmio.

Gismondi, A., G. Di Marco, F. Martini, L. Sarti, M. Crespan, C. Martínez-Labarga, O. Rickards and A. Canini (2016), 'Grapevine Carpological Remains Revealed the Existence of a Neolithic Domesticated Vitis vinifera L. Specimen Containing Ancient DNA Partially Preserved in Modern Ecotypes', *JAS*, 69: 75–84.

Gladstones, J. (2011), *Wine, Terroir and Climate Change*, Mile End: Wakefield Press.

Glicksman, K. (2005), 'Internal and external trade in the Roman province of Dalmatia', *Opvsvla Archaeologica*, 29 (1): 189–230.

Glicksman, K. (2007), 'Olive and Vine Cultivation in the Roman Province of Dalmatia', *Histria Antiqua*, 15: 43–50.

Glonti, T. and Z. Glonti (2013), *The Remarkable Qvevri Wine (Why Is the Qvevri Wine Better than One Made without It?)*, Tbilisi: Qvevri Foundation.

Gnirs, A. (1915), 'Forschungen über antiken Villenbau in Südistrien, I. Die Grabung in der antiken Villenanlage von Val Catena, II. Eine villa rustica am strand der Bucht Olmo grande', *Jahreshefte des Österreichischen archäelogischen Instituts*, 18: 99–163.

Godino, Y., Lebole C. and G. Di Gangi (2020), 'Fornire la pratica che sostiene la teoria: una riflessione sull'Archeologia Sperimentale', *Archeologie Sperimentali: Temi, Metodi, Ricerche*, 1: 1–27.

Goodchild, H. (2009), 'Modelling the Productive Landscape of the Middle Tiber Valley', in F. Coarelli and H. Patterson (eds), *Mercator placidissimus: the Tiber Valley in Antiquity: New Research in the Upper and Middle River Valley: Rome, 27–28 February 2004*, 769–96, Roma.

Goodchild, H. (2013), 'Agriculture and the Environment of Republican Italy', in J. DeRose Evans (ed.), *A Companion to the Archaeology of the Roman Republic*, 198–213, Oxford: Wiley-Blackwell.

Goodchild, H. (2019), 'De Agri Cultura Experientia: From Modern Agronomy to Roman Economic Analysis', in J. Remesal Rodríguez, V. Revilla Calvo, D.-J. Martin-Arroyo Sánchez and A. Martin I Oliveras (eds), *Productive Landscapes and Trade Networks in the Roman Empire*, 155–78, Barcelona: University of Barcelona.

Goodman, D. and S. Piro (2013), *GPR Remote Sensing in Archaeology*, Geotechnologies and the Environment 9, New York: Springer.

Goode, J. (2006), *The Science of Wine: From Vine to Glass*, Berkeley: University of California Press.

Grace, V. (1979), *Amphoras and the Ancient Wine Trade*, Princeton.

Grassi, F. and G. De Lorenzis (2021), 'Back to the Origins: Background and Perspective of Grapevine Domestication', *Int J Mol Sci*, 22 (9): 4518.

Grasso, A. M. and G. Fiorentino (2012), 'Archeologia e storia della vite e del vino nel Medioevo italiano: Il contributo dell'archeobotanica e di nuove metodologie di analisi integrate per la caratterizzazione varietale applicate ai contesti archeologici della Puglia meridionale', in F. Redi and A. Forgione (eds), *VI Congresso Nazionale di Archeologia Medievale (L'Aquila, 12–15.9.2012)*, 688–92, Firenze: All'Insegna del Giglio.

Grave, P. and L. Kealhofer (1999), 'Assessing Bioturbation in Archaeological Sediments using Soil Morphology and Phytolith Analysis', *JAS*, 26: 1239–48.

Greig, J. (2015), 'Pollen', in M. Atkinson and S. J. Preston Heybridge (eds), *A Late Iron Age and Roman Settlement: Excavations at Elms Farm 1993–5*, Internet Archaeology, 40, http://dx.doi.org/10.11141/ia.40.1.greig

Grigalashvili, Z. A. and M. B. Khomasuridze (2021), 'Impact of Oak Barrel and Its Alternatives on Sensory Characteristics and Phenolic Compounds of Qvevri Wine', *Georgian Engineering News*, 92: 159–66.

Grisso, R., M. M. Alley, D. Holshouser and W. Thomason (2009), *Precision Farming Tools: Soil Electrical Conductivity*, Petersburg: Virginia Polytechnic Institute and State University.

Guasch-Jané, M. R. (2019), 'Grape Archaeology and Ancient DNA Sequencing', in D. Cantu and M. Andrew Walker (eds), *The Grape Genome*, 57–75, Cham: Springer.

Guasch-Jané, M. R., M. Iberno-Gómez, C. Andrés-Lacueva, O. Jáuregui, R. M. and Lamuela-Raventós (2004), 'Liquid Chromatography with Mass Spectrometry in Tandem Mode Applied for the Identification of Wine Markers in Residues from Ancient Egyptian Vessels', *Analytical Chemistry*, 76: 1672–77.

Guasch-Jané, M. R., C. Andrés-Lacueva, O. Jáuregui and R. M. Lamuela-Raventós (2006), 'First Evidence for White Wine in Ancient Egypt from Tutankhamun's Tomb', *Journal of Archaeological Science*, 33: 1075–80.

Guido, M. A., M. Mariotti Lippi, B. I. Menozzi, S. Placerani and C. Montanari (2004), 'Il paesaggio vegetale montano della Liguria centro-occidentale nell'età del Ferro: area del Monte Beigua (Savona)', in R. C. De Marinis and G. Spadea (eds), *I Liguri*, 91–5, Milano: Skira.

Guiot, J. and D. Kaniewski (2015), 'The Mediterranean Basin and Southern Europe in a warmer world: what can we learn from the past?', *Frontiers in Earth Science*, 3: 28.

Guiot, J., N. Bernigaud, A. Bondeau, L. Bouby and W. Cramer (2023), 'Viticulture Extension in Response to Global Climate Change Drivers – Lessons from the Past and Future Projections', *Climate of the Past*, 19 (6): 1219–44.

Hadjisavvas, S. (2020), *Archaeology of Wine in Cyprus*, Nicosia: Napaphos.

Hanson, J. W. (2016), *Cities database (OXREP databases). Version 1.0*, accessed here: http://oxrep.classics.ox.ac.uk/databases/cities/. DOI: https://doi.org/10.5287/bodleian:eqapevAn8

Harper, K. and M. McCormick (2019), 'Reconstructing the Roman Climate', in W. Scheidel (ed.), *The Science of Roman History: Biology, Climate, and the Future of the Past*, 11–52, Princeton: Princeton University Press.

Harrison, A., U. Rajala, S. Stoddart, R. E. Witcher and E. Zubrow (2004), 'The Enhancement of the South Etruria Survey: Phase 1', in H. Patterson (ed.), *Bridging the Tiber: Approaches to Regional Archaeology in the Middle Tiber Valley*, 29–35, London: British School at Rome.

Harutyunyan, M. and M. Malfeito-Ferreira (2022), 'Historical and Heritage Sustainability for the Revival of Ancient Wine-Making Techniques and Wine Styles', *Beverages*, 8 (1): 10.

Harutyunyan, M., R. Viana, J. Granja-Soares, M. Martins, H. Ribeiro and M. Malfeito-Ferreira (2022), 'Adaptation of Ancient Techniques to Recreate "Wines" and "Beverages" using Withered Grapes of Muscat of Alexandria', *Fermentation*, 8 (2): 85.

Harutyunyan, M., R. Viana, J. Granja-Soares, A. Asryan, J. C. Marques and M. Malfeito-Ferreira (2023), 'Consumer Acceptance of Sweet Wines and Piquettes Obtained by the Adaptation of Ancient Winemaking Techniques ("AWT")', *Journal of Sensory Studies*, 38 (3), e12823.

Harvey, S. M. (2010), 'Iron Tools from a Roman Villa at Boscoreale, Italy, in the Field Museum and the Kelsey Museum of Archaeology', *AJA*, 114: 697–714.

Hegyi, A., D. Diaconescu, P. Urdea, A. Sarris, M. Pisz and A. Onaca (2021), 'Using Geophysics to Characterize a Prehistoric Burial Mound in Romania', *Remote Sensing*, 13: 842.

Hendy, C. H. (1971), 'The Isotopic Geochemistry of Speleothems in the Calculation of the Effects of Different Modes of Formation on Isotopic Composition of Speleothems and their Applicability as Palaeoclimatic Indicators', *Geochimica Cosmochimica Acta*, 35: 801–24.

Hendy, C. H. and A. T. Wilson (1968), 'Palaeoclimate Data from Speleothems', *Nature*, 219: 48–51.

Heron, C. and A. M. Pollard (1987), 'The analysis of natural resinous materials from amphorae', in *Science and Archaeology*, Glasgow, Oxford: British Archaeological Reports.

Hine, H. M. (2011), '"Discite . . . Agricolae": Modes of instruction in Latin Prose Agricultural Writing from Cato to Pliny the Elder', *CQ*, 61: 624–54.

Hirschfeld, Y. (1983), 'Ancient Wine Presses in the Park of Aijalon', *Israel Exploration Journal*, 33 (3–4): 207–18.

Hjelle, K. L., I. K. Mehl, S. Sugita and G. L. Andersen (2015), 'From Pollen Percentage to Vegetation Cover: Evaluation of the Landscape Reconstruction Algorithm in Western Norway', *Journal of Quaternary Science*, 30 (4): 312–24.

Hodges, R. (2010), 'Adriatic Sea Trade in a European Perspective', *Scottish Archaeological Journal*, 32 (2): 95–119.

Hoffelinck, A. and F. Vermeulen (2021), 'Ubi sunt macella? The Contribution of Non-Invasive Archaeological Survey to the Identification and Study of Roman Food Markets', *Oxford Journal of Archaeology*, 40: 105–33.

Holloway, B. (2019), 'Archaeological Trial-Trenching Evaluation: Land North of Elmstead Road, Colchester, Essex, CAT Project Ref.: 14/02h, NGR: TM 0221 2442', Colchester and Ipswich Museum accession code: 2014.19.

Horden, P. and N. Purcell (2000), *The Corrupting Sea: A Study of Mediterranean History*, Oxford: Wiley.

Hudson, N. F. (2010), 'Changing Places: The Archaeology of the Roman Convivium', *American Journal of Archaeology*, 114 (4): 663–95.

Iglesias, J. (1979), *El fogatge de 1553*, Barcelona: Fundació Salvador Vives Casajuana.

Imazio, S. and M. Labra (2005), 'I marcatori molecolari e il loro impiego nello studio della domesticazione della vite, attraverso il confronto con la vite selvatica', A. Ciacci, A. Zifferero (eds), *VINUM: Un progetto per il riconoscimento della vite silvestre nel paesaggio archeologico della Toscana e del Lazio settentrionale*, 57–66, Siena: Ci.Vin.

Indelicato, M. (2017), 'Coltivare la vite, produrre il vino: un approccio sperimentale e multidisciplinare allo studio della viticoltura della Sicilia romana', unpublished tesi di Specializzazione in Archeologia Classica, Università degli Studi di Catania.

Indelicato, M. (2020a), 'Archeologia Sperimentale e alimentazione: il panorama italiano', *Archeologie Sperimentali: Temi, Metodi, Ricerche*, 1: 36–55.

Indelicato, M. (2020b), 'Columella's Wine: a Roman Enology Experiment', *EXARC*, 1: 1–8.

Indelicato, M., D. Malfitana and G. Cacciaguerra (2017), 'The Archaeology of Wine in Italy: A Sicilian Experiment', in R. Alonso, J. Baena and D. Canales (eds), *Playing with the Time: Experimental Archaeology and the Study of the Past*, 321–8, Madrid: Universidad Autónoma de Madrid.

Inserra, F., A. Pecci, M. Á. Cau Ontiveros and J. Roig Buxó (2015), 'Organic Residue Analysis of Late Antique Pottery from Plaça Major-Horts de Can Torras (Castellar del Vallés, Catalonia, Spain)', *Periodico di Mineralogia*, 84 (1): 123–38.

Jackson, R. S. (2008), *Wine Science: Principles and Applications*, London: Academic Press.

James, T. G. H. (1995), 'The Earliest History of Wine and Its Importance in Ancient Egypt', in P. E. McGovern, S. J. Fleming and S. Katz (eds), *The Origins and Ancient History of Wine*, 197–213, New York: Gordon and Breach.

Jashemski, W. (1968), 'Excavations in the "Foro Boario" at Pompeii: A Preliminary Report', *AJA*, 72: 69–73.

Jashemski, W. (1973), 'The Discovery of a Large Vineyard at Pompeii: University of Maryland Excavation, 1970', *AJA*, 77: 27–41.

Jashemski, W. (1979), *The Gardens of Pompeii, Herculaneum and the Villas Destroyed by Vesuvius*, Vol. 1, New Rochelle: Caratzas.

Jashemski, W. (1993), *The Gardens of Pompeii: Herculaneum and the Villas Destroyed by Vesuvius*, Vol. 2: Appendices, New York: Caratzas.

Jerkovíc, I., Z. Marijanović, M. Gugić and M. Roje (2011), 'Chemical Profile of the Organics Residue from Ancient Amphora found in the Adriatic Sea Determined by Direct GC and GC-MS Analysis', *Molecules*, 16: 7936–48.

Jochum, K. P., D. Scholz, B. Stoll, U. Weis, S. A. Wilson, Q. Yang, A. Schwalb, N. Börner, D. E. Jacob and M. O. Andreae (2012), 'Accurate Trace Element Analysis of Speleothems and Biogenic Calcium Carbonates by LA-ICP-MS', *Chemical Geology*, 318–19: 31–44.

Johnson, H. and J. Robinson (2019), *The World Atlas of Wine*, 8th edn, London: Octopusbooks.

Jones, D. W. (2014), *Economic Theory and the Ancient Mediterranean*, Malden: Wiley-Blackwell.

Jones, G. V. (2007), 'Climate Change: Observations, Projections and General Implications for Viticulture and Wine Production', in *Climate and Viticulture Congress, Zaragoza, April 10–14, 2007/8th International Symposium Innovations in Enology, Stuttgart April 20–23, 2007*, 1–13, accessed here: https://www.infowine.com/intranet/libretti/libretto4594-01-1.pdf

Jones, G. V. and R. Davis (2000), 'Climate Influences on Grapevine Phenology, Grape Composition and Wine Production and Quality for Bordeaux, France', *Am J Enol Vitic* 51 (3): 249–61.

Jones, G. V., É. Duchêne, D. Tomasi, J. Yuste, O. Braslavská, H. Schuktz, C. Martínez, S. Boso, F. Langellier, C. Perruchot and G. Guimberteau (2005), 'Changes in European Winegrape Phenology and Relationships with Climate', in *XIV International GESCO Viticulture Congress, Geisenheim, Germany, 23–27 August, 2005*, 54–61.

Jones, G., R. Reid and A. Vilks (2012), 'Climate, Grapes and Wine: Structure and Suitability in a Variable and Changing Climate', in P. H. Dougherty (ed.), *The Geography of Wine: Regions, Terroir and Techniques*, 109–33, Dordrecht: Springer.

Jones, G. and L. B. Webb (2010), 'Climate Change, Viticulture and Wine: Challenges and Opportunities', *Journal of Wine Research*, 21 (2): 103–6.

Jongman, W. (1988), *The Economy and Society of Pompeii*, Dutch Monographs on Ancient History and Archaeology 4, Amsterdam: Gieben.

Jouanna, J. (2012), 'Wine and Medicine in Ancient Greece', in P. van der Eijk (ed.), *Greek Medicine from Hippocrates to Galen: Selected Papers*, Studies in Ancient Medicine 40, 173–93, Boston: Brill.

Joyce, J. (2019), 'Modelling Agricultural Strategies in the Dutch Roman Limes via Agent-Based Modelling (ROMFARMS)', in P. Verhagen, J. Joyce and M. R. Groenhuijzen (eds), *Finding the Limits of the Limes: Modelling Demography, Economy and Transport on the Edge of the Roman Empire*, 109–27, Cham: Springer.

Jung, C., H. Pomarèdes, M. Compan, I. Figueiral, S. Martin, O. Ginouvez and C. Tardy (2013), 'Pratiques culturales et système agraire gallo-romain: L'exemple de la vallée de l'Hérault et du Biterrois (Hérault)', *RAN*, 46: 159–77.

Kaal, J., F. Criado-Boado, M. Costa-Casis, J. A. López-Sáez, L. López-Merino, T. Mighall, Y. Carrión, N. S. Sánchez and A. M. Cortizas (2013), 'Prehistoric Land Use at an Archaeological Hot-Spot (the Rock Art Park of Campo Lameiro, NW Spain) Inferred from Charcoal, Synanthropic Pollen and Non-Pollen Palynomorph Proxies', *Journal of Archaeological Science*, 40: 1518–27.

Kaniewski, D., N. Marriner C. Morhange, D. Rius M.-B. Carre S. Faivre and E. Van Campo (2018), 'Croatia's Mid-Late Holocene (5200–3200 BP) Coastal Vegetation Shaped by Human Societies', *Quaternary Science Reviews*, 200: 334–50.

Kaplan, J. O., I. C. Prentice and N. Buchman (2002), 'The Stable Carbon Isotope Composition of the Terrestrial Biosphere: Modelling at Scales from the Leaf to the Globe', *Global Biogeochemical Cycles*, 16 (4): 8–11.

Karaman, L. (1930), *Iz kolijevke hrvatske prošlosti*, Zagreb: Matica hrvatska.

Karlioğlu, N., H. Caner, N. K. Rauh, Ü. Akkemik and N. Kose (2014), 'Palynological Evidence for Human Occupance in Western Rough Cilicia (Antalya, Gazipaşa)', *ANMED 2014–12 (News of Archaeology from Anatolia's Mediterranean Areas)*, 158–62, Antalya: Suna and Inan Kirac Research Institute on Mediterranean Civilizations.

Karlioğlu, N., H. Caner, N. K. Rauh, U. Akkemik and E. M. Connor (2016), 'Palynological Evidence for Human Occupation in Western Rough Cilicia (Southwest Turkey)', *Quaternary International*, 401: 109–22.

Kay, J. E., C. Deser, A. Phillips, A. Mai, C. Hannay, G. Strand, J. M. Arblaster, S. C. Bates, G. Danabasoglu, J. Edwards, M. Holland, P. Kushner, J. F. Lamarque, D. Lawrence, K. Lindsay, A. Middleton, E. Munoz, R. Neale, K. Oleson, L. Polvani and M. Vertenstein (2014), 'The Community Earth System Model (CESM) Large Ensemble Project: A Community Resource for Studying Climate Change in the Presence of Internal Climate Variability', *Bull. Am. Meteorol. Soc*, 96 (8): 1333–49.

Kimura, P., G. Okamoto and K. Hitano (1998), 'The Mode of Pollination and Stigma Receptivity in Vitis Coignetiae Pulliat', *American Journal of Enology and Viticulture*, 49: 1–5.

Kliewer, W. M. (1967), 'Concentration of Tartrates, Malates, Glucose and Fructose in the Fruits of the Genus Vitis', *American Journal of Enology and Viticulture*, 18: 87–96.

Kolendo, J. (1980), *L'agricoltura nell'italia romana: tecniche agrarie e progresso economico dalla tarda repubblica al principato*, Rome: Editori Riuniti.

Kolendo, J. (1985), 'Le attività agricole degli abitanti di Pompei e gli attrezzi agricoli ritrovati all'interno della città', *Opus*, 4: 111–24.

Komar, P. (2016), 'Wine from Cyprus and Cilicia: Taste and Trade', *Electrum*, 23: 155–85.

Komar, P. (2020), 'In Vino Sanitas: Medical Qualities of Greek Wines', *Mélanges de l'Ecole Française de Rome*, 132: 429–47.

Komar, P. (2021), 'Wine Imports and Economic Growth in Rome Between the Late Republic and Early Empire', *Historia*, 70 (4): 437–62.

Kvavadze, E., M. Chichinadze and I. Martkoplishvili (2010), 'The Pollen Production and Yield in Vineyard of Kakheti According to the Results of Pollen Monitoring', in *33rd World Congress of Vine and Wine*, Georgian National Museum.

Labate, D. (2010), 'Il popolamento in età romana della collina modenese: l'insediamento e gli impianti produttivi', in F. Guandalini and D. Labate (eds), *Insediamento di Montegibbio: una ricerca interdisciplinare per l'archeologia: Atti del convegno, Sassuolo (Modena), 7 febbraio 2009*, 21–30, Borgo S. Lorenzo: All'Insegna del Giglio.

Labra, M., O. Failla, T. Fossati, S. Castiglione, A. Scienza and F. Sala (1999), 'Philogenetic Analysis of Grapevine cv. Ansonica Growing on the Island of Giglio, Italy, by AFLP and SSR Markers', *Vitis*, 38: 161–6.

Lagóstena, L. (1996), *Alfarería romana en la Bahía de Cádiz*, Cádiz: Universidad de Cádiz.

Lágostena, L. (2016), 'El Lacus Ligustinus como agente de articulación urbana y conectividad territorial: De las riberas de Hasta Regia a Carissa Aurelia y el acceso a la serranía', *Studia Historica*, 34: 63–86.

Lagóstena, L. (2021), 'Estructura de la ocupación rural romana en las campiñas Hastenses y Gaditanas: aportación de la exploración GPR a su conocimiento', in V. Mayoral Herrera, I. Grau Mira and J. P. Bellón Ruiz (eds), *Arqueología y sociedad de los espacios agrarios: en busca de la gente invisible en la materialidad del paisaje*, 109–17, Mérida: Mytra.

Lagóstena, L. and J. Molina (2019), 'La figlina como ámbito de investigación cuantitativa de la economía romana: Aportación y potencial de la exploración georradar', in V. Revilla Calvo, A. Aguilera Martín, L. Pons Pujol and M. García Sánchez (eds), *Ex Baetica Romam: Homenaje a José Remesal Rodríguez*, Collecció Homenatges 58, 923–46, Barcelona: University of Barcelona.

Lagóstena, L. and P. Trapero (2019), 'La localización de los viñedos de Marco Columela: literatura agronómica y análisis GIS en el interfluvio Guadalquivir-Guadalete', in J. Remesal, V. Revilla, D. J. Martín-Arroyo and A. Martín i Oliveras (eds), *Paisajes productivos y redes comerciales en el imperio romano*, Instrumenta 65, 89–114, Barcelona.

Lake, M. (2020), 'Spatial Agent-Based Modelling', in M. Gillings, P. Hacigüzeller and G. Lock (eds), *Archaeological Spatial Analysis: A Methodological Guide*, 247–72, New York: Routledge.

Lamboglia, N. (1950), *Gli Scavi di Albintimilium e la cronologia della ceramica romana I (campagne di scavo 1938–1940)*, Bordighera: Istituto Internazionale di Studi Liguri.

Lamboglia, N. (1952), 'La nave romana di Albegna', *Rivista di Studi Liguri*, 28: 131–2.

Lamboglia, N. (1955), 'Sulla cronologia delle anfore romane di età repubblicana', *Rivista di Studi Liguri*, 31: 241–70.

Landa, V., Y. Shapira, M. David, A. Karasik, E. Weiss, Y. Reuveni and E. Drori (2021), 'Accurate Classification of Fresh and Charred Grape Seeds to the Varietal Level, Using Machine Learning Based Classification Method', *Scientific Reports*, 11 (13577).

Lasfargues, J. (1985), 'Informations archéologiques: Rhône-Alpes', *Gallia*, 43: 542–3.

Lavee, S. (2000), 'Grapevine (Vitis Vinifera) Growth and Performance in Warm Climates', in A. Erez (ed.), *Temperate Fruit Crops in Warm Climates*, 343–66, Dordrecht: Springer.

Laville, P. (1990), 'Le terroir, un concept indispensable à l'élaboration et à la protection des appellations d'origine comme à la gestation des vignobles: le cas de la France', *Bulletin de l'OIV*, 63: 217–41.

Lechterback, J. and C. E. Jensen (2020), 'Exploring the Potential of Palynology in Archaeological Contexts: Proceedings of the Session Held at the 24th Annual Meeting of the European Association of Archaeologists in Barcelona 2018', *Vegetation History and Archaeobotany*, 29: 111–12.

Leckebusch, J. (2003), 'Ground-Penetrating Radar: A Modern Three-Dimensional Prospection Method', *Archaeological Prospection*, 10: 213–41.

Lee, A. (2022), 'People of Roman Lincoln – Marcus Aurelius Lunaris. The Collection Art and Archaeology in Lincolnshire', accessed at: https://www.thecollectionmuseum.com/?/blog/view/people-of-roman-lincoln-marcus-aurelius-lunaris Last accessed 8/11/2022

Leidwanger, J. (2020), *Roman Seas: A Maritime Archaeology of the Eastern Mediterranean*, Oxford: Oxford University Press.

Lelli, E. (2010), *L'agricoltura Antica: I Geoponica di Cassiano Basso*, Soveria Mannelli: Rubbettino Editore.

Lentjes, D. (2016), *Planting the Seeds of Change: Landscape and Land Use in First Millennium* BC *Southeast Italy*, Amsterdam: Amsterdam University Press.

Lentjes, D. and G. S. Semerari (2016), 'Big Debates over Small Fruits: Wine and Oil Production in Protohistoric Southern Italy (ca. 1350–750 BC)', *BABesch*, 91: 1–16.

Leveau, Ph. and J. C. Golvin (1979), 'L'amphithéâtre et le théâtre-amphithéâtre de Cherchel: Monuments à spectacles et histoire urbaine à Caesarea de Maurétanie', *MÉFRA*, 91 (2): 817–43.

Leveau, Ph. (2011), 'Entre faits archéologiques et concepts: la recherche sur les aglomérations protohistóriques et gallo-romaines', in *Les Rutènes: Du peuple à la cité*, Supp. Aquitania 25, 591–603, Bordeaux: Aquitania.

Lewis, Ch. and T. Short (1879), *A Latin Dictionary*, London: Clarendon Press.

Lewit, T. (2020), 'Invention, Tinkering, or Transfer? Innovation in Oil and Wine Presses in the Roman Empire', in P. Erdkamp, K. Verboven, A. Zuiderhoek (eds), *Capital, Investment, and Innovation in the Roman World*, 307–53, Oxford: OUP.

Lewit, T. and P. Burton (2019), 'Wine and Oil Presses in the Roman to Late Antique Near East and Mediterranean: Balancing Textual and Archaeological Evidence', in A. Squitieri and D. Eitam (eds), *Stone Tools in the Ancient Near East and Egypt: Ground Stone Tools, Rock-Cut Installations and Stone Vessels from Prehistory to Late Antiquity*, 97–110, Oxford: Archaeopress.

Li, X. J., A. Bevan, M. Martinón-Torres, T. Rehren, W. Cao, Y. Xia and K. Zhao (2014), 'Crossbows and Imperial Craft Organisation: The Bronze Triggers of China's Terracotta Army', *Antiquity*, 88 (339): 126–40.

Lindhagen, A. (2009), 'The Transport Amphoras Lamboglia 2 and Dressel 6A: A Central Dalmatian Origin?', *Journal of Roman Archaeology*, 22: 83–108.

Lindhagen, A. (2013), 'The Freedman Milieus at Delos and Narona: New Perspectives on the Lamboglia 2 Wine Trade', in A.-L. Schallin (ed.), *Perspectives on Ancient Greece: Papers in celebration of the 60th anniversary of the Swedish Institute at Athens*, 231–50, Stockholm: Paul Astroms.

Lindhagen, A. (2020), 'A Central Dalmatian Origin of the Adriatic Wine Amphorae? New Evidence from Xrf-analyses', in I. Kamenjarin and M. Ugarkovic (eds), *Exploring the Neighborhood: The Role of Ceramics in Understanding Place in the Hellenistic World, Proceedings of the 3rd conference of IARPotHP, Kastela, June 2017, 1st–4th*, 343–52, Wien: Phoibos Verlag.

Linford, N. (2006), 'The Application of Geophysical Methods to Archaeological Prospection', *Reports on Progress in Physics*, 69: 2205–57.

Lipovac Vrkljan, G. (2009), 'L'officina ceramica di Crikvenica', in S. Pesavento Mattioli and M.-B. Carre (eds), *Olio e pesce in epoca romana: Produzione e commercio nelle regioni dell'alto Adriatico, Atti del Convegno (Padova, 16 febbraio 2007)*, 309–14, Rome: Quasar.

Lipovac Vrkljan, G., I. Radic Rossi and A. Konestra, eds (2017), *AdriAmphorae: Amphorae as a resource for the reconstruction of economic development in the Adriatic region in Antiquity: local production, Proceedings of the workshop, Zagreb 21st April 2016*, Zagreb: Institut za Arheologiju.

Lipovac Vrkljan, G. and A. Konestra, eds (2019), *Pottery Production, Landscape and Economy of Roman Dalmatia*, Oxford: Archaeopress.

Lissarague, F. (1989), *L'immaginario del simposio Greco*; Roma; Bari: Editori Laterza.

Liverani, P. (1987), 'Termini muti di centuriazione o contrappesi di torchi?', *MEFRA*, 99: 111–27.

Liverani, P. (1988), 'Resti di torchi di tradizione romana in Valpolicella', *Annuario storico della Valpolicella*, 6: 53–62.

Lodwick, L. and E. Rowan (2022), 'Archaeobotanical Research in Classical Archaeology', *American Journal of Archaeology*, 126 (4): 593–623.

López Amador, J. J. and J. A. Ruiz Gil (2007), 'Arqueología de la vid y el vino en El Puerto de Santa María', *Revista de Historia del Puerto*, 38: 11–36.

López Amador, J. J. and J. A. Antonio Ruiz Gil (2005), 'Arqueología de la repoblación Alfonsí: Pocito Chico en La Laguna del Gallo de El Puerto de Santa María (Cádiz, España)', *Revista de Historia de El Puerto*, 35: 11–51.

Lyne, M. (2006), *Roman Wight*, Isle of Wight History Centre, accessed here: https://www.iwhistory.org.uk/resourceassessment/iow%20Roman.pdf

Macias-Fauria, M. and K. J. Willis (2013), 'Landscape Planning for the Future: Using Fossil Records to Independently Validate Bioclimatic Envelope Models for Economically Valuable Tree Species in Europe', *Global Ecology and Biogeography*, 22 (3): 318–33.

Madrigal, J., A. K. W. Runge, L. Bouby, T. Lacombe, J. A. Samaniego, A.-F. Adam-Bondon, I. Figueiral, C. Hallavant, J. M. Martinez-Zapater, C. Schaal, R. Töpfer, B. Patersen, T. Sicheritz-Pontén, P. This, R. Bacilieri, M. Gilbert and N. Whales (2019), 'Palaeogenomic Insights into the Origins of French Grapevine Diversity', *Nature Plants*, 5: 595–603.

Maggi, P. and Y. Marion (2011), 'Le produzioni di anfore e di terra sigillata a Loron e la loro diffusione', in G. Lipovac Vrkljan, I. Radić Rossi and B. Siljeg (eds), *Officine per la produzione di ceramica e vetro in epoca romana: Produzione e commercio nella regione adriatica. Atti del convegno (Crikvenica, 23–24 octobre 2008)*, 175–87, Crikvenica.

Magri, D. and L. Sadori (1999), 'Late Pleistocene and Holocene Pollen Stratigraphy at Lago di Vico, Central Italy', *Vegetation History and Archaeobotany*, 8: 247–60.

Magris, G., I. Jurman, A. Fornasiero, E. Paparelli, R. Schwope, F. Marroni, G. di Gaspero and M. Morgante (2021), 'The Genomes of 204 Vitis Vinifera Accessions Reveal the Origin of European Wine Grapes', *Nature Communications*, 12: 7240.

Malaguti, C., M. Marchesini, L. Casagrande, V. Cobianchi, I. Gobbo, S. Marvelli, L. Mura and E. Rizzoli (2011), 'Il pozzo di Badia Polesine (Rovigo): Archeologia e tecnica dei pozzi per acqua dalla pre-protostoria all'età moderna' in, S. Cipriano and E. Pettenò (eds), *Archeologia e tecnica dei pozzi per acqua dalla pre-protostoria all'età moderna*, Antichità Altoadriatiche LXX, 85–114, Rome: Quasar.

Malavolti, F. (1948), 'Rapporti tra alluvioni ed antichi insediamenti umani nella pianura emiliana', *Emilia Preromana*, 1: 79–86.

Malheiro, A. C., J. A. Santos, H. Fraga and J. G. Pinto (2010), 'Climate Change Scenarios Applied to Viticultural Zoning in Europe', *Climate Research*, 43: 163–77.

Manataki, M., A. Sarris, J. C. Donati, C. Cuenca and T. Kalayci (2015), 'GPR: Theory and Practice in Archaeological Prospection', in A. Sarris (ed.), *Best Practices of Geoinformatic Technologies for the Mapping of Archaeolandscapes*, 13–24, Oxford: Archaeopress.

Manca, R., L. Pagliantini, E. Pecchioni, A. P. Santo, F. Cambi, L. Chiarantini, A. Corretti, P. Costagliola, A. Orlando and M. Benvenuti (2016), 'The Island of Elba (Tuscany, Italy) at the Crossroads of Ancient Trade Routes: An Archaeometric Investigation of Dolia Defossa from the Archaeological Site of San Giovanni', *Miner Petrol*, 110: 693–711.

Mangafa, M. and K. Kotsakis (1996), 'A New Method for the Identification of Wild and Cultivated Charred Grape Seeds', *Journal Archaeological Science*, 23: 409–18.

Mange, M. and T. Bezeczky (2006), 'Petrography and Provenance of Laecanius Amphorae from Istria, Northern Adriatic Region, Croatia', *Geoarchaeology*, 21 (5): 429–60.

Manning, J. G. (2018), *The Open Sea: The Economic Life of the Ancient Mediterranean World from the Iron Age to the Rise of Rome*, Princeton and Oxford: Princeton University Press.

Manzi, L. (1883), *La Viticoltura e l'Enologia presso I Romani*, Rome: Quasar.

Manzoni, L. (1965), 'Esame istologico dei reperti di fusti di vite a Torcello', *Memorie di Biogeografia Adriatica*, 6: 147–56.

Marcello, A. (1965), 'Testimonianze di una antica ortofrutticoltura nell'isola di Torcello', *Memorie di Biogeografia Adriatica*, 6: 111–45.

Marchesini, M. (1998), *Il paesaggio vegetale nella pianura Bolognese in età romana sulla base di analisi archeopalinologiche ed archeocarpologiche*, PhD thesis, Università degli Studi di Firenze.

Marchesini, M. and L. Forlani (2002), 'I legni del pozzo di Cognento (Modena) dal periodo tardo romano all'età moderna', *Archeologia dell'Emilia Romagna*, 3: 229–42.

Marchesini, M., L. Forlani and N. Giordani (1998), 'Reperti antraco/xilologici del pozzo di Rubiera (53 m s.l.m., Reggio Emilia-Nord Italia): età tardo antica; fine VI-metà VII sec. d.C.' in *Studi in ricordo di Daria Bertolani Marchetti: Atti della Giornata di Studi, Formigine, 18 maggio 1996*, 375–83.

Marchesini, M. and S. Marvelli (2010), 'Ricostruzione del paesaggio vegetale e antropico nelle aree centuriate dell'Emilia Romagna attraverso le indagini archeobotaniche: Sistemi e opere di assetto agrario tra età romana primo medioevo', *Agri Centuriati: An International Journal of Landscape Archaeology*, 6: 313–23.

Marchesini, M. and S. Marvelli (2017), 'Paesaggio vegetale e agricoltura nella pianura padana in età romana', in E. Lo Cascio and M. Maiuro (eds), *Popolazione e risorse nell'Italia del nord dalla romanizzazione ai longobardi*, 289–304, Roma; Bari: Edipuglia.

Marchesini, M., S. Marvelli and E. Rizzoli (2021), 'La diffusione della viticoltura nell'età del bronzo in Italia', in I. Damiani, A. Cazzella and V. Copat (eds), *Studi di Preistoria e Protostoria – 6, Preistoria del cibo: L'alimentazione nella preistoria e nella protostoria*, 535–42, Firenze.

Marchesini, M., A. C. Muscogiuri, E. Rizzoli and S. Marvelli (forthcoming), 'Paesaggio e alimentazione: il contributo delle analisi archeobotaniche', in P. Basso, G. Falezza, M. Migliavacca and F. Saggioro (eds), *Veronensium mensa: Food and Wine in Ancient Verona*, Rome: Quasar.

Marcone, A. (1997), *Storia dell'agricoltura romana*, Rome: Nuova Italia Scientifiche.

Marengo, S. M. (2002), 'Laterizi con marchio C. HER. DIOG.', in G. Paci (ed.), *Antiqua Frustula, Macerata*, 116, no. 17, Pollenza: Tipografia S. Giuseppe.

Marescalchi, A. and G. Dalmasso (1931), *Storia della Vite e del Vino in Italia*, Milan: Editrice Unione Italiana Vini.

Marinval, P., D. Maréchal and D. Labadie (2002), 'Arbres fruitiers et cultures jardinées gallo-romains à Longueil-Sainte-Marie (Oise)', *Gallia*, 59: 253–71.

Mariotti Lippi, M., M. Guido, B. I. Menozzi, C. Bellini and C. Montanari (2007), 'The Massaciuccoli Holocene Pollen Sequence and the Vegetation History of the Coastal Plains by the Mar Ligure (Tuscany and Liguria, Italy)', *Vegetation History and Archaeobotany*, 16 (4): 267–77.

Mariotti Lippi, M., M. Mori Secci and C. Bellini (2012), 'L'archeobotanica e lo studio della vite nella preistoria della Toscana', in A. Ciacci, P. Rendini and A. Zifferero (eds), *Archeologia della vite e del vino in Toscana e nel Lazio: Dalle tecniche dell'indagine archeologica alle prospettive della biologia molecolare*, 119–24, Borgo San Lorenzo: All'Insegna del Giglio.

Mariottini, V. and F. Salvadori (2012), 'Archeologia dell'alimentazione, dallo scavo open air allo studio dei reperti mobili, riflessioni metodologiche', in Redi F. and A. Forgione (eds), *VI Congresso Nazionale di Archeologia Medievale: Atti del convegno, L'Aquila, 12–15 settembre 2002*, 69–72, Firenze.

Marlière, E. (2001), 'Le tonneau en Gaule romaine', *Gallia*, 58: 181–201.

Marlière, E. (2002), *L'outre et le tonneau dans l'Occident romain*, Montagnac: Editions Mergol.

Marlière, E. and J. Costa (2005), 'Tonneaux et amphores a Vindolanda: contribution à la connaissance de l'approvisionnement des troupes stationnées sur le mur d'Hadrien (II)', in A. Birley and J, Blacke (eds), *Vindolanda: The Excavations of 2003–2004*, 214–36, Durham: Vindolanda Trust.

Marret, F., J. O'Keefe, P. Osterloff, M. Pound and L. Shumilovskikh, eds (2021), *Applications of Non-Pollen Palymorphs: From Palaeoenvironmental Reconstructions to Biostratigraphy*, Geological Society Special Publication 511, London: Geological Society of London.

Marsh, E. J. and J. R. Ferguson (2010), 'Introduction', in J. R. Ferguson (ed.), *Designing Experimental Research in Archaeology: Examining Technology Through Production and Use*, 1–12, Boulder: University of Colorado.

Marston, J. M., C. Warinner and J. D'Alpoim Guedes (2014), *Method and Theory in Paleoethnobotany*, Boulder: University of Colorado.

Martín, R. (1971), *Recherches sur les agronomes latins et leurs conceptions économiques et sociales*, Paris: Societé d'édition les belles lettres.

Martín i Oliveras, A. (2013), 'The CELLA VINARIA Project and Archaeological Park (Teià, Maremse, Barcelona): A Great Experimental Archaeology Laboratory', in F. W. F. Foulds (ed.), *Experimental Archaeology and Theory: Recent Approaches to Archaeological Hypotheses*, 67–100, Oxford: Oxbow Books.

Martín i Oliveras, A. (2015a), 'Arqueología del Vino en Época Romana: Teoría económica, lógica productiva y comercial aplicada al envasado, la expedición, el transporte y la distribución de ánforas vinarias del nordeste peninsular (s. I a.C.-I d.C.)', in V. Martínez Ferreras (ed.), *La difusión Comercial de las Ánforas Vinarias de Hispania Citerior Tarraconensis (s.I aC-I dC)*, Roman Archaeology 4, 19–38, Oxford: Archaeopress.

Martín i Oliveras, A. (2015b), *Arqueologia del vi a l'època romana. Del cultiu al consum: Marc teòric i epistemològic*, Premi d'Arqueologia Memorial Josep Barberà i Farràs (12a edició), Barcelona: Societat Catalana d'Arqueologia.

Martín i Oliveras, A. (2016), 'Aspectos transversales de lógica económica, productiva y comercial aplicada al envasado, la expedición, el transporte y la distribución de ánforas vinarias del nordeste peninsular (siglos I a.C. - III d.C.): Algunas reflexiones', in R. Járrega and P. Berni (eds), *Amphorae ex Hispania: paisajes de producción y consumo. Actas del III Congreso de la SECAH-Sociedad de Estudios de Cerámica Antigua-Ex Officina Hispana, celebrado en Tarragona el 10–13 de diciembre de 2014*, 34–54, Tarragona: Institut Català d'Arqueologia Clàssica.

Martín i Oliveras, A. and F. Bayés (2009), 'Cella Vinaria de Vallmora (Teià-Maresme-Barcelona): Estudi per la reconstrucció de dues premses romanes', in M. Prevosti and A. Martín i Oliveras (eds), *El vi tarraconense i laietà: ahir i avui, Actes del Simpòsium, Tarragona-Teià 2007*, 215–48, Tarragona: Institut Cataà d'Arqueologia.

Martín i Oliveras, A., D. Martín-Arroyo Sánchez and V. Revilla Calvo (2017), 'The Wine Economy in Roman Hispania: Archaeological Data and Modellization', in J. Remesal Rodriguez (ed.), *The Roman Economy: New Perspectives*, 189–237, Barcelona: University of Barcelona.

Martín i Oliveras, A. and B. Parés (2020), 'Modelización 3D de un *dolium* romano a partir de técnicas combinadas de metrología dimensional y fotogrametría digital', in V. Revilla, A. Aguilera, Ll. Pons and M. García (eds), *'Ex Baetica Romam': homenaje a José Remesal Rodríguez*, 947–74, Barcelona: Universitat de Barcelona.

Martín i Oliveras, A. and V. Revilla (2019), 'The Economy of Laetanian Wine: A Conceptual Framework to Analyse an Intensive/Specialized Winegrowing Production System and Trade (First Century BC to Third Century AD)', in P. Verhagen, J. Joyce and M. Groenhuijzen (eds), *Finding the Limits of the Limes*, 129–63, Cham: Springer.

Martín i Oliveras, A. and V. Revilla (2020), 'Quantifying Laetanian Roman Wine Production Function (1st century BC–3rd century AD): A Microeconomic Approach to Calculating Vineyard's Crop and Winemaking Processing Facilities Yields', in A. Marzano (ed.), *Archaeology and Economy in the Ancient World 17: Villas, Peasant Agriculture, and the Roman Rural Economy. Proceedings of the 19th International Congress of Classical Archaeology, Panel 3.15*, 89–112, Cologne–Bonn: Propylaeum.

Martín i Oliveras, A., V. Revilla, C. Carreras and J. Remesal (2020), 'Viticulture and Demography in the Laetanian Region (Hispania Citerior Tarraconensis), 1st c. BC – 3rd c. AD', in D. Van Limbergen, S. Maréchal and W. De Clercq (eds), *The Resilience of the Roman Empire: Regional Case Studies on the Relationship between Population and Food Resources*, BAR International Series 3000, 47–69, Oxford: BAR.

Martín-Arroyo, D. and J. Remesal (2018), 'Modelado GIS de la agricultura romana en un marco colonial teórico. Vegetación de ribera y viticultura en Hasta Regia', *SPAL*, 27 (1): 211–35.

Mathieu, J. R. (2002), 'Introduction', in J. R. Mathieu (ed.), *Experimental Archaeology: Replicating Past Objects, Behaviours and Processes*, 1–12, Oxford: Archaeopress.

Maune, S. (2003), 'La villa gallo-romaine de "Vareilles" à Paulhan (Hérault; fouille de l'autoroute A75). Un centre domanial du Haut-Empire spécialisé dans la viticulture?', *Revue archéologique de Picardie*, 1 (1): 309–37.

Marvelli, S. (1999), 'Testimonianze archeobotaniche: Reperti di vite in Emilia-Romagna', in M. Marchesini, S. Marvelli and P. Pancaldi (eds), *La vite maritata: Storia, cultura, coltivazione, ecologia della piantata nella pianura padana*, 35–50, San Giovanni in Persiceto: Comune di S. Giovanni in Persiceto.

Marvelli, S., S. De Siena, E. Rizzoli and M. Marchesini (2013), 'The Origin of Grapevine Cultivation in Italy: The Archaeobotanical Evidence', *Annali di Botanica*, 3: 155–63.
Marzano, A. (2007), *Roman Villas in Central Italy: A Social and Economic History*, Boston: Brill.
Marzano, A. (2013), 'Agricultural Production in the Hinterland of Rome: Wine and Olive Oil', in A. Bowman and A. Wilson (eds), *The Roman Agricultural Economy: Organisation, Investment, and Production*, 85–106, Oxford: Oxford University Press.
Matijašić, R. (1993), 'Oil and Wine Production in Istria and Dalmatia in Classical Antiquity and the Early Middle Ages', in M.-C. Amouretti and J.-P. Brun (eds), *La production du vin et de l'huile en Méditerranée*, 247–61, BCH Supp. XXVI, Athens: École Française d'Athènes.
Matijašić, R. (1998), *Gospodarstvo antičke Istre*, Pula: Zavičajna naklada 'Zakan Juri'.
Matijašić, R. and F. Tassaux (2000), 'Liber et Silvanus', in C. Delplace and F. Tassaux (eds), *Les cultes polythéistes dans l'Adriatique romaine*, 65–117, Bordeaux: Ausonius Éditions.
Matijašić, R. (2007), 'O Nalazu Kasniantićkih Tijesaka u Poreću', *Opuscvla Archaeologica*, 31: 265–81.
Matijašić, R. (2012), 'Le Città Romane Dell'Istria: Nuovi Particolari', in G. de Marinis, G. M. Fabrini, G. Paci, R. Perna and M. Silvestrini (eds), *I processi formativi ed evolutivi della città in area adriatica*, BAR International Series 2419, 445–52, Oxford: BAR.
Mazzetti, G. (1892), 'Per lo scavo di un nuovo pozzo in Modena: Cenno intorno alla fauna e alla flora del sottosuolo di Modena', in *Atti della Società dei Naturalisti di Modena*, 59–73, Modena.
McCormick, M., U. Büntgen, M. A. Cane, E. R. Cook, K. Harper, P. J. Huybers, T. Litt, S. W. Manning, P. A. Mayewski, A. F. M. More, K. Nicolussi and W. Tegel (2012), 'Climate Change During and After the Roman Empire: Reconstructing the Past from Scientific and Historical Evidence', *Journal of Interdisciplinary History*, 43 (2): 169–220.
McDermott, F. (2004), 'Palaeo-climate Reconstruction from Stable Isotope Variations in Speleothems: A Review', *Quaternary Science Reviews*, 23: 901–18.
McDermott, F., H. Schwarcz and P. J. Rowe, (2005), 'Isotopes in Speleothems', in M. J. Leng (ed.), *Isotopes in Paleoenvironmental Research*, 186–226, Dordrecht: Springer.
McGovern, P. E. (1995), 'Science in Archaeology: A Review', *American Journal of Archaeology*, 99: 79–142.
McGovern, P. E. and H. M. Rudolph (1996), 'The Analytical and Archaeological Challenge of Detecting Ancient Wine: Two Case Studies from the Ancient Near East', in P. E. McGovern, S. J. Fleming and S. H. Katz (eds), *The Origins and Ancient History of Wine*, 57–67, New York.
McGovern, P. E. (2001), 'The Origins of the Tomb U-j Syro-Palestinian Type Jars as Determined by Neutron Activation Analysis', in U. Hartung (ed.), *Umm El-Qaab II: Importkeramik aus dem Friedhof U in Abydos (Umm el-Qaab) und die Beziehungen Ägyptens zu Vorderasien im 4. Jahrtausend v. Chr*, Archäologische Veröffentlichungen 92, 407–16, Mainz: P. von Zabern.
McGovern, P. E. (2003), *Ancient Wine: The Search for The Origins of Viniculture*, Princeton: Princeton University Press.
McGovern, P. E. (2004), *L'archeologo e l'uva*, Rome: Carocci.
McGovern, P. E. (2009), *Uncorking the Past: The Quest for Wine, Beer, and Other Alcoholic Beverages*, Berkeley: University of California.
McGovern, P. E. (2011), 'Two Luxury Items of the Canaanites and Phoenicians: Royal Purple and Wine', in *Forum II: Ligue des cités Cananéennes, Phéniciennes et Puniques, 30 October 2009*, 184–90, Beirut: Fondation Tyr.
McGovern, P. E. (2013), 'From East to West: The Ancient Near Eastern "Wine Culture" Travels Land and Sea', in S. Perez and J. Perez (eds), *Patrimonio Cultural de la Vid y el Vino: Vine and Wine Cultural Heritage*, 233–41, Madrid: Universidad Autónoma de Madrid.
McGovern, P. E. (2017a), *Ancient Brews Rediscovered and Re-created*, New York: W.W. Norton.
McGovern, P. E. (2019a), *Ancient Wine: The Search for the Origins of Viniculture* (with updated Afterword), Princeton: Princeton University Press.

McGovern, P. E. (2019c), 'Uncorking the Past: Alcoholic Fermentation as Humankind's First Biotechnology', in K. Hockings and R. I. M. Dunbar (eds), *Alcohol and Humans: A Long and Social Affair*, 81–92, Oxford: Oxford University Press.

McGovern, P. E. (2022), 'Vessel Contents as Revealed by Organic Residue Analysis: The Scorpion I Tomb as a Case Study', in R. David (ed.), *Concise Manual for Ceramic Studies / دليل موجز لدراسات الفخار from the Nile Valley to the Middle East*, 109–13, Paris: Soleb.

McGovern, P. E., M. P. Callahan, G. R. Hall, W. C. Petersen, D. Cavalieri, D. L. Hartl, O. Jáuregui and R. M. Lamuela-Raventós (2021), 'A Response to Léa Drieu et al., 2020, "Is it possible to identify ancient wine production using biomolecular approaches?"' *STAR: Science & Technology of Archaeological Research*, 7: 43–48.

McGovern, P. E. and G. R. Hall (2015), 'Charting a Future Course for Organic Residue Analysis in Archaeology', *Journal of Archaeological Method and Theory*, 23: 592–622.

McGovern, P. E., M. Jalabadze, S. Batiuk, M. P. Callahan, K. E. Smith, G. R. Hall, E. Kvavadze, D. Maghradze, N. Rusishvili, L. Bouby, O. Failla, G. Cola, L. Mariani, E. Boaretto, R. Bacilieri, P. This, N. Wales and D. Lordkipanidze (2017), 'Early Neolithic Wine of Georgia in the South Caucasus', *The Proceedings of the National Academy of Sciences USA*, 114: E10309–E10318.

McGovern, P. E., B. P. Luley, N. Rovira, A. Mirzoian, M. P. Callahan, K. E. Smith, G. R. Hall, T. Davidson and J. M. Henkin (2013), 'Beginning of Viniculture in France', *The Proceedings of the National Academy of Sciences USA*, 110: 10147–52.

McGovern, P. E., A. Mirzoian and G. R. Hall (2009), 'Ancient Egyptian Herbal Wines', *The Proceedings of the National Academy of Sciences USA*, 106: 7361–6.

McGovern, P. E., S. J. Fleming and S. H. Katz, eds (1996), *The Origins and Ancient History of Wine*, Amsterdam: Routledge.

McGovern, P. E., J. Zhang, J. Tang, Z. Zhang, G. R. Hall, R. A. Moreau, A. Nuñez, E. D. Butrym, M. P. Richards, C. S. Wang, G. Cheng, Z. Zhao and C. Wang (2004), 'Fermented Beverages of Pre- and Proto-Historic China', *Proceedings of the National Academy of Sciences USA*, 101: 17593–8.

McLean, A. and X. Rubio-Campillo (2022), 'Beyond Least Cost Paths: Circuit Theory, Maritime Mobility and Patterns of Urbanism in the Roman Adriatic', *Journal of Archaeological Science*, 138: 105534. https://doi.org/10.1016/j.jas.2021.105534.

Meadows, I. (1996), 'Wollaston: The Nene Valley, a British Moselle?', *Current Archaeology*, 150: 212–15.

Menotti, E. M. (1999), *Archeologia dal territorio mantovano*, Catalogo della mostra, Mantova.

Menozzi, B. I., A. Fichera, M. A. Guido, M. Mariotti Lippi, C. Montanari, G. Zanchetta, F. P. Bonadonna and F. Garbari (2002), 'Lineamenti paleoambientali del bacino del Lago di Massaciuccoli (Toscana nord-occidentale)', *Atti della Società Toscana di Scienze Naturali Serie B*, 109: 177–87.

Mentzer, S. M., D. G. Romano and M. E. Voyatzis (2017), 'Micromorphological Contributions to the Study of Ritual Behavior at the Ash Altar to Zeus on Mt. Lykaion, Greece', *Archaeological Anthropological Sciences*, 9: 1017–43.

Mercati, F., G. De Lorenzis, A. Mauceri, M. Zerbo, L. Brancadoro, C. D'Onofrio, C. Morcia, M. G. Barbagallo, C. Bignami, M. Gardiman, L. de Palma, P. Ruffa, V. Novello, M. Crespan and F. Sunseri (2021), 'Integrated Bayesian Approaches Shed Light on the Dissemination Routes of the Eurasian Grapevine Germplasm', *Frontiers in Plant Science*, 5, August 2021.

Mercuri, A. M. (2014), 'Genesis and Evolution of the Cultural Landscape in Central Mediterranean: The "Where, When and How" Through the Palynological Approach', *Landscape Ecology*, 29: 1799–1810.

Mercuri, A. M., C. A. Accorsi, M. Bandini Mazzanti, G. Bosi, A. Cardarelli, D. Labate, M. Marchesini and G. T. Grandi (2006), 'Economy and Environment of Bronze Age Settlements – Terramaras – in the Po Plain (Northern Italy): First Results of the Archaeobotanical Research at the Terramara di Montale', *Vegetation History and Archaeobotany*, 16: 43–60.

Mercuri, A. M., C. A. Accorsi, M. Bandini Mazzanti, P. Bigi, G. Bottazzi, G. Bosi, M. Marchesini, M. C. Montecchi and D. Pedini (2009), 'From the "Treasure of Domagnano" to the Archaeobotany of a Roman and Gothic Settlement in the Republic of San Marino', in J.-P. Morel and A. M. Mercuri (eds), *Plants and Culture: Seeds of the Cultural Heritage of Europe*, 69–91, Bari: Edipuglia.

Mercuri, A. M., P. Torri, A. Florenzano, E. Clò, M. Mariotti Lippi, E. Sgarbi and C. Bignami (2021), 'Sharing the Agrarian Knowledge with Archaeology: First Evidence of the Dimorphism of Vitis Pollen from the Middle Bronze Age of N Italy (Terramara Santa Rosa di Poviglio)', *Sustainability*, 13 (4): 2287.

Mezquíriz, M. Á. (2004), 'La producción de vino en época romana a través de los hallazgos en territorio navarro', *Trabajos de Arqueología de Navarra*, 17: 133–60.

Middleton, R. and M. J. Bunting (2004), 'Mosaic v1.1: Landscape Scenario Creation Software for Simulation of Pollen Dispersal and Deposition', *Review of Palaeobotany and Palynology*, 132: 61–6.

Middleton, W. D., L. Barba, A. Pecci, J. H. Burton, A. Ortiz, L. Salvini and R. Rodriguez-Suárez (2010), 'The Study of Archaeological Floors: Methodological Proposal for the Analysis of Anthropogenic Residues by Spot Tests, ICP-OES, and GC-MS', *Journal of Archaeological Method and Theory*, 17: 183–208.

Milanesi, C., F. Antonucci, P. Menesatti, C. Costa, C. Faleri and M. Cresti (2011), 'Morphology and Molecular Analysis of Ancient Grape Seeds', *Interdisciplinaria Archaeologica*, 2: 95–100.

Mille, P. and Ph. Rollet (2020), 'Étude de trois grands tonneaux mis au jour à Reims/*Durocortorum* (Marne): le savoir-faire des tonneliers antique', *Gallia: Archéologie de la France antique*, 77 (2): 123–55.

Miniero, P. (1991–2), 'Stabiae. Attività dell'ufficio Scavi 1991', *RStPom*, 5: 221–8.

Montana, G., L. Randazzo, D. Barca and M. Carroll (2021), 'Archaeometric Analysis of Building Materials and Dolia Defossa from the Roman Imperial Estate of Vagnari', *Journal of Archaeological Science: Reports*, 38: 103057.

Moore, P. D., J. A. Webb and M. E. Collinson (1991), *Pollen Analysis*, 2nd edn, London: Blackwell Scientific.

Mori Secci, M. (2005), 'Archeobotanica della vite', in A. Ciacci and A. Zifferero (eds), *VINUM: Un progetto per il riconoscimento della vite silvestre nel paesaggio archeologico della Toscana e del Lazio settentrionale*, 67–74, Siena: Ci.Vin.

Moricca, C., L. Bouby, V. Bonhomme, S. Ivorra, G. Pérez-Jordà, L. Nigro, F. Spagnoli, L. Peña-Chocarro, P. Van Dommelen and L. Sadori (2021), 'Grapes and Vines of the Phoenicians: Morphometric Analyses of Pips from Modern Varieties and Iron Age Archaeological Sites in the Western Mediterranean', *Journal of Archaeological Science: Reports*, 37: 102991.

Morris, D. (2017), *The Naked Ape: A Zoologist's Study of the Human Animal*, first published 1967, London: Jonathan Cape Publishing.

Motta, L. and K. Beydler (2020), 'Agriculture in Iron Age and Archaic Italy', in D. Hollander and T. Howe (eds), *A Companion to Ancient Agriculture*, 399–415, Hoboken: John Wiley & Sons.

Mulholland, S. C. and G. Rapp Jr (1992), 'Phytolith Systematics: An Introduction', in G. Rapp Jr. and S. C. Mulholland (eds), *Phytolith Systematics: Emerging Issues*, 1–14, New York: Plenum Press.

Musti, D. (2005), *Il Simposio nel suo sviluppo storico*, Roma; Bari: Laterza.

Myles, S., A. R. Boyko, C. L. Owens, P. J. Brown, F. Grassi, M. K. Aradhya, B. Prins, A. Reynolds, J. M. Chia, D. Ware, C. D. Bustamante and E. S. Buckler (2011), 'Genetic Structure and Domestication History of the Grape', *PNAS*, 108: 3530–5.

Naqinezhad, A., R. Ramezani, M. Djamali, A. Schnitzler and C. Arnold (2018), 'Wild Grapevine (Vitis vinifera subsp. Sylvestris) in the Hyrcanian Relict Forests of Northern Iran: An Overview of Current Taxonomy, Ecology and Paleorecords', *Journal of Forestry Research*, 29: 1757–68.

Naso, A. (2012), 'Sui primi contatti dell'Etruria con il mondo greco', in A. Mandolesi and M. Sannibale (eds), *Etruschi: L'ideale eroico e il vino lucente* (Catalogo della Mostra, Asti 2012), 48–55, Milan: Electa.

Naveh, Z. and J. Dan (1973), 'The Human Degradation of Mediterranean Landscapes in Israel', in F. Di Castri and H. A. Mooney (eds), *Mediterranean Type Ecosystems*, 373–90, Berlin: Springer.

Nencini, P. (2009), *Ubriachezza e sobrietà nel mondo antico: Alle radici del bere moderno*, Milano: GEM Edizioni.

Nesbitt, A., S. Dorling and A. Lovett (2018), 'A Suitability Model for Viticulture in England and Wales: Opportunities for Investment, Sector Growth and Increased Climate Resilience', *Journal of Land Use Science*, 13 (4): 414–38.

Notarstefano, F. (2012), *Ceramica e alimentazione: L'analisi chimica dei residui organici nelle ceramiche applicata ai contesti archeologici*, Quaderno 10, Bari: B.A.C.T.

Nowak, A. (2018), 'Symmetry Park, Bicester', *Oxfordshire Archaeological Recording Action; Phase 2*, Report No. MK 141/18.

Odiot, T. (1984), 'Un grand domaine viticole à Donzère (Drôme)', in *Archéologie en Rhône-Alpes: Protohistoire et monde galloromaine: dix ans de recherches. Catalogue de l'exposition (Lyon 1983–1984)*, 28–9, Lyon.

OPEL IV (2002), Onomasticon Provinciarum Europae Latinarum (*OPEL*) 4, Quadratia-Zures, Wien: Forschungsgesellschaft Wiener Stadtarchäologie.

Orengo, H. A., F. C. Conesa, A. Garcia-Molsosa, A. Lobo, A. S. Green, M. Madella and C. A. Petrie (2020), 'Automated Detection of Archaeological Mounds using Machine-Learning Classification of Multisensor and Multitemporal Satellite Data', *Proceedings of the National Academy of Sciences*, 117 (31): 18240–50.

Orfila, M., A. Rodríguez-Antón, E. Chávez-Álvarez, A. C. González-García and J. A. Belmonte (2020), 'Determinación de orientaciones de ciudades romanas por medio de la "varatio" / Orientation layout of Roman towns by using the "varatio"', *Archivo Español de Arqueología*, 93: 127–46.

Orrù, M., O. Grillo, G. Lovicu, G. Venora and G. Bachetta (2013), 'Morphological Characterization of Vitis vinifera L. Seeds by Image Analysis and Comparison with Archaeological Remains', *Vegetation History and Archaeobotany*, 22: 231–42.

Orsingher, A., S. Amicone. J. Kamlah, H. Sader H. and C. Berthold (2020), 'Phoenician Lime for Phoenician Wine: Iron Age Plaster from a Wine Press at Tell el-Burak, Lebanon', *Antiquity*, 94: 1224–44.

Paci, G. (2009), 'Una Nuova Testimonianza sul Consumo di Olive Picene in Ambito Renano', *Picus*, 29: 179–86.

Paci, G. and R. Perna (2016), 'Una villa romana nel territorio di Pollentia-Urbs Salvia: note preliminari sulle indagini archeologiche condotte presso Villamagna (Urbisaglia-MC)', *The Journal of Fasti Online*, 371: 1–14.

Pagnoux, C., L. Bouby, S. Ivorra, C. Petit, S.-M. Valamoti and J.-F. Terral (2015), 'Inferring the Ancient Grapevine (Vitis Vinifera L.) Agrobiodiversity in Greece by a Comparative Shape Analysis of Archaeological and Modern Seeds', *Vegetation History and Archaeobotany*, 24: 75–84.

Pagnoux, C., L. Bouby, S.-M. Valamoti, V. Bonhomme, S. Ivorra, E. Gkatzogia, A. Karathanou, D. Kotsachristou, H. Kroll and J.-F. Terral (2021), 'Local Domestication or Diffusion? Insights into Viticulture in Greece from Neolithic to Archaic Times, using Geometric Morphometric Analyses of Archaeological Grape Seeds', *Journal of Archaeological Science*, 125: 105263.

Palet, J. M. (1997), *Estudi territorial del Pla de Barcelona. Estructuració i evolució del territori entre l'època íberoromana i l'altmedieval segles II-I aC. / X-XI dC.*, Estudis i Memòries d'Arqueologia de Barcelona 1, Barcelona.

Palet, J. M., Fiz, J. I. and H. A. Orengo (2009), 'Centuriació i estructuració del l'ager de la colònia Barcino: anàlisi arqueomorfològica i modelació del paisatge', *Quaderns d'Història i Arqueologia de la Ciutat de Barcelona*, 5: 103–26.

Palet, J. M., Orengo H. A. and S. Riera (2010), 'Centuriación del territorio y modelación del paisaje en los llanos litorales de Barcino (Barcelona) y Tarraco (Tarragona): Una investigación interdisciplinar a través de la integración de datos arqueomorfológicos y paleoambientales', *Agri Centuriati: An International Journal of Landscape Archaeology*, 7: 113–29.

Palet, J. M., Orengo H. A. and S. Riera (2011), 'Centuriación del territorio y modelación del paisaje en los llanos litorales de Barcino y Tarraco: Una investigación interdisciplinar a través de la integración de datos arqueomorfológicos y paleoambientales', in P. L. Dall'aglio and G. Rosada (eds), *Sistemi Centuriali e Opere di Assetto Agrario tra Età Romana e primo Medioevo: Atti del convegno borgoricco (Padova)-Lugo (Ravenna) 10–12 Settembre 2009*, 113–29, Pisa: Fabrizio Serra.

Palet, J. M., M. J. Ortega and C. Miró (2021), 'The Territory of Roman Barcino: Methodological Advances Applied to the Study of a Centuriated Landscape', in M. Prevosti and J. Guitart (eds), *Proceedings of the 1st TIR-FOR Symposium: From Territory Studies to Digital Cartography*, 173–88, Barcelona.

Panella, C. (1989), 'Le anfore italiche del II sec. d.C.', in *Amphores romaines et histoire économique: dix ans de recherche: Actes du Colloque de Sienne (22–24 mai 1986)*, 139–78, Rome: École Française de Rome.

Parducci, L., K. Bennett, G. F. Ficetola, I. G. Alsos, Y. Suyama, J. Wood and M. Winther Pedersen (2017), 'Ancient Plant DNA in Lake Sediments', *New Phytologist*, 214: 924–42.

Parker, A. J. (1990), 'The Wines of Roman Italy', *JRA*, 3: 325–31.

Parker, A. J. (1992a), *Ancient Shipwrecks of the Mediterranean and the Roman Provinces*, Oxford: BAR.

Parker, A. J. (1992b), 'Cargoes, Containers and Stowage: The Ancient Mediterranean', *IJNA*, 21: 89–100.

Parker, A. J. (1996), 'Sea Transport and Trade in the Ancient Mediterranean', in E. E. Rice (ed.), *The Sea and History*, 97–109, Stroud: Sutton.

Parker, R. M. (2008), *Wine Buyers Guide*, 7th edn, New York: Simon and Schuster.

Parsons, R. W. and I. C. Prentice (1981), 'Statistical Approaches to R-values and the Pollen Vegetation Relationship', *Rev Palaeobot Palynol*, 32: 127–52.

Pasqui, A. (1897), 'La villa pompeiana della Pissanella presso Boscoreale', *MonAL*, 7: 397–554.

Pasquinucci, M. and S. Menchelli (2017), 'Rural, Urban and Suburban Communities and their Economic Interconnectivity in Coastal North Etruria (2nd Century BC–2nd Century AD)', in T. de Haas and G. Tol (eds), *The Economic Integration of Roman Italy*, 322–41, London: Brill.

Peacock, D. P. S. and D. F. Williams (1986), *Amphorae and the Roman Economy: An Introductory Guide*, London: Longman.

Pearsall, D. M. and D. R. Piperno, eds (1993), *Current Research in Phytolith Analysis: Applications in Archaeology and Paleoecology*, MASCA Research Papers in Science and Archaeology, Vol. 10, Philadelphia: University of Pennsylvania.

Pecci, A (2005), 'Per una definizione funzionale degli spazi e delle ceramiche all'interno degli insediamenti in corso di scavo: un progetto archeometrico', unpublished PhD thesis in Medieval Archaeology, University of Siena.

Pecci, A. (2007), 'Potenzialità delle analisi chimiche applicate all'archeologia dei consumi alimentari: un bilancio delle conoscenze', in A. Ciacci. P. Rendini and A. Zifferero (eds), *Archeologia della vite e del vino in Etruria*, 123–31, Siena: Ci.Vin.

Pecci, A. (2009), 'Analisi funzionali della ceramica e alimentazione medievale', *Archeologia Medievale*, 36: 21–42.

Pecci, A. (2013), 'Almost Ten Years of Plasters Residue Analysis in Italy: Activity Areas and the Function of Structures', *Periodico di Mineralogia*, 82 (3): 393–410.

Pecci, A. (2016), 'Appendice: Analisi dei residui in tre dolia rinvenuti nella UT 88', in G. Cordiano (ed.), *Carta archeologica del litorale ionico aspromontano*, 163–6, Pisa: Edizioni Ets.

Pecci, A. (2018), 'Analisi dei residui organici nelle ceramiche: Contenuti, rivestimenti, uso e funzione', in M. Giorgio (ed.), *Storie [di] Ceramiche 4: Ceramiche e Archeometria*, 13–20, Firenze: All'Insegna del Giglio.

Pecci, A. (2018–19), 'La Producción y consumo de vino a través del análisis de residuos químicos en materiales arqueológicos', *Tribuna d'Arqueologia*, 1130–7781: 322–33.

Pecci, A. (2019), 'The Life Cycle of Wine: Examples from the Bronze Age to the Middle Ages on the Italian Peninsula', in P. Stockhammer and J. Fries-Knoblach (eds), *Was Tranked die Fruhen Kelten? Bedeutungen und Funktionen mediterraner Importe im früheisenzeitlichen Mitteleuropa (BEFIM 1)*, 203–9, Leiden: Sidestone Press.

Pecci, A., E. Borgna. S. Mileto, E. Dalla Longa, G. Bosi, A. Florenzano, A. M. Mercuri, S. Corazza, M. Marchesini and M. Vidale (2020), 'Wine Consumption in Bronze Age Italy: Combining Organic Residue Analysis, Botanical Data and Ceramic Variability', *Journal of Archaeological Science*, 123: 105256.

Pecci, A. and M. Á. Cau (2010), 'Análisis de residuos orgánicos en ánforas. El problema de la resina y del aceite', in J. M. Blázquez and J. Remesal J. (eds), *Estudios sobre el Monte Testaccio (Roma)* V, Instrumenta 35, 593–600, Barcelona: Universitat de Barcelona.

Pecci, A., M. Á. Cau and N. Garnier (2013), 'Identifying Wine and Oil Production: Analysis of Residues from Roman and Late Antique Plastered Vats', *Journal of Archaeological Science*, 40 (12): 4491–8.

Pecci, A., J. Clarke, M. Thomas, J. Muslin, I. van der Graaff, L. Toniolo, D. Miriello, G. M. Crisci, M. Buonincontri and G. Di Pasquale (2017), 'Use and Reuse of Amphorae. Wine Residues in Dressel 2–4 Amphorae from Oplontis Villa B (Torre Annunziata, Italy)', *Journal of Archaeological Science: Reports*, 12: 515–21.

Pecci, A., A. Contino, S. Mileto, L. Toniolo and P. Reynolds (2021b), 'Anfore africane antiche a Pompei: uso e riuso in base all'analisi dei contenuti', *Rivista di Studi Pompeiani*, 32: 87–102.

Pecci, A., S. Domínguez-Bella, M. Buonincontri, D. Miriello, R. De Luca, G. Di Pasquale, D. Cottica and D. Bernal-Casasola (2018), 'Combining Residue Analysis of Floors and Ceramics for the Study of Activity Areas at the Garum Shop at Pompeii', *Archaeological and Anthropological Sciences*, 10: 485–502.

Pecci, A. and G. Giorgi (forthcoming), 'L'analisi chimica dei pavimenti e dei contenitori per derrate della cosiddetta Casa delle Anfore – UT 161', in A. Zifferero (ed.), *Materiali per Marsiliana d'Albegna 3. Ricerche nell'area suburbana.*

Pecci, A., G. Giorgi, L. Salvini and M. Á. Cau Ontiveros (2013), 'Identifying Wine Markers in Ceramics and Plasters Using Gas Chromatography–Mass Spectrometry: Experimental and Archaeological Materials', *JAS*, 40: 109–15.

Pecci, A., P. Reynolds, S. Mileto, J. M. Vargas Girón and D. Bernal-Casasola (2021a), 'Production and Transport of Goods in the Roman Period: Residue Analysis and Wine Derivatives in Late Republican Baetican Ovoid Amphorae', *Environmental Archaeology*, 1–13. https://doi.org/10.1080/14614103.2020.1867291

Pecci, A., L. Salvini, E. Cirelli and A. Augenti (2010), 'Castor Oil at Classe (Ravenna, Italy): Residue Analysis of some Late Roman Amphorae coming from the Port', in S. Menchelli, S. Santoro, M. Pasquinucci and G. Guiducci (eds), *LRCW3: Late Roman Coarse Wares, Cooking Wares and Amphorae in the Mediterranean. Archaeology and Archaeometry*, 617–22, BAR International Series 2185, Oxford: BAR.

Pecci, A., M. Vidale, S. Mileto, S. Bergamini and M. Cupitò (2021), 'Il primo bicchiere non si scorda mai,' *Archeo: Attualità del Passato*, 439: 106–11.

Peña Cervantes, Y. (2010), *Torcularia: la producción de vino y aceite en Hispania*, Documenta 14, Tarragona: ICAC.

Peña Cervantes, Y. (2020), 'Wine Making in the Iberian Peninsula During the Roman Period: Archaeology, Archaeobotany and Biochemical Analysis', in J.-P. Brun, N. Garnier and G. Olcese (eds), *A. Making Wine in Western-Mediterranean. B. Production and the Trade of*

Amphorae: Some New Data from Italy. Panel 3.5, Archaeology and Economy in the Ancient World 9, 73–81, Heidelberg: Propylaeum.

Peña Cervantes, Y. (2023), *De tornos y tornillos: Tecnologías de prensado de la uva y la aceituna en elm undo romano y tardoantiguo*, Granada: Comares.

Pereira, P. and T. Silvino (2015), 'Chemical Analysis about Roman Wine on the Douro valley – The Site of Prazo (Freixo de Numao, Portugal)', in C. Oliveira, R. Morais and A. Morillo (eds), *Archaeoanalytics: Chromatography and DNA Analysis in Archaeology. International Symposium*, 187–91, Porto: University of Porto.

Perez-Arantegui, J., J. A. Paz Peralta and E. Ortiz Palomar (1996), 'Analysis of Products Container in Two Roman Glass Unguentaria from the Colony of Selsa (Spain)', *JAS*, 23: 649–55.

Pérez-Jordà, G., N. Alonso, N. Rovira, I. Figueiral, D. López-Reyes, P. Marinval, E. Montes, L. Peña-Chocarro, R. Pinaud-Querrac'h, J. Ros, M. Tarongi, M. Tillier and L. Bouby (2021), 'The Emergence of Arboriculture in the 1st Millennium BC along the Mediterranean's "Far West"', *Agronomy*, 11 (5): 1–32.

Perna, R. (2006), *Urbs Salvia: Forma e Urbanistica*, Rome: L'Erma di Bretschneider.

Perret, M. (1997), 'Polymorphisme des génotypes sauvages et cultivés de Vitis vinifera l. détecté à l'aide de marqueurs RAPD', *Bulletin de la société Neuchâteloise des sciences naturelles*, 120: 45–54.

Pesavento Mattioli, S. (2003), 'Produzione e commercio del vino: un percorso di ricerca nella Valpolicella di età romana', *Annuario Storico della Valpolicella*, 19: 103–16.

Pesavento Mattioli, S. and M. Mongardi (2018), 'Anfore Vinarie dalla Cisalpina in età augustea: un aggiornamento su alcune serie bollate', in M. Cavalieri and C. Boschetti (eds), *MVLTA PER AEQUORA. Il polisemico significato della moderna ricerca archeologica: Omaggio a Sara Santoro*, 321–45, Louvain: Presses universitaires de Louvain.

Pfunter, L. (2017), 'Review of An Urban Geography of the Roman World, 100 BC to AD 300. Archaeopress Roman Archaeology, 18', *Bryn Mawr Classsical Review*, 2017.05.12.

Pignatti, S. (2017–19), *Flora d'Italia*, I–IV, Bologna: Edagricole Calderini.

Pini, R., C. Ravazzi, R. Comolli, R. Perego, L. Castellano, C. Croci, M. De Amicis, D. A. El Khair, G. Furlanetto and D. Marsetti (2021), 'Life on a Hilltop: Vegetation History, Plant Husbandry and Pastoralism at the Dawn of Bergamo-Bergomum (Northern Italy, 15th to 7th century BC)', *Vegetation History and Archaeobotany*, 30: 525–53.

Piperno, D. R. (1988), *Phytolith Analysis: An Archaeological and Geological Perspective*, San Diego: Academic Press.

Piperno, D. R. (2006a), *Phytoliths: A comprehensive guide for archaeologists and paleoecologists*, Lanham: AltaMira Press.

Piperno, D. R. (2006b), *Phytolith Analysis in Archaeology and Environmental History*, Walnut Grove: AltaMira Press.

Pinto, G., C. Poni and U. Tucci, eds (2002), *Storia dell'Agricoltura Italiana: Il Medioevo e l'Età Moderna*, Firenze: Edizioni Polistampa.

Poglajen, S. and T. Zerjal (2007), 'The Potential of Slovenian Istria for Roman Viticulture and Oleoculture', *Histria Antiqua*, 15: 267–80.

Pollard, A. M. and C. Heron (2008), *Archaeological Chemistry*, Cambridge: Royal Society of Chemistry Publishing.

Porat, R., Y. Kalman, R. Chachy, S. Terem, R. Bar-Natan, E. Weiss, A. Ecker, T. Ben-Gedalya and E. Drori (2018), 'Herod's Royal Winery and Wine Storage Facility in the Outer Structure of the Mountain Palace-Fortress at Herodium Antiquities', *Qadmoniot*, 156: 106–14.

Poux, M. (2011), 'Le "vin du Triumvir" à Lyon: Témoignages archéologiques et littéraires d'une production de vin sur le territoire colonial de Lugdunum', *Gallia*, 68 (1): 13–91.

Poux, M., J.-P. Brun and M.-L. Hervé-Monteil, eds (2011), 'La Vigne et le Vin dans les Trois Gaules', *Gallia*, 68: 1–285.

Prentice, I. C. (1985), 'Pollen Representation, Source Area, and Basin Size: Toward a Unified Theory of Pollen Analysis', *Quatern Res*, 23: 76–86.

Prentice, I. C. and R. W. Parsons (1983), 'Maximum Likelihood Linear Calibration of Pollen Spectra in Terms of Forest Composition', *Biometrics*, 39: 1051–7.

Preston, D., Pearman, D. A. and A. R. Hall (2004), 'Archaeophytes in Britain', *Botanical Journal of the Linnean Society*, 145: 257–94.

Pullen, J., J. D. Doyle, T. Haack, C. Dorman, R. P. Signell and C. M. Lee (2007), 'Bora Event Variability and the Role of Air-Sea Feedback', *Journal of Geophysical Research*, 112: C03S18.

Punt, W., A. Marks and P. P. Hoen (2003), 'VITACEAE', *Review of Palaeobotany and Palynology*, 123: 67–70.

Purcell, N. (1985), 'Wine and Wealth in Ancient Italy', *JRS*, 75: 1–19.

Purcell, N. (1990), 'Review of: Jongman, the Economy and Society of Pompeii', *The Classical Review*, 40: 111–16.

Py, M. (1985), 'Les amphores etrusques de Gaule meridionale', in M. Cristofani (ed.), *Il commercio etrusco arcaico*: *Atti dell'Incontro di Studio, Roma 1983*, 73–94, Rome: Consiglio Nazionale delle Ricerche.

Raimondi, S., G. Tumino, P. Ruffa, P. Boccacci, G. Gambino and A. Schneider (2020), 'DNA-Based Genealogy Reconstruction of Nebbiolo, Barbera and Other Ancient Grapevine Cultivars from Northwestern Italy', *Scientific Reports*, 10: 1–16.

Ramón, J. (2010), 'La cerámica fenicia del mediterráneo extremooccidental y del Atlántico (s. VIII – 1R. 1/3 del VI AC)', in L. Nigro (ed.), *Problemas y perspectivas actuales: Motya and the Phoenician Ceramic Repertoire between the Levant and the West 9th–6th Century* BC, 211–53, Rome: Missione Archeologica a Mozia.

Ramos-Madrigal, J., A. K. Wiborg Runge, L. Bouby, T. Lacombe, J. A. Samaniego Castruita, A.-F. Adam-Blondon, I. Rigueiral, C. Hallavant, J. M. Martínez-Zapater, C. Schaal, R. Töpfer, B. Petersen, T. Sicheritz-Pontén, P. This, R. Bacilieri, M. T. P. Gilbert and N. Wales (2019), 'Paleogenomic Insights into the Origins of French Grapevine Diversity', *Nature Plants*, 5: 595–603.

Rapp, G. and S. C. Mulholland, eds (1992), *Phytolith Systematics: Emerging Issues*. New York: Plenum.

Rauh, N. K. (2006), 'The Archaeology of Deforestation in Ancient Rough Cilicia', *ANMED 2006-4 (News of Archaeology from Anatolia's Mediterranean Areas)*, 83–4, Antalya: Suna & Inan Kirac Research Institute on Mediterranean Civilizations.

Rauh, N. K. (2008), 'The Rough Cilicia Archaeological Survey Project: Paleo-Environmental Research of the 2007 Season', *ANMED 2008–6 (News of Archaeology from Anatolia's Mediterranean Areas)*, 117–19, Antalya: Suna & Inan Kirac Research Institute on Mediterranean Civilizations.

Rauh, N. K. (2012), 'Dağlık Kilikiya Yüksek Arazi Arkeoloji Araştıması: 2011 Yılı Raporu, 30', *Araştırma Sonuçları Toplantısı*, 30 (2): 223–34.

Rauh, N. K., M. J. Dillon, C. Dore, R. M. Rothaus and M. Korsholm (2006), 'Viticulture, Oleoculture, and Economic Development in Roman Rough Cilicia', *Münster Beiträge fur Antike Handelsgeschichte*, 25: 49–98.

Ravegnani, G. (2004), *I Bizantini in Italia*, Bologna: Il Mulino.

Ravenna 2007 = D. Vitali, ed. (2007), *Le fornaci e le anfore di Albinia: Primi dati su produzioni e scambi dalla costa tirrenica al mondo gallico, Atti del Seminario Internazionale, Ravenna 2006*, Bologna: Università di Bologna.

Reill, D. K. (2012), *Nationalists Who Feared the Nation: Adriatic Multi-Nationalism in Habsburg Dalmatia, Trieste, and Venice*, Stanford: Stanford University Press.

Remola, J. A., J. Lozano, I. Ruisanchez, M. S. Larrechi, F. X. Rius and J. Zupan (1996), 'New Chemometric Tools to Study the Origin of Amphorae Produced in the Roman Empire', *Trend Anal. Chem.*, 15: 137.

Revilla Calvo, V. (1995), *Producción cerámica, viticultura y propiedad rural en Hispania Tarraconensis (Siglos I a.C. – III d.C.)*, Barcelona: L'Estaquirot-Barcelona.

Reynolds, J. M. (2011), *An Introduction to Applied and Environmental Geophysics*, 2nd edn, Oxford: Wiley-Blackwell.
Ribechini, E., F. Modugno, M. P. Colombini and R. P. Evershed (2008a), 'Gas Chromatographic and Mass Spectrometric Investigations of Organic Residues from Roman Glass Unguentaria', *Journal of Chromatography A*, 1182: 158–69.
Ribechini, E., F. Modugno, C. Baraldi, P. Baraldi and M. P. Colombini (2008b), 'An Integrated Analytical Approach for Characterizing an Organic Residue from an Archaeological Glass Bottle Recovered in Pompeii', *Talanta*, 74 (4): 555–61.
Rice, P. (1987), *Pottery Analysis*, First Edition, Chicago: University of Chicago Press.
Rice, C. (2016), 'Mercantile Specialization and Trading Communities: Economic Strategies in Roman Maritime Trade', in A. Wilson and M. Flohr (eds), *Urban Craftsmen and Traders in the Roman World*, 97–114, Oxford: OUP.
Rinaldi, F. (2003), 'Motivi geometrici e temi figurati nelle pavimentazioni musive della villa romana di Negrar', in A. Buonopane and A. Brugnoli (eds), *La Valpolicella in età romana: Atti del II Convegno (Verona, 11.5.2002)*, 133–60, Verona: Centro di Documentazione per la storia della Valpolicella.
Rinaldi, R., M. Bandini Mazzanti and G. Bosi (2013), 'Archaeobotany in Urban Sites: The Case of Mutina', *Annali di Botanica*, 3: 217–30.
Roaf, M. (1992), *Atlante della Mesopotamia e dell'Antico Vicino Oriente*, Milan: De Agostini.
Robinson, J. (2015), *The Oxford Companion to Wine*, 4th edn, Oxford: Oxford University Press.
Roffet-Salque, M., P. I. Bogucki, J. Pyzel, I. Sobkowiak-Tabaka, R. Grygiel, M. Szmyt and R. P. Evershed (2013), 'Earliest Evidence for Cheese Making in the Sixth Millennium B.C. in Northern Europe.' *Nature* 493: 522–5.
Rojo Muñoz, S. (2021), 'Il vino etrusco: una concorrenza tra grand crus?', *Archeo: Attualità del Passato*, 439: 20–2.
Romanowska, I., C. D. Wren and S. A. Crabtree (2021), *Agent-Based Modelling for Archaeology: Simulating the Complexity of Societies*, Santa Fe: The Santa Fe Institute Press.
Romanus, K., J. Baeten, J. Poblome, S. Accardo, P. Degryse, P. Jacobs, D. De Vos and M. Waelkens (2009), 'Wine and Olive Oil Permeation in Pitched and Non-Pitched Ceramics: Relation with Results from Archaeological Amphorae from Sagalassos, Turkey', *JAS*, 36 (3): 900–9.
Rosada, G. and J. Turchetto (2008), 'La questione dell'assetto agrario e dello sfruttamento delle risorse nella Val Belluna romana', in P. Basso, A. Buonopane, A. Cavarzere and S. P. Mattioli (eds), *Est enim ille flos Italiae: Vita economica e sociale nella Cisalpina Romana, Atti delle Giornate di Studi in onore di Ezio Buchi (Verone, 30 novembre–1 dicembre 2006)*, 501–19, Verona: QuiEdit.
Rosen, A. M. (1999), 'Phytolith Analysis in Near Eastern Archaeology', in S. Pike and S. Gitin (eds), *The Practical Impact of Science on Aegean and Near Eastern Archaeology*, 86–92, London: Archetype Press.
Rosen, A. M. (2008), 'Phytolith Analysis', in D. M. Pearsall (ed.), *Encyclopedia of Archaeology*, 1818–22, New York: Academic Press.
Rossiter, J. J. (1981), 'Wine and Oil Processing at Roman Farms in Italy', *Phoenix*, 35: 345–61.
Rossiter, J. J. (1998), 'Pressing Issues: Wine and Oil Production', *Journal of Roman Archaeology*, 11: 597–602.
Rostovtzeff, M. I. (1957), *The Social and Economic History of the Roman Empire*, 2nd edn, edited by P. M. Fraser, Oxford: Clarendon Press.
Roth-Congés, A. (1996), 'Modalités pratiques d'implantation de cadastres romains: quelques aspects', in *MEFRA* 108, 299–422.
Rotshild-Boros, M. C. (1981), 'The Determination of Amphora Contents', in G. Barker and R. Hodges (eds), *Archaeology and Italian Society*, BAR International Series 102, 79–89, Oxford: Archaeopress.
Rottoli, M. and E. Castiglioni (2011), 'Plant Offering from Roman Cremations in Northern Italy: A Review', *Vegetation History and Archaeobotany*, 20: 495–506.

Roushannafas, T., A. Bogaard and M. Charles (2022), 'Geometric Morphometrics Sheds New Light on the Identification and Domestication Status of "New Glume Wheat" at Neolithic Çatalhöyük', *Journal of Archaeological Science*, 142: 105599.

Rousse, C., D. Munda, G. Benčić, O. Bourgeon, P. Maggi, V. Dumas, K. Bartolić, E. Botte and F. Welc (2020), 'Loron / Santa Marina (Tar-Vabriga, Poreč, Croatie). La villa de Santa Marina. Campagne 2019', *Chronique des activités archéologiques de l'École française de Rome*, https://doi.org/10.4000/cefr.4862

Rousse, C., G. Benčić, D. Munda, N. Garnier, K. Bartolić Sirotić, V. Dumas, N. Basuau and P. Maggi (2022), 'La villa de Santa Marina (Tar Vabriga – Torre Abrega, Croatie)', *Bulletin archéologique des Écoles françaises à l'étranger*, https://doi.org/10.4000/baefe.5098

Rowlett, L. M. and D. M. Pearsall (1993), 'Archaeological Age Determinations derived from Opal Phytoliths by Thermoluminescence', in D. M. Pearsall and D. R. Piperno (eds), *Current Research in Phytolith Analysis: Applications in Archaeology and Paleoecology*, MASCA Research Papers in Science and Archaeology 10, 25–30, Philadelphia: University of Pennsylvania.

Rubio-Campillo, X. (2016), 'Model Selection in Historical Research Using Approximate Bayesian Computation', *PLOS ONE*, 11 (1): e0146491.

Rubio-Campillo, X., M. Coto-Sarmiento, J. Pérez-Gonzalez and J. Remesal Rodríguez (2017), 'Bayesian Analysis and Free Market Trade within the Roman Empire', *Antiquity*, 91 (359): 1241–52.

Ruffing, K., J. Renger and A. Gutsfeld (2002), 'Wein', in *Der Neue Pauly* 12/2, 424–36, Stuttgart: Brill.

Ruiz, M., I. Rondán, J. Catalán, L. Lagóstena and J. Remesal (2021), 'Lectura de Arva desde la investigación no invasiva de la urbs y el suburbium', in *Congreso Internacional Small Town: Una realidad urbana en la Hispania romana*, Alicante.

Ruiz Gil, J. A., L. Lagóstena, J. Pérez Marrero, D. Martín, P. Trapero and J. Catalán (2019), 'Villae y Figlinae a orillas del Lacus Ligustinus. Análisis GIS y prospección geofísica en el territorium ribereño de Hasta Regia', in J. Remesal, V. Revilla, D. J. Martín-Arroyo and A. Martín i Oliveras (eds), *Paisajes productivos y redes comerciales en el imperio romano*, Instrumenta 65, 115–38, Barcelona: Universitat de Barcelona.

Ruíz Hernández, M. (2021), 'Densidad del vino', *La Rioja 25*, 11 June. Available online: https://www.larioja.com/opinion/densidad-vino-20210611214244-nt.html (accessed 30 April 2022).

Rylatt, K. (2004), *Report on a Programme of Archaeological Trial Trenching: Lincoln Eastern Bypass, Lincolnshire*, Lincoln: Pre-Construct Archaeology.

Saltini, A. (1979), *Storia delle scienze agrarie: venticinque secoli di pensiero agronomico*, Bologna: Edagricole.

Salviat, F. and A. Tchernia (2013), *Vins, Vignerons et Buveurs de l'Antiquité*, Rome: L'Erma Di Bretschneider.

San Quirico d'Orcia 2010 = G. Barbieri, A. Ciacci and A. Zifferero (eds), *Eleiva Oleum Olio. Le origini dell'olivicoltura in Toscana: nuovi percorsi di ricerca tra archeologia, botanica e biologia molecolare, Atti della Giornata di Studi, San Quirico d'Orcia 2007*, San Quirico d'Orcia: Edizioni Don Chisciotte.

Sands, R. and E. Marlière (2020), 'Produce, Repair, Reuse, Adapt, and Recycle: The Multiple Biographies of a Roman Barrel', *European Journal of Archaeology*, 23 (3): 356–80.

Santangeli Valenzani R. and R. Volpe (2012), 'Paesaggi agrari della viticoltura a Roma e nel suburbio', in A. Ciacci, P. Rendini and A. Zifferero (eds), *Archeologia della vite e del vino in Toscana e nel Lazio: Dalle tecniche dell'indagine archeologica alle prospettive della biologia molecolare*, 61–70, Borgo San Lorenzo: All'Insegna del Giglio.

Santos, J. A., A. C. Malheiro, J. G. Pinto and G. V. Jones (2012), 'Macroclimate and Viticultural Zoning in Europe: Observed Trends and Atmospheric Forcing', *Climate Research*, 51: 89–103.

Sarpaki, A. (2012), 'Revisiting the Visibility of the Grape, Grape-Products, By-Products and Some Insights of Its Organization from the Prehistoric Aegean, as Guided by New Evidence from Monastiraki, Crete', *IANSA*, 3: 211–20.

Scali, M., A. Zifferero and R. Vignani (2018), 'Distribution and Characterization of the Vitis vinifera L. subsp. sylvestris in Southern Tuscany', *Recent Patents on Biotechnology*, 12: 208–20.

Scansano 2007 = A. Ciacci, P. Rendini and A. Zifferero (eds), *Archeologia della Vite e del Vino in Etruria: Atti del Convegno Internazionale di Studi, Scansano 2005*, Siena: Ci.Vin.

Scheidel, W. (1992), 'Neuen Wein in leere Schläuche: Jongman's Pompeii, Modelle und die kampanische Landwirtschaft', *Athenaeum*, 80: 207–13.

Schmidt, A. (2009), 'Electric and Magnetic Methods in Archaeological Prospection', in S. Campana and S. Piro (eds), *Seeing the Unseen: Geophysics and Landscape Archaeology*, 67–81, London: Routledge.

Schmitt, A. (1998), 'Amphorae from Lyons: Petrographic and Chemical Arguments', *Archaeometry*, 40: 293–310.

Schneider, V. and S. Troxell (2018), *Acidity Management in Must and Wine*, San Francisco: Board and Bench.

Scholz, D. and D. L. Hoffmann (2011), 'StalAge – An Algorithm Designed for Construction of Speleothem Age Models', *Quaternary Geochronology*, 6 (3–4): 369–82.

Scienza, A. (2004), 'Il terzo anello: storia di un viaggio', in F. Del Zan, O. Failla and A. Scienza (eds), *La vite e l'uomo: Dal rompicapo delle origini al salvataggio delle reliquie. Evidenze storico-ampelografiche per ripercorrere il viaggio della vite da Oriente alle soglie dell'Occidente*, 99–148, Gorizia: ERSA.

Scienza, A., ed. (2006), *Vitigni tradizionali ed antichi italiani: La storia, il paesaggio, la ricerca*, Siena: Ci.Vin.

Scienza, A. (2010), 'L'origine dei vitigni coltivati: una storia interdisciplinare della cultura europea', in G. Di Pasquale (ed.), *Vinum nostrum: Arte, scienza e miti del vino nelle civiltà del Mediterraneo antico*, Catalogo della Mostra, Firenze 2010, 23–41, Firenze: Giunti Editore.

Scienza, A. and O. Failla (1996), 'La circolazione dei vitigni in ambito padano-veneto ed atesino: le fonti storico-letterarie e l'approccio biologico-molecolare', in G. Forni and A. Scienza (eds), *2500 anni di cultura della vite nell'ambito alpino e cisalpino*, 185–268, Trento: Istituto Trentino del Vino.

Scienza, A. and O. Failla (2016), 'La circolazione varietale della vite nel Mediterraneo: lo stato della ricerca', *Rivista di Storia dell'Agricoltura*, 56 (1–2): 31–48.

Scienza, A. and S. Imazio (2018), *La stirpe del vino*, Milan: Sperling and Kupfer.

Scollar, A., A. Tabbagh, A. Hesse and I. Herzog (1990), *Archaeological Prospecting and Remote Sensing*, Cambridge: Cambridge University Press.

Seiler, F. (2023), 'Der Fluss Sarno: Schnittstelle in der Entwicklung der Küstenregion Pompejis zwischen Umweltbedingungen und Siedlungsdynamiken', *Archäologischer Anzeiger*, 2: 1–37. doi: 10.34780/78ah-0gx3.

Seiler, F., M. Märker, P. Kastenmeier, S. Vogel, D. Esposito, U. Heussner, M. Boni, G. Balassone, G. Di Maio and M. Joachimski (2011), 'Interdisciplinary Approach on the Reconstruction of the Ancient Cultural Landscape of the Sarno River Plain Before the Eruption of Somma-Vesuvius A. D. 79', in H.-R. Bork, H. Meller and R. Gerlach (eds), *Umweltarchäologie – Naturkatastrophen und Umweltwandel im archäologischen Befund: 3. Mitteldeutscher Archäologentag vom 07. bis 09. Oktober 2010*, 145–54, Halle (Saale): Landesmuseum für Vorgeschichte.

Seiler, F., S. Vogel and D. Esposito (2019), 'Ancient Rural Settlement and Land Use in the Sarno River Plain (Campania, Italy): Predictive Models and Quantitative Analyses', in J. R. Rodríguez, V. Revilla Calvo, D. J. Martín-Arroyo Sánchez and A. Martín i Oliveras (eds), *Productives Landscapes and Trade Networks in the Roman Empire*, 179–99, Barcelona: Universitat de Barcelona Edicions.

Senatore, F. (1998), '"Ager Pompeianus": viticoltura e territorio nella piana del Sarno nel I secolo d.C.' in F. Senatore (ed.), *Pompei il Sarno e la Penisola Sorrentina: Atti del primo ciclo di conferenze di geologia, storia e archeologia. Pompei, aprile-giugno 1997*, 135–66, Rome: Bardi Editore.

Senior, L. M. and D. P. Birnie (1995), 'Accurately Estimating Vessel Volume from Profile Illustrations', *American Antiquity*, 60: 319–34.
Sereni, E. (1964), 'Per la storia delle più antiche tecniche e della nomenclatura della vite e del vino in Italia', *Atti e Memorie dell'Accademia Toscana di Scienze e Lettere La Colombaria*, 29: 75–204.
Sereni, E. (1987), *Storia del paesaggio agrario italiano*, Roma; Bari: Editori Laterza.
Shumilovskikh, L. S. and B. van Geel (2020), 'Non-Pollen Palynomorphs', in A. G. Henry (ed.), *Handbook for the Analysis of Micro-Particles in Archaeological Samples*, 65–94, Cham: Springer.
Shumilovskikh, L. S., E. S. Shumilovskikh, F. Schlütz and B. van Geel (2022), 'NPP-ID - Non-Pollen Palynomorphs Image Database as a Research and Educational Platform', *Vegetation History and Archaeobotany*, 31: 323–8.
Singleton, V. L. (1996), 'An Enologist's Commentary on Ancient Wines', in P. E. McGovern, S. J. Fleming and S. H. Katz (eds), *The Origins and Ancient History of Wine*, 67–77, New York: Gordon and Breach.
Slovenec, D., I. Sondi and B. Crnkovic (1991), 'Mineraloske karakteristike amfora s podrucja srednjeg Jadrana', *Geoloski Vjesnik*, 44: 129–37.
Smekalova, T. N., B. W. Bevan, A. V. Chudin and A. S. Garipov (2016), 'The Discovery of an Ancient Greek Vineyard', *Archaeological Prospection*, 23: 15–26.
Smith, W. and C. Anthon (1854), *A New Classical Dictionary of Greek and Roman Biography, Mythology and Geography, Partly Based Upon the Dictionary of Greek and Roman Biography and Mythology*, New York: Harper and Brothers.
Smith, H. and G. Jones (1990), 'Experiments on the Effects of Charring on Cultivated Grape Seeds', *JAS*, 17: 317–27.
Smith, W., W. Wayte and G. E. Marindin, eds (1890), *A Dictionary of Greek and Roman Antiquities*, London: John Murray.
Sondi, I. and D. Slovenic (2003), 'The Mineralogical Characteristics of the Lamboglia 2 Roman-Age Amphorae from the Central Adriatic (Croatia)', *Archaeometry*, 45: 251–62.
Sourisseau, J-C. (2011), 'La diffusion des vins grecs d'Occident du VIIIe au IVe s. av. J.-C., sources écrites et documents archéologiques', in M. Lombardo, A. Siciliano and A. Alessio (eds), *La vigna di Dioniso: vite, vino e culti in Magna Grecia. Atti del Quarantanovesimo Convegno di Studi sulla Magna Grecia (Taranto 2009)*, 143–52, Taranto: Istituto per la Storia e l'Archeologia della Magna Grecia.
Spallanzani, L. (1798), *Travels in the Two Sicilies and Some Parts of the Apennines*, Eng. trans. by G. G. Robinson and J. Robinson, London: Paternoster-Row.
Spötl, C. and D. Mattey, (2012), 'Scientific Drilling of Speleothems – A Technical Note', *Int. J. Speleology*, 41: 29–34.
Šprem, K. (2021), 'All Aboard! Quarries and Transport in Roman Istria', *Proceedings from the 7th Scientific Conference Methodology and Archaeometry*, 67–85, Zagreb: University of Zagreb.
Stančič, Z., ed. (1999), *The Archaeological Heritage of the Island of Brač, Croatia: The Adriatic Islands Project Contact, Commerce and Colonialism: 6000 BC–AD 600*, Oxford: Archaeopress.
Steinhübel, L. (2013), 'Palynological Analysis of a Pompeiian Excavation Site in Scafati, Italy, to Reconstruct Vegetation During Roman Times', unpublished thesis, Faculty of Biology and Psychology Georg-August-University Göttingen, Göttingen.
Stevenson, A. C. and P. D. Moore (1988), 'Studies in the Vegetational History of S.W. Spain IV. Palynological investigations of a valley Mire at El Acebron, Huelva', *Journal of Biogeography*, 15 (2): 339–61.
Stromberg, C. A. E. (2002), 'The Origin and Spread of Grass-Dominated Ecosystems in the Late Tertiary of North America: Preliminary Results Concerning the Evolution of Hypsodonty', *Palaeogeography Palaeoclimatology Palaeoecology*, 177 (1–2): 59–75.
Stromberg, C. A. E. (2004), 'Using Phytolith Assemblages to Reconstruct the Origin and Spread of Grass-Dominated Habitats in the Great Plains of North America during the Late Eocene to Early Miocene', *Palaeogeography Palaeoclimatology Palaeoecology*, 207 (3–4): 239–75.

Stubert, L., A. Martín i Oliveras, M. Märker, H. Schernthanner and S. Vogel (2020), 'Viticulture in the Laetanian Region (Spain) during the Roman Period: Predictive Modelling and Geomatic Analysis', *Geosciences*, 10 (6): 206.

Stummer, A. (1911), 'Zur Urgeschichte der Reben und des Weinbaues', *Mitteilungen der Anthropologischen Gesellschaft in Wien*, 41: 283–96.

Sugita, S. (1994), 'Pollen Representation of Vegetation in Quaternary Sediments: Theory and Method in Patchy Vegetation', *J Ecol*, 82: 881–97.

Tang, P., F. Chen, A. Jiang, W. Zhou, H. Wang, G. Leucci, L. de Giorgi, M. Sileo, R. Luo, R. Lasaponara and N. Masini (2018), 'Multi-Frequency Electromagnetic Induction Survey for Archaeological Prospection: Approach and Results in Han Hangu Pass and Xishan Yang in China', *Surveys in Geophysics*, 39: 1285–1302.

Tassaux, F. (2005), 'Patrimoines sénatoriaux de la Decima Regio', *Cahiers du Centre Gustave Glotz*, 26: 139–64.

Tassaux, F., R. Matijašić and V. Kovačić (2001), *Loron (Croatie): Un grand centre de production d'amphores à huile istriennes (Ier-IVe s. P.C.)*, Mémoires 6, Bordeaux: Institut Ausonius.

Tchernia, A. (1968–1970), 'Premiers résultats des fouilles de juin 1968 sur l'épave de Planier', *EtCl*, 3: 51–82.

Tchernia, A. (1969), 'Recherches archéologiques sous-marines', *Gallia*, 27: 465–99.

Tchernia, A. (1986), *Le vin de l'Italie romaine: Essai d'histoire économique d'après les amphores*, Bibliothèque des Écoles Françaises d'Athènes et de Rome 261, Rome: De Boccard.

Tchernia, A. (1990), 'La vinification des Romains', in G. Garrier (ed.), *Le Vin des Historiens: Actes du Ier Symposium Vin et Histoire, 19–21 mai 1989*, 65–74, Suze-la-Rousse: Université du Vin.

Tchernia, A. (1998), 'Archéologie expérimentale et goût du vin romain', in *El Vi a l'antiguitat: economia, producció i comerç al Mediterrani occidental: II Colloqui Internacional d'Arqueologia Romana, actes (Barcelona 6-9 de maig de 1998)*, 503–9, Badalona: Museu de Badalona.

Tchernia, A. (2013), 'Les dimensions de quelques vignobles Romains', in F. Salviat and A. Tchernia (eds), *Vins, Vignerons et Buveurs de l'Antiquité*, 153–66, Rome: L'Erma di Bretschneider.

Tchernia, A. and J.-P. Brun (1999), *Le vin romain antique*, Grenoble: Glénat.

Terral, J.-F. (2002), 'Quantitative Anatomical Criteria for Discriminating Wild Grapevine (Vitis vinifera ssp. sylvestris) from Cultivated Vines (Vitis vinifera ssp. vinifera)', in S. Thiébault (ed.), *Charcoal Analysis: Methodological Approaches, Paleoecological Results and Wood Uses*, BAR International Series 1063, 59–64, Oxford: BAR.

Terral, J.-F., E. Tabard, L. Bouby, S. Ivorra, T. Pastor, I. Figueiral, S. Picq, J. B. Chevance, C. Jung, L. Fabre, C. Tardy, M. Compan, R. Bacilieri and T. Lacombe (2010), 'Evolution and History of Grapevine (Vitis vinifera) under Domestication: New Morphometric Perspectives to Understand Seed Domestication Syndrome and Reveal Origins of Ancient European Cultivars', *Annals of Botany*, 105: 443–55.

Terrenato N. and J. A. Becker, eds (2012), *Roman Republican Villas: Architecture, Context, and Ideology*, University of Michigan Press, Papers and Monographs of the American Academy in Rome.

Thomas, R. and A. Wilson (1994), 'Water Supply for Roman Farms in Latium and South Etruria', *Papers of the British School at Rome*, 62: 139–96.

Thurmond, D. L. (2017), *From Vines to Wines in Classical Rome: A Handbook of Viticulture and Oenology in Rome and the Roman West*, Leiden: Brill.

Tirone, L. (1970), 'Le vigne dans l'exploitation agricole in Italie', *Méditerranée*, 1 (4): 339–62.

Tobler, W. R. (1979), 'Cellular Geography', in S. Gale and G. Olsson (eds), *Philosophy in Geography*, 379–86, Dordrecht: Springer.

Tol, G., T. De Haas, J. Sevink, M. Schepers, B. Ullrich and W. De Neef (2020), '"There is More than Meets the Eye": Developing an Integrated Archaeological Approach to Reconstruct Human-Environment Dynamics in the Pontine Marshes (Lazio, Central Italy)', *Geoarchaeology*, 36 (1): 109–29.

Toniolo, L. and A. Pecci (2020), 'Il ciclo di vita del vino: Casi studio dall'area Vesuviana', in J.-P. Brun, N. Garnier and G. Olcese (eds), *A. Making Wine in Western-Mediterranean. B. Production and the Trade of Amphorae: Some New Data from Italy, Panel 3.5*, 131–42, Heidelberg: Propylaeum.

Tosi, G. (1983–4), 'La villa romana di Negrar di Valpolicella', *Annuario Storico della Valpolicella*, 2: 91–102.

Toupet, C. and P. Lemaître (2003a), 'Une plantation de vignes gallo-romaine, dans le Nord de la Gaule, à Bruyères-sur-Oise (Val d'Oise)', *Actualitéde la Recherche en Histoire et Archéologie agraires: Actes du Colloque international AGER V, Septembre 2000*, Annales Littéraires 764, Série 'Environnement, Sociétés et Archéologie' 5, 209–23, Besançon: Presses Universitaires Franc-Comtoise.

Toupet, C. and P. Lemaître (2003b), 'Vignobles et modes d'exploitation viticoles antiques dans le Nord de la Gaule: L'exemple de Bruyères-sur-Oise (Val-d'Oise): une relecture', *Revue archéologique de Picardie*, 1–2: 209–26.

Trapero Fernández, P. (2016), 'Roman Viticulture Analysis based on Latin Agronomists and the Application of a Geographic Information System in Lower Guadalquivir', *Virtual Archaeology Review*, 7 (14): 53–60.

Trapero Fernández, P. (2018), 'Ways to Make History and Archaeology with Non-Invasive Techniques: The Application of GIS, RPA and GPR in Hasta Regia Roman City', in M. J. Mosquera and M. L. Almoraima Gil (eds), *Conserving Cultural Heritage*, 129–31, London: CRC Press.

Trapero Fernández, P. (2021a), 'Condicionantes para el cultivo de la vid en época romana a través de Columela y los SIG', *Revista de Estudios Andaluces*, 41: 25–46.

Trapero Fernádez, P. (2021b), *La viticultura romana en el estuario del Guadalquivir: Las practicas de cultivo, producción distribución y modelado SIG enn la colonia Hasta Regia*, BAR International Series 3022, Oxford: BAR.

Trentacoste, A. and L. Lodwick (2023), 'Agriculture and Animal Husbandry in First-Millennium BC Italy: Towards an Agroecology of the Roman Expansion', in D.-E. Padilla Peralta, S. Bernard and L. Mignone (eds), *New Approaches to Middle Republican Rome and Italy*, Cambridge: Cambridge University Press.

Trevizam, M. (2013), *Prosa Técnica: Catão, Varrão, Vitrúvio e Columela*, Campinas: Editora da Uni.

Trocchi, T., M. Marchesini, S. Marvelli and L. Lambertini, eds (2014), *La villa nel pozzo: Un insediamento rustico romano a Sant'Agata Bolognese*, San Giovanni in Persiceto: Museo Archeologico Ambientale.

Trojsi, G. (2017), 'Villa romana di Cottanello: Indagini archeometriche su alcuni campioni di Dolia e di ceramiche comuni', in P. Pensabene and C. Sfameni (eds), *Villa romana di Cottanello: Ricerche 2010–2016*, 28–292, Bari: Edipuglia.

Tsartsidou, G., S. Lev-Yadun, R. M. Albert, A. Miller Rosen, N. Efstratiou and S. Weiner (2007), 'The Phytolith Archaeological Record: Strengths and Weaknesses Evaluated Based on a Quantitative Modern Reference Collection from Greece', *JAS*, 34 (8): 1262–75.

Turner, S. D. and A. G. Brown (2004), 'Vitis Pollen Dispersal in and from Organic Vineyards I: Pollen Trap Data', *Review of Palaeobotany and Palynology*, 129: 117–32.

Turner, P. and G. L. Greasy (2003), 'Terroir: Competing Definitions and Applications', *Australian and New Zealand Wine Industry Journal*, 18 (6): 48–55.

Tyree, E. L. (1994), 'Phytolith Analysis of Olive Oil and Wine Sediments for Possible Identification in Archaeology', *Canadian Journal of Botany*, 72: 499–504.

Valamoti, S. M. (1998), 'Η εκμετάλλευση της αμπέλου στο χώρο της μακεδονίας και της ?ρ ?κης κατά τη ?εολιθική και την ρρώιμη εποχή του ?αλκού (The Exploitation of the Grapevine in Macedonia and Thrace during the Neolithic and the Early Bronze Age)', in *Αμπελοοινική Ιστορία στο Χώρο της Μακεδονίας και της Θράκης (History of Viticulture and Wine in Macedonia and Thrace)*, 137–49, Athens: ETBA.

Valamoti, S. M. (2004), *Plants and People in Late Neolithic and Bronze Age Northern Greece: An Archaeobotanical Investigation*, BAR International Series 1258, Oxford: Archaeopress.
Valamoti, S. M. (2009), *Η Αρχαιοβοτανική Έρευνα της Διατροφής στην Προϊστορική Ελλάδα (An Archaeobotanical Investigation of Prehistoric Diet in Greece)*, Thessaloniki: University Studio Press.
Valamoti, S. M. (2015), 'Harvesting the "wild"? Exploring the Context of Fruit and Nuts Exploitation at Neolithic Dikili Tash, with Special Reference to Wine', *Vegetation History and Archaeobotany*, 24: 35–46.
Valamoti, S. M., M. Mangafa, H. Koukouli-Chrysanthaki and D. Malamidou (2007), 'Grape-Pressings from Northern Greece: The Earliest Wine in the Aegean?', *Antiquity*, 81: 54–61.
Valamoti, S. M., P. Darcque, C. Koukouli-Chrysanthaki, D. Malamidou and Z. Tsirtsoni (2015), 'An Archaeobotanical Investigation of Prehistoric Grape Vine Exploitation and Wine Making in Northern Greece: Recent Finds from Dikili Tash', in A. Diler, K. Senol and Ü. Aydinoğlu (eds), *Olive Oil and Wine Production in Eastern Mediterranean during Antiquity: International Symposium Proceedings 17–19 November 2011, Urla, Turkey*, 125–39, Izmir: Ege Üniversitesi.
Valamoti, S. M., E. Gkatzogia, I. Hristova and E. Marinova (2018), 'Iron Age Cultural Interactions, Plant Subsistence and Land Use in SE Europe Inferred from Archaeobotanical Evidence of Greece and Bulgaria', in S. Gimatzidis, M. Pieniazek and S. Mangalağlu-Votruba (eds), *Archaeology Across Frontiers and Borderlands: Fragmentation and Connectivity in the North Aegean and the Central Balkans from Bronze to Iron Age*, Oriental and Europaean Archaeology 9, 269–90, Wien: Institute for Oriental and European Archaeology.
Valamoti, S. M., C. Pagnoux, M. Ntinou, L. Bouby, V. Bonhomme and J. F. Terral (2020), 'More than Meets the Eye: New Archaeobotanical Evidence on Bronze Age Viticulture and Wine Making in the Peloponnese, Greece', *Vegetation History and Archaeobotany*, 29: 35–50.
Valdambrini, C., A. Pecci and L. Salvini (2007), 'Ceramiche da mensa e da dispensa da alcuni siti della provincia di Grosseto: rapporto contenuto/contenitore', in *Atti del Convegno Internazionale della ceramica: La ceramica da fuoco e da dispensa nel basso medioevo e nella prima età moderna (secoli XI-XVI), Savona, 26–27 maggio 2006*, 120–8, Firenze: All'Insegna del Giglio.
Vallelonga, F. and A. Zifferero (2010), 'Archeologia dell'olivo e dell'olio nel Lazio settentrionale', in G. Barbieri, A. Ciacci and A. Zifferero (eds), *Eleiva Oleum Olio: Le origini dell'olivicoltura in Toscana: nuovi percorsi di ricerca tra archeologia, botanica e biologia molecolare, Atti della Giornata di Studi, San Quirico d'Orcia 2007*, 49–70, San Quirico d'Orcia: Edizioni Don Chisciotte.
Van der Mersch, C. (1996), 'Vigne, vin et économie dans l'Italie du Sud grecque à l'époque archaïque', *Ostraka*, 1: 155–85.
Van der Veen, M., A. Livarda and A. Hill (2008), 'New Plant Foods in Roman Britain – Dispersal and Social Access', *Environmental Archaeology*, 13: 11–36.
van Joolen, E. (2003), 'Archaeological Land Evaluation: A Reconstruction of the Suitability of Ancient Landscapes for Various Land Uses in Italy Focused on the First Millennium BC', unpublished PhD thesis, University of Groningen.
Van Leeuwen, C. and P. Darriet (2016), 'The Impact of Climate Change on Viticulture and Wine Quality', *Journal of Wine Economics*, 11 (1): 150–67.
Van Limbergen, D. (2011), 'Vinum Picenum and Oliva Picena: Wine and Oil Presses in Central Adriatic Italy between the Late Republic and the Early Empire: Evidence and Problems', *BABSCH*, 86: 71–94.
Van Limbergen, D. (2015), 'Pots, Presses, People and Land: The Role of Overseas Export and Local Consumption Demand in the Development of Viticulture and Oleoculture in Central Adriatic Italy (250 BC – AD 200)', unpublished PhD thesis, Ghent University.
Van Limbergen, D. (2016), 'A Note on Olives and Olive Oil from Picenum (Marche, Northern Abruzzo): An Obscured Food Product Within the Economy of Central Adriatic Italy in Roman Times?', *PICUS*, 36: 171–82.

Van Limbergen, D. (2017), 'Changing Perspectives on Roller Presses in Northern Syria', *Syria*, 94: 307–23.

Van Limbergen, D. (2018), 'The Central Adriatic Wine Trade of Italy Revisited: The Central Adriatic Wine Trade of Italy', *Oxford Journal of Archaeology*, 37 (2): 201–26.

Van Limbergen, D. (2019), 'Vinum Picenum and Oliva Picena II: Further Thoughts on Wine and Oil Presses in Central Adriatic Italy', *BABSCH*, 94: 1–31.

Van Limbergen, D. (2020a), 'Wine, Greek and Roman,' in T. Whitmarsh and N. Morley (eds), *The Oxford Classical Dictionary*, Oxford: Oxford University Press.

Van Limbergen, D. (2020b), 'Growing grapes in populous landscapes: demography, food, land and vine agroforestry in central Adriatic Italy', in D. Van Limbergen, S. Maréchal and W. De Clercq (eds), *The Resilience of the Roman Empire: Regional case studies on the relationship between population and food resources* (BAR IntSeries 3000), Oxford: BAR Publishing: 73–111.

Van Limbergen, D. (2022), 'Debating the Yield Potential of Vineyards in Roman Italy', *Revue Belge de Philologie et d'Histoire*, 100: 69–110.

Van Limbergen, D. and W. De Clercq (2021), 'Viticulture as a Climate Proxy for the Roman World? Global Warming as a Comparative Framework for Interpreting the Ancient Source Material in Italy and the West (ca. 200 BC–200 AD)', in P. Erdkamp, J. G. Manning and K. Verboven (eds), *Climate Change and Ancient Societies in Europe and the Near East: Diversity in Collapse and Resilience*, 443–84, London: Palgrave Macmillan.

Van Limbergen, D. and P. Komar (forthcoming), 'Making Wine in Big Clay Jars: Comparative Perspectives on Roman Winemaking', *Antiquity*.

Van Limbergen, D., P. Monsieur and F. Vermeulen (2017), 'The Role of Overseas Export and Local Consumption Demand in the Development of Viticulture in Central-Adriatic Italy (200BC–AD150): The Case of the Ager Potentinus and the Wider Potenza Valley', in T. de Haas and G. Tol (eds), *The Economic Integration of Roman Italy: Rural Communities in a Globalizing World*, 342–66, Leiden: Brill.

Van Limbergen, D. and F. Vermeulen (2021), 'A Method for Estimating Roman Population Sizes from Urban Survey Contexts: An Application in Central Adriatic Italy', in K. Verboven (ed.), *Complexity Economics: Building a New Approach to Ancient Economic History*, 203–49, Cham: Palgrave Macmillan.

Van Pool, T. L. and R. D. Leonard (2011), *Quantitative Analysis in Archaeology*, London: Blackwell.

Van Valkenburgh, P. and J. A. Dufton (2020), 'Big Archaeology: Horizons and Blindspots', *Journal of Field Archaeology*, 45: S1–S7.

Vaschalde, C., C. Rousse, B. Brossier and G. Benčić (2021), 'Production and Management of Fuel for the Imperial Ceramic Workshop of Loron (Croatia): An Archaeological and Anthracological Approach', in D. Van Limbergen and D. Taelman (eds), *The Exploitation of Raw Materials in the Roman World: A Closer Look at Producer-Resource Dynamics, Panel 4.4, Archaeology and Economy in the Ancient World – Proceedings of the 19th International Congress of Classical Archaeology (Cologne/Bonn 2018, Band 27)*, 31–45, Heidelberg, Propylaeum.

Vaschalde, C., M. Tillier, N. Rovira, F. Guibal, D. Kaniewski, C. De Bruxelles, M.-B. Carre, C. Rousse, G. Benčić, V. Kovačić and D. Munda (forthcoming), 'Landscape, Resources Management and Vegetal Economy in the Parentium Area (Croatia) between the Imperial Period and the Beginning of the Middle Ages. First Results of Archaeobotanical Investigation', in I. Borzić et al. (eds), *Transformations of Adriatic Europe (IIe - IXe s. A.D.). Proceeding of the international Conference (Zadar 2016)*, Oxford: Archaeopress.

Venable, W. N. and B. D. Ripley (2002), *Modern Applied Statistics with S*, 4th edn, New York: Springer.

Verdonck, L., A. Launaro and F. Vermeulen (2020), 'Ground-Penetrating Radar Survey at Falerii Novi: A New Approach to the Study of Roman Cities', *Antiquity*, 94: 705–23.

Verhagen, P. (2008), 'Testing Archaeological Predictive Models: A Rough Guide', in A. Posluschny, K. Lambers and I. Herzog (eds), *Layers of Perception: Proceedings of the 35th International Conference on Computer Applications and Quantitative Methods in Archaeology (CAA), Berlin, 2.-6. April 2007*, 285–91, Bonn: Koll.

Verhagen, P., M. Kleijn and J. Joyce (2021), 'Different Models, Different Outcomes? A Comparison of Approaches to Land Use Modeling in the Dutch Limes', *Heritage*, 4: 2081–104.

Verhagen, P. and T. G. Whitley (2020), 'Predictive Spatial Modelling', in M. Gillings, P. Hacigüzeller and G. Lock (eds), *Archaeological Spatial Analysis: A Methodological Guide*, 231–46, Oxon and New York: Routledge.

Verheyden, S. (2004), 'Trace Elements in Speleothems: A Short Review of the State of the Art', *Int. J. Speleol.*, 33 (1/4): 95–101.

Vermeulen, F., G.-J. Burgers, S. Keay and C. Corsi, eds (2012), *Urban Landscape Survey in Italy and the Mediterranean*, Oxford: Oxbow Books.

Vermeulen, F., D. Van Limbergen, P. Monsieur and D. Taelman, eds (2017), *The Potenza Valley Survey (Marche, Italy): Settlement Dynamics and Material Culture in an Adriatic Valley between Iron Age and Late Antiquity*, Academia Belgica Studia Archaeologica 1, Rome: Fondazione Dià Cultura.

Vidal, J. M. and D. M. Corredor (2018), 'The Roman Amphorae Average Capacity (AC)', *Oxford Journal of Archaeology*, 37 (3): 299–311.

Vidale, M. (2000), 's.v. *Archeologia sperimentale*', in R. Francovich and D. Manacorda (eds), *Dizionario di Archeologia: Temi, concetti e metodi*, 280–2, Bari; Roma: Editori Laterza.

Vignani, R., E. Paolucci, M. Scali, J. Bigliazzi, M. Cresti and V. Zorzi (2012), 'Il "Progetto ArcheoVino": caratteri e genoma della vite silvestre in Maremma', in A. Ciacci, P. Rendini and A. Zifferero (eds), *Archeologia della Vite e del Vino in Toscana e nel Lazio: Dalle tecniche dell'indagine archeologica alle prospettive della biologia molecolare*, 653–62, Borgo San Lorenzo: All'Insegna del Giglio.

Vita-Finzi, C., E. Higgs, D. Sturdy, J. Harriss, A. Legge and H. Tippett (1970), 'Prehistoric Economy in the Mount Carmel Area of Palestine: Site Catchment Analysis', *Proceedings of the Prehistoric Society*, 36: 1–37.

Vitali, D., ed. (2007), *Le fornaci e le anfore di Albinia: Primi dati su produzione e scambi della costa tirrenica al mondo gallico*, Bologna: Dip. Archeo.

Vogel, S. and M. Märker, M. (2012), 'Comparison of Pre-AD 79 Roman Paleosols in Two Contrasting Paleo-Topographical Situations Around Pompeii (Italy)', *Geografia Fisica e Dinamica Quaternaria*, 35: 199–209.

Vogel, S, and M. Märker (2014), 'Analysis of Post-Burial Soil Developments of Pre-AD 79 Roman Paleosols Near Pompeii (Italy)', *Open Journal of Soil Science*, 4 (10): 337–56.

Vogel, S., M. Märker, D. Esposito and F. Seiler (2016a), 'The Ancient Rural Settlement Structure in the Hinterland of Pompeii Inferred from Spatial Analysis and Predictive Modeling of Villae Rusticae', *Geoarchaeology*, 31 (2): 121–39.

Vogel, S., M. Märker, I. Rellini, P. Hoelzmann, S. Wulf, M. Robinson, L. Steinhübel, G. Di Maio, C. Imperatore, P. Kastenmeier, L. Liebmann, D. Esposito and F. Seiler (2016b), 'From a Stratigraphic Sequence to a Landscape Evolution Model: Late Pleistocene and Holocene Volcanism, Soil Formation and Land Use in the Shade of Mount Vesuvius (Italy)', *Quaternary International*, 394: 155–79.

Vogel, S., M. Märker and F. Seiler (2011), 'Revised Modelling of the Post-AD 79 Volcanic Deposits of Somma-Vesuvius to Reconstruct the Pre-AD 79 Topography of the Sarno River Plain (Italy)', *Geologica Carpathica*, 62 (1): 5–16.

Volpe, R. (2009), 'Vino, vigneti ed anfore in Roma repubblicana', in V. Jolivet (ed.), *Suburbium II: Il suburbio di Roma dalla fine dell'età monarchica alla nascita del sistema delle ville, V–II secolo a.C.*, 369–92, Rome: École française de Rome.

Vouillamoz, J. F. and S. Grando (2006), 'Genealogy of Wine Grape Cultivars: "Pinot" Is Related to "Syrah"', *Heredity*, 97: 102–10.

Vouillamoz, J. F., A. Monaco, L. Costantini, M. Stefanini, A. Scienza and M. S. Grando (2007a), 'The Parentage of "Sangiovese": The Most Important Italian Wine Grape', *Vitis*, 46: 19–22.
Vouillamoz, J. F., A. Schneider and M. S. Grando (2007b), 'Microsatellite Analysis of Alpine Grape Cultivars (*Vitis vinifera* L.): Alleged Descendants of Pliny the Elder's *Raetica* are Genetically Related', *Genet Resour Crop Evol*, 54: 1095–1104.
Vouillamoz, J. F., A. Monaco, L. Costantini, J. Zambanini, M. Stefanini, A. Scienza, M. S. Grando (2008), 'Il Sangiovese è per metà figlio del Calabrese di Montenuovo', *L'Informatore Agrario*, 5: 59–62.
Wales, N., C. Caroe, M. Sandoval-Velasco, C. Gamba, R. Barnett, J. A. Samaniego, J. R. Madrigal, L. Orlando, M. Thomas and P. Gilbert (2015), 'New Insights on Single-Stranded Versus Double-Stranded DNA Library Preparation for Ancient DNA', *Biotechniques*, 59: 368–71.
Wales, N., J. R. Madrigal, E. Cappellini, A. C. Baez, J. A. S. Castruita, J. A. Romero-Navarro, C. Carøe, M. C. Ávila-Arcos, F. Peñaloza, J. V. Moreno-Mayar, B. Gasparyan, D. Zardaryan, T. Bagoyan, A. Smith, R. Pinhasi, G. Bosi, G. Fiorentino, A. M. Grasso, A. Celant, G. Bar-Oz, Y. Tepper, A. Hall, S. Scalabrin, M. Miculan, M. Morgante, G. Di Gaspero and M. T. P. Gilbert (2016), 'The Limits and Potential of Paleogenomic Techniques for Reconstructing Grapevine Domestication', *Journal of Archaeological Science*, 72: 57–70.
Waliszewski, T. (2014), *Elaion: Olive oil production in Roman and Byzantine Syria-Palestine*, PAM Monograph 6, Warsaw: Warsaw University Press.
Ward-Perkins, J. B. (1962), 'Etruscan Towns, Roman Roads and Medieval Villages: The Historical Geography of Southern Etruria', *The Geographical Journal*, 128 (4): 389–404.
Waterhouse, A. L., G. L. Sacks and D. W. Jeffery (2016), *Understanding Wine Chemistry*, Hoboken: Wiley.
Webb, D. A. (1968), 'Vitacee', in T. G. Tutin, V. H. Heywood, N. A. Burges, D. M. Moore, D. H. Valentine, S. M. Walters D. A. Webb (eds), *Flora Europaea* 2, 246–47, Cambridge: Cambridge University Press.
Webster, D., H. Webster and D. F. Petch (1967), 'A Possible Vineyard of the Romano-British Period at North Thoresby, Lincolnshire', *Lincolnshire History and Archaeology*, 2: 55–61.
Welc, F., G. Lipovac Vrkljan, A. Konestra and T. Rosic (2017), 'Remote Sensing of a Roman Pottery Workshop: Report on a Geophysical Survey Carried out in Crikvenica (Ancient Ad Turres, Croatia)', *Studia Quaternaria*, 34 (2): 119–30.
Welc, F., A. Konestra, A. Dugonjic, P. Andrioc Gracanin, K. Rabiega and B. Nowacki (2019), 'Multidisciplinary Insights into Late Roman Rural Settlement on the Northeastern Adriatic Coast of Croatia: Island of Rab Case Study', *Polish Archaeology in the Mediterranean*, 28 (2): 433–54.
Welc, F., C. Rousse and G. Bencic (2020), 'Results of Geophysical Scanning of a Roman Senatorial Villa in the Santa Marina Bay (Croatia, Istria) using the Amplitude Data Comparison Method (ADCM)', *Studia Quaternaria*, 37 (2): 79–90.
Wendt, K. A., X. Li and R. Lawrence Edwards (2021), 'Uranium–Thorium Dating of Speleothems', *Elements*, 17 (2): 87–92.
Whelton, H. L., S. Hammann, L. J. E. Cramp, J. Dunne, M. Roffet-Salque and R. P. Evershed (2021), 'A Call for Caution in the Analysis of Lipids and other Small Biomolecules from Archaeological Contexts', *Journal of Archaeological Science*, 132: 105397.
White, K. D. (1967), *Agricultural Implements of the Roman World*, Cambridge: Cambridge University Press.
White, K. D. (1970a), *Roman Farming*, London: Thames & Hudson Ltd.
White, K. D. (1970b), *A Bibliography of Roman Agriculture*, Reading: University of Reading.
White, K. D. (1975), *Farm Equipment of the Roman World*, Cambridge: Cambridge University Press.
Willcox G. (1999), 'Archaeobotanical Significance of Growing Near Eastern Progenitors of Domestic Plants at Jalès, France', in P. Anderson (ed.), *Prehistory of Agriculture*, Monograph 40, 103–17, Los Angeles: University of California.
Wilkes, J. J. (1969), *Dalmatia*, Harvard: Harvard University Press.

Wilkins, B. and B. Roberts (2003), *Land to the South of the A120, Takerley, Essex (Barkers Tank Site)*, Archaeological Excavation Interim Site Narrative. Hertfordshire Archaeological Trust, Report No. 1301.

Williams, D. (1977), 'A Consideration of the Sub-Fossil Remains of Vitis vinifera L. as Evidence for Viticulture in Roman Britain', *Britannia*, 8: 327–34.

Wilson, A. (2009), 'Approaches to Quantifying Roman Trade', in A. K. Bowman and A. I. Wilson (eds), *Quantifying the Roman Economy: Methods and Problems*, 213–49, Oxford: Oxford University Press.

Wilson, A. (2011), 'Developments in Mediterranean Shipping and Maritime Trade from the Hellenistic Period for AD 1000', in D. Robinson and A. Wilson (eds), *Maritime Archaeology and Ancient Trade in the Mediterranean*, 33–60, Oxford: Oxford University Press.

Wilson, A. (2014), 'Quantifying Roman Economic Performance by Means of Proxies: Pitfalls and Potential', in F. de Callataÿ (ed.), *Quantifying the Greco-Roman Economy and Beyond*, 147–67, Bari: Edipuglia.

Wiltshire, P. E. J. (1997), 'Evidence for the Environment and Economy: The Easterton Brook Channel. Palynological Analysis of the Palaeochannel Sediments', in P. Williams and R. Newman (eds), *Market Lavington, Wiltshire: An Anglo-Saxon cemetery and settlement. Excavations at Grove Farm 1986–90*, Wessex Archaeology Report No. 19, 121–37, Wessex Archaeology.

Winkler, A. J., J. A. Cook, W. M. Kliewer and L. A. Lider (1992), *General Viticulture*, 2nd edn, Berkeley: University of California Press.

Wiseman, T. P. (1987), *Roman Studies: Literary and Historical*, Liverpool: Francis Cairns.

Wiseman, R., E. Brewer, R. Luxford, J. Losh, R. Fosberry, M. Roberts, C. Jackson-Slater and A. Boulton (2020), *Roman Planting Trenches in the East of England: Archaeology on Furlough*, Cambridge: Cambridge Archaeological Unit.

Witcher, R. (2013), 'On Rome's Ecological Contribution to British Flora and Fauna: Landscape, Legacy and Identity', *Landscape History*, 34: 5–26.

Witcher, R. (2016), 'Agricultural Production in Roman Italy', in A. E. Cooley (ed.), *A Companion to Roman Italy*, 459–82, Chichester: Wiley-Blackwell.

Won, I. J., D. Keiswetter, D. Hanson, E. Novikova and T. Hall (1997), 'GEM 3: A Monostatic Broadband Electromagnetic Induction Sensor', *J. Environ. Eng. Geophys.*, 2 (1): 53–64.

Wong, C. J. and D. O. Breecker (2015), 'Advancements in the Use of Speleothems as Climate Archives', *Quat. Sci. Rev.*, 127: 1–18.

Woodfield, C. and C. Johnson (1989), 'A Roman Site at Stanton Low, on the Great Ouse, Buckinghamshire. Excavated by Margaret Jones, 1957–58', *Archaeological Journal*, 146, 135–278.

Woodworth, M., D. Bernal, M. Bonifay, N. B. Garnier, S. Keay, A. Pecci, J. Poblome, M. Pollard, F. Richez and A. Wilson (2015), 'The Content of African Keay 25/Africana Amphorae: Initial Results of the CORONAM Project', in C. Oliveira, R. Morais and A. Morillo (eds), *Archaeoanalytics. Chromatography and DNA analysis in archaeology: International Symposium*, 41–57, Porto: University of Porto.

Wynn, J. C. (1986), 'A Review of Geophysical Methods Used in Archaeology', *Geoarchaeology: An International Journal*, 1 (3): 245–57.

Zamora, J.-A. (2000), *La Vid y el Vino en Ugarit*, Madrid: Consejo Superior de Investigacions Cientificas.

Zech-Matterne, V. and L. Bouby (2011), 'Viticulture et viniculture dans le nord du bassin parisien d'après les données archéobotaniques', *Gallia*, 68 (1): 257–62.

Zevi, F. (1966), 'Appunti sulle anfore romane. I – la tavola tipologica del Dressel', *AC*, 18: 208–47.

Zifferero, A. (2012), 'Parchi per l'archeologia e il paesaggio: Uno sviluppo possibile per Archeo-Vino', in A. Ciacci, P. Rendini and A. Zifferero (eds), *Archeologia della vite e del vino in Toscana e nel Lazio: Dalle tecniche dell'indagine archeologica alle prospettive della biologia molecolare*, 683–704, Borgo San Lorenzo: All'Insegna del Giglio.

Zifferero, A. (2016), 'Archeologia e circolazione varietale: prospettive di ricerca e valorizzazione del paesaggio agrario in Italia centrale', *Rivista di Storia dell'Agricoltura*, 56 (1-2): 13–30.

Zifferero, A. (2017a), 'Le attività artigianali nel territorio vulcente: la Valle dell'Albegna e Marsiliana', *Scienze dell'Antichità*, 23 (2): 311–29.

Zifferero, A. (2017b), 'Il paesaggio del vino a *Mutina*: circolazione varietale, produzione e consumo', in L. Malnati, S. Pellegrini, F. Piccinini and C. Stefani (eds), *Mutina splendidissima: la città romana e la sua eredità*, Catalogo della Mostra, Modena 2017–18, 275–84, Rome: De Luca Editori d'Arte.

Zifferero, A. (2021), 'Produzione e consumo del vino in Etruria nel periodo orientalizzante: metodi di analisi e prospettive di ricerca', *Mediterranea*, Supplementi 1: 479–500.

Zlateva, B. and M. Rangelov (2015), 'Chemical Analysis of Organic Residues Found in Hellenistic Time Amphorae from SE Bulgaria', *Journal of Applied Spectroscopy*, 82: 221–7.

Zohary, D. and M. Hopf (2000), *Domestication of Plants in the Old World*, 3rd edn, New York.

Zohary, M. (1973), *Geobotanical Foundations of the Middle East*, Stuttgart: G. Fischer.

INDEX

www.ingramcontent.com/pod-product-compliance
Ingram Content Group UK Ltd.
Pitfield, Milton Keynes, MK11 3LW, UK
UKHW022214051026
15624UKWH00001B/154

9 781350 346642